VOLUME ONE HUNDRED AND FORTY FIVE

PROGRESS IN MOLECULAR BIOLOGY AND TRANSLATIONAL SCIENCE

Molecular Biology of Placental Development and Disease

VOLUME ONE HUNDRED AND FORTY FIVE

Progress in MOLECULAR BIOLOGY AND TRANSLATIONAL SCIENCE

Molecular Biology of Placental Development and Disease

Edited by

WILLIAM R. HUCKLE
Virginia-Maryland College of Veterinary Medicine,
Virginia Polytechnic Institute & State University,
Blacksburg, VA, United States

ELSEVIER

ACADEMIC PRESS
An imprint of Elsevier
elsevier.com

Academic Press is an imprint of Elsevier
50 Hampshire Street, 5th Floor, Cambridge, MA 02139, United States
525 B Street, Suite 1800, San Diego, CA 92101-4495, United States
The Boulevard, Langford Lane, Kidlington, Oxford OX5 1GB, United Kingdom
125 London Wall, London EC2Y 5AS, United Kingdom

First edition 2017

Notices
Knowledge and best practice in this field are constantly changing. As new research and experience broaden our understanding, changes in research methods, professional practices, or medical treatment may become necessary.

Practitioners and researchers must always rely on their own experience and knowledge in evaluating and using any information, methods, compounds, or experiments described herein. In using such information or methods they should be mindful of their own safety and the safety of others, including parties for whom they have a professional responsibility.

To the fullest extent of the law, neither the Publisher nor the authors, contributors, or editors, assume any liability for any injury and/or damage to persons or property as a matter of products liability, negligence or otherwise, or from any use or operation of any methods, products, instructions, or ideas contained in the material herein.

ISBN: 978-0-12-809327-6
ISSN: 1877-1173

For information on all Academic Press publications visit our website at https://www.elsevier.com/

Publisher: Zoe Kruze
Acquisition Editor: Alex White
Editorial Project Manager: Helene Kabes
Production Project Manager: Magesh Mahalingam
Cover Designer: Vicky Pearson Esser

Typeset by SPi Global, India

CONTENTS

CONTRIBUTORS

K.J. Baines
The University of Western Ontario, London, ON, Canada

P.-A. Bolze
University of Lyon 1; French Reference Center for Gestational Trophoblastic Diseases; Joint Unit Hospices Civils de Lyon-bioMérieux, University Hospital Lyon Sud, Pierre Bénite, France

N.C. Clark
Center for Reproductive Biology, Washington State University, Pullman, WA, United States

S. Hafez
Virginia-Maryland College of Veterinary Medicine, Virginia Polytechnic Institute and State University, Blacksburg; Virginia Tech Carilion School of Medicine and Research Institute, Roanoke, VA, United States

W.R. Huckle
Virginia-Maryland College of Veterinary Medicine, Virginia Polytechnic Institute & State University, Blacksburg, VA, United States

K. Imakawa
Animal Resource Science Center, Graduate School of Agricultural and Life Science, The University of Tokyo, Kasama, Japan

T. Jansson
University of Colorado Anschutz Medical Campus, Aurora, CO, United States

F. Mallet
Joint Unit Hospices Civils de Lyon-bioMérieux, University Hospital Lyon Sud, Pierre Bénite; EA 7526 Pathophysiology of Injury-Induced Immunosuppression, University of Lyon1-Hospices Civils de Lyon-bioMérieux, Hôpital Edouard Herriot, Lyon, France

M. Mommert
Joint Unit Hospices Civils de Lyon-bioMérieux, University Hospital Lyon Sud, Pierre Bénite; EA 7526 Pathophysiology of Injury-Induced Immunosuppression, University of Lyon1-Hospices Civils de Lyon-bioMérieux, Hôpital Edouard Herriot, Lyon, France

S. Nakagawa
Biomedical Informatics Laboratory, Tokai University School of Medicine, Isehara, Japan

T.L. Powell
University of Colorado Anschutz Medical Campus, Aurora, CO, United States

C.A. Pru
Center for Reproductive Biology, Washington State University, Pullman, WA, United States

J.K. Pru
Center for Reproductive Biology, Washington State University, Pullman, WA, United States

S.J. Renaud
The University of Western Ontario; Children's Health Research Institute, The University of Western Ontario, London, ON, Canada

G.E. Rice
Exosome Biology Laboratory, Centre for Clinical Diagnostics, UQ Centre for Clinical Research, Royal Brisbane and Women's Hospital, The University of Queensland, Brisbane, QLD, Australia; Ochsner Baptist Hospital, New Orleans, LA, United States

F.J. Rosario
University of Colorado Anschutz Medical Campus, Aurora, CO, United States

C. Salomon
Exosome Biology Laboratory, Centre for Clinical Diagnostics, UQ Centre for Clinical Research, Royal Brisbane and Women's Hospital, The University of Queensland, Brisbane, QLD, Australia; Ochsner Baptist Hospital, New Orleans, LA, United States

O.R. Vaughan
University of Colorado Anschutz Medical Campus, Aurora, CO, United States

PREFACE

What an extraordinary tissue is the placenta—that developmental project undertaken jointly by mother and her unborn, its extraembryonic offshoots invading an accommodating endometrium, ultimately serving both parties as cellular gatekeeper, trafficker of nutrients and molecular signals, immune protector, vascular network, endocrine organ, and more—only to disappear at birth. Little wonder that the placenta is subject after birth to a host of traditional practices around the world, many being a ritual acknowledgment of the vital roles that we understand the tissue to play biologically. Particularly appealing is the belief, held in one form or another by several cultures, that the placenta is able to serve as an ethereal, twinned guardian throughout the child's life, and so must be accorded due reverence and proper handling at the time of birth.

While the folklore surrounding the placenta undoubtedly dates back many centuries, this organ presently receives what some might consider long overdue reverence of a scientific nature. In 2011, the Eunice Kennedy Shriver National Institute of Child Health and Human Development (NICHD) at the US National Institutes of Health began a series of workshops, involving hundreds of scientists and other stakeholders and aiming to set research priorities for the years to follow. Emerging from these discussions, and stemming in part from the broad recognition of the potential lifelong consequences of an individual's experience in utero, was the launch of the "Human Placenta Project" by the NICHD in 2014. To date, three rounds of Requests for Applications based directly on the stated goals of the HPP have been issued, including calls titled "Novel Tools to Assess Human Placental Structure and Function, Using Omics to Define Human Placental Development and Function Across Pregnancy, and Assessing Human Placental Development and Function Using Existing Data." Close to 40 grants have been awarded, additional workshops have been held to exchange findings and refine goals, and interested readers invited to follow the progress of the HPP in detail at http://www.nichd.nih.gov. This endeavor promises to bring to bear the powerful new tools of ultrasensitive transcriptional profiling, epigenetics, noninvasive imaging, and computational modeling in service of improving the prospects for prevention, diagnosis, and treatment of the numerous pregnancy complications that involve the placenta. It is reasonable to expect that the HPP will do justice to the vision articulated by NICHD leaders Drs. Guttmacher and Spong, writing in an October 2015 supplement to the *American Journal of Obstetrics & Gynecology* devoted to expert reviews on placental biology and disorders: "The HPP is designed to rectify the long-neglected need to understand the human placenta across gestation. If successful, it should change how we understand and manage pregnancy and all that grows from pregnancy."

The success of the HPP will of course be built upon long-standing and diversified expertise in placental research. The aforementioned journal, together with *Placenta* (the *Official Journal of the International Federation of Placenta Associations*), other serials, and the professional societies and conferences that serve the reproductive biology, theriogenology, fertility, endocrine, immunology, and vascular biology communities, are rich with basic and clinical investigations that define our state of the art. I am pleased that many of the investigators responsible for building this knowledge base have joined me in the assembly of Volume 145 of *Progress in Molecular Biology and Translational Sciences (PMBTS), Molecular Biology of Placental Development and Disease.*

The volume begins with two chapters of an introductory nature, the first reviewing the comparative developmental anatomy of placenta in species most relevant to clinical medicine and basic research (Hafez), and the second giving an overview of experimental models used to investigate placenta formation and function (Huckle). Next comes a set of four chapters that address molecular mechanisms by which the cells of the extraembryonic membranes proliferate, differentiate, form syncytia, and invade and remodel the uterine wall: "Transcription factors that regulate trophoblast development and function" by Baines and Renaud, "The phylogeny of placental evolution through dynamic integrations of retrotransposons" by Imakawa and Nakagawa, "Contribution of syncytins and other endogenous retroviral envelopes to human placenta pathologies" by Bolze, Mommert, and Mallet, and "Role of exosomes in placental homeostasis and pregnancy disorders" by Salomon and Rice. These chapters describe the remarkably complex and precise feats of transcriptional coordination that ultimately produce a fully developed, functional placenta. Even more startling is the contribution made to that process by the expressed products of retroviruses incorporated into mammalian genomes in the dim evolutionary past. The volume concludes with a pair of chapters that address the placental function at the fetal–maternal interface in addition to developmental questions: "Novel regulators of hemodynamics in the pregnant uterus" by Clark, Pru, and Pru and "Regulation of placental amino acid transport and fetal growth" by Vaughn, Rosario, Powell, and Jansson.

It is my hope that readers will find in this volume new reasons to be intrigued by the placenta as a biological phenomenon, as well as an appreciation for the ways that cutting-edge molecular biology is being engaged to ameliorate serious and frequently occurring complications of pregnancy. I am grateful to Professor P. Michael Conn, editor of the PMBTS series, for the opportunity to assemble this volume, to the contributing authors for sharing their expertise, and to Ms. Helene Kabes, Mr. Alex White, Mr. Magesh Mahalingam and their colleagues at Elsevier for unfailing professionalism and patient guidance through the editorial process.

W.R. Huckle
Blacksburg, Virginia

CHAPTER ONE

Comparative Placental Anatomy: Divergent Structures Serving a Common Purpose

S. Hafez[*,†,1]
*Virginia-Maryland College of Veterinary Medicine, Virginia Polytechnic Institute and State University, Blacksburg, VA, United States
†Virginia Tech Carilion School of Medicine and Research Institute, Roanoke, VA, United States
[1]Corresponding author: e-mail address: shafez@vt.edu

Contents

Abstract

The placenta, one of the most important transient organs, forms by the apposition of fetal membranes and maternal tissues. Its role is to mediate physiological exchanges between mother and fetus. The word "apposition" covers a wide range of structural variations. It includes approximation, adhesion, interdigitation, or actual fusion between

Progress in Molecular Biology and Translational Science, Volume 145
ISSN 1877-1173
http://dx.doi.org/10.1016/bs.pmbts.2016.12.001

fetal and maternal tissues.[1] Formation of the placenta establishes hemotropic nutrition for the fetus: essential metabolites must be provided to maintain the growing fetus, and these must come to it via the maternal circulatory system.[2,3] Equally important, the placenta also provides oxygen and removes metabolic waste products from fetal blood. Nutritive and excretory roles of the placenta are not its only functions: it also has immune and endocrine activities.[4] Nutrient and gas transport, waste removal, immunological protection of the fetus, and hormonal secretion influencing the maternal metabolism are all complex functions. They may also to some extent be conflicting purposes; hence, the placenta is a complex fetal organ. It is structurally adapted to perform its roles somewhat differently in different species, but the set of functions remain the same. Understandably, the placenta has been the subject of extensive research, and it will continue be an important topic thanks to its complexity. The intent of this chapter is to provide a simple description of placental anatomy using classic categories and to describe anatomical species variations in humans, important domestic animals, and the major laboratory species.

THE PLACENTA

The placentae of eutherian mammals have features in common, ones that facilitate essential functions, but there are of course unique species-associated configurations. While the placentae of different species have many structural variations, the overriding necessity of fulfilling its essential functions means that different systems are designed to achieve the same purpose.

The process of placentation starts with a small area of maternofetal apposition, which in time increases in size, in response to the growth of the fetus. The development of the placenta begins with the implantation of the blastocyst into the wall of the uterus.

1. FETAL MEMBRANES

Understanding the development of fetal membranes is necessary before discussing the anatomy of the completed placental organ. Fetal membranes provide the basis for the formation of structures essential to the physiological maintenance and protection of the embryo. Fetal membranes (also referred to as extraembryonic membranes, despite the fact that they are really embryonic in origin) are formed from and are continuous with the three embryonic layers: ectoderm, mesoderm, and endoderm[5]. The prefix "extra" is used in context of "outside of" the embryo proper. Fetal membranes are formed from somatic or splanchnic mesoderm plus ectoderm or endoderm. For details, see Steven, 1975; Perry, 1981; Noden and De Lahunta, 1985; Mossman, 1987; Leiser and Kaufmann, 1994; Wooding and Burton, 2008.[3,5–9]

There are four fetal membranes: the chorion, the amnion, the yolk sac, and the allantois (Fig. 1). The chorion and amnion are derived from the somatopleure (i.e., trophoblastic ectoderm and extraembryonic somatic mesoderm). The yolk sac and allantois are derived from splanchnopleure (the endoderm and extraembryonic splanchnic mesoderm).

The following general description applies to human and domestic animals, including the horse, pig, ruminant, and carnivores, but species variations will be noted and specified. The development of fetal membranes in rodents is unique to those species, so the development of fetal membranes in mice and rats will be discussed in the section of the placenta of rodents.

A placenta is formed when fetal tissues acquire contact or fusion with maternal tissue for physiological exchange. In mammals this always involves the chorion and either the yolk sac or the allantois. The amnion remains avascular, and its function is chiefly mechanical.[7]

1.1 Yolk Sac

The yolk sac is formed from hypoblast endoderm and extraembryonic mesoderm. The hypoblast separates from the inner surface of the embryonic disc in early blastocyst stage, forming an endodermal tube within the trophoblast

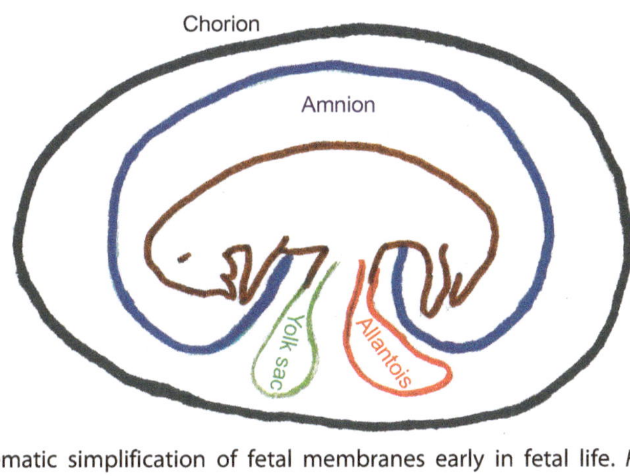

Fig. 1 Schematic simplification of fetal membranes early in fetal life. *Redrawn after Leiser R, Kaufmann P. Placental structure: in a comparative aspect.* Exp Clin Endocrinol. *1994;102:122–134.*

tube. The hypoblast tube is invested with splanchnic mesoderm after its formation and splitting.

The yolk sac is the part of the tube outside the embryo. The yolk sac is one of the components of a choriovitelline placenta; the other component is the chorion. It is the connection between the yolk sac and the chorion on the abembryonic side that forms the choriovitelline placenta, that is, the apposition of yolk sac endoderm, fused somatic and splanchnic mesoderm, and the trophoblast. This combination of embryonic structures is termed a "trilaminar omphalopleure." This connection is formed early in gestation in the horse and in carnivores; it remains functioning in the horse for a longer period than in any other mammal (for the first quarter of the total gestation period), and it is the primary source of nutrients during that period. The yolk sac/chorion connection is broken down later in carnivores, except in the extremities where it remains functioning well after establishment of the chorioallantoic placenta. The apposition between the yolk sac and the chorion is transitory in ruminants and pigs, but it is nevertheless functional for a short period.

In humans, the primary yolk sac is formed in a similar fashion as in domestic animals. With fetal growth, however, it is displaced to the abembryonic pole and ultimately degenerates. The space that constituted the primary yolk sac becomes the definitive yolk sac. It is small to begin with, provides very limited nutritive function, and regresses early, but it is still important in respect to other functions. The yolk sac mesoderm is a major site of hematopoiesis, and the yolk sac endoderm is the source of primordial germ cells.

The yolk sac and its vitelline vessels provide temporary nourishment early in embryonic life. The nutritive role of the yolk sac is later taken over by the allantois, after the latter has developed. In most species, the yolk sac's degeneration leaves no visible remnant at birth. The attachment between chorionic and yolk sac mesoderm at the extremities in carnivores persists until birth and can be seen as a tubular structure extending throughout the length of the fetal membranes.

1.2 Allantois

The allantois is derived from splanchnopleure (endoderm and splanchnic mesoderm). It arises as a diverticulum of the hindgut and gradually fills the entire extraembryonic coelom (exocoelom) in most species. The allantois does not extend to the area where the connection of yolk sac and chorion exists in the horse and carnivores, nor where the mesamnion is located in the pig and ruminants. In humans, the allantois is vestigial, but in a functional sense, the human placenta is a chorioallantoic type (see later). The vessels of the allantois vascularize the chorion and amnion, with allantoic arteries as branches of the two dorsal aortae. Allantoic veins or umbilical veins drain into the caudal (inferior) vena cava through the sinus venosus.

1.3 Chorion

The chorion is derived from trophoblastic ectoderm and extraembryonic mesoderm (somatopleure). There is an intimate association between the forming chorion and amnion. These form by folding in domestic animals and by so-called cavitation in humans, mice, and rats.

In domestic animals, the chorion and amnion are the products of bilateral folding of the extraembryonic somatopleure. This arches dorsal to the embryo and continues to grow. Fusion of the chorioamniotic folds occurs at the mesamnion or chorioamniotic raphe. Dorsal fusion results in formation of two layers of somatopleure separated by the exocoelom: the outer somatopleure becomes the chorion and the inner somatopleure the amnion. Thus, the chorion is lined by mesoderm from inside, and amnion is lined by mesoderm from outside. This ensures that the chorionic trophoblasts face the endometrium.

When complete separation of the chorion and amnion occurs, the exocoelom fully surrounds the amnion. When this happens, the fetus is born covered with the amnion, as in the case of the horse. If the dorsal connection (mesamnion) persists, which is the case in pigs and ruminants, the fetus is born without being covered by the amnion.

In humans, the chorion is simply the original trophoblast, which becomes lined by somatic mesoderm. The chorion is relatively avascular; blood perfusion is achieved instead by the allantoic vessels. But the chorion (vascularized by the allantois) is the essential component of chorionic villi. The chorioallantoic placenta is the permanent functional placenta in domestic mammals and humans, taking the place of the transitory choriovitelline placenta.

1.4 Amnion

The formation of the amnion is associated with the formation of the chorion as described earlier. The amnion is the outer membrane, created by bilateral folding of the extraembryonic somatopleure and fusion of the chorioamniotic folds dorsal to the embryo in domestic animals.

In humans, the amniotic cavity develops within the inner cell mass, which has two layers; the epiblast and the hypoblast. These two layers are separated by a basement membrane that is in contact with the hypoblastic layer. The epiblast cells form the primitive ectoderm; the hypoblast cells form the primitive endoderm. The epiblast cells closest to the primitive endoderm become polarized and form a columnar epiblast epithelium separated from the "roof" of amniotic ectodermal cells, a process that takes place by apoptosis of the cells in the center. The internal "space" resulting from apoptotic death of the center cells is the proamniotic cavity that becomes the mature amniotic cavity, lined with mesoderm when fully formed. This mesoderm populates the amniotic membrane from the outside. The amnion surrounds the embryo until term; in humans, the placenta is, structurally, a "chorioamniotic" type because fusion occurs between the amniotic and chorionic somatic mesodermal layers.

2. CLASSIFICATION OF PLACENTAE

Classical placenta types are categorized based on the membranes of origin, gross morphological features, histologic nature of the maternofetal interface, or the relative directions of the maternal and fetal blood flow. Unfortunately, there is no single classification that provides a complete picture of the anatomy of the placenta of any given species. Understanding all the various classes is important to understanding the architecture of both the maternal and fetal sides of the placenta. For details, see Amoroso, 1952; Ramsey, 1975; Noden and De Lahunta, 1985; Kaufmann and Burton, 1994; Leiser and Kaufmann, 1994; Wooding and Burton, 2008.[1,3,5,9–11]

2.1 Classification Based on Origin

This approach to classification is based on the nature of the maternofetal interdigitation areas, which can be described as *choriovitelline* or *chorioallantoic*. A true chorionic placenta only exists in very early stages of development, before the invasion of the vitelline or allantoic vessels into the chorion.[1]

In most species, regions of maternofetal interdigitation are composed of chorioallantoic placental tissues (i.e., vascularized allantois plus chorion). When the allantoic splanchnic mesoderm fuses with the somatic mesoderm of the chorion, its vessels vascularize it. Therefore, the chorioallantoic placenta is the permanent functional placenta in most domestic mammals, taking over from the transitory choriovitelline placenta. The trilaminar omphalopleure remains functional for a longer period in the horse than in any other mammal. In carnivores, the extremities remain functioning after the establishment of the chorioallantoic placenta.

2.2 Classification Based on Morphology

Placenta can be classified according to the scope and arrangement of the apposition area between the maternal and fetal surfaces (Fig. 2). The maternofetal exchange surface area can be vastly increased by elaboration of villi or folds. Trophoblasts proliferate, forming trophoblastic bud-like outgrowths, which are merely villous stems without a mesenchymal core. These are termed primary villi. Those primary villi acquire a core of

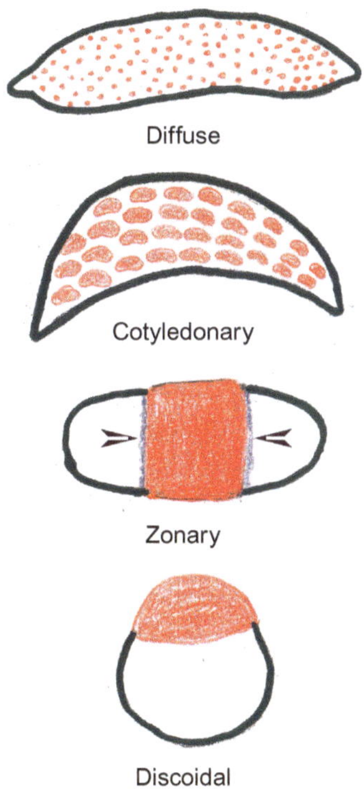

Fig. 2 Schematic representation of the scope of the apposition area of the maternal and fetal surfaces. The *arrows* on the zonary placenta indicate the marginal hematomas in carnivores. *Redrawn after Noden D, De Lahunta A.* The Embryology of Domestic Animals Developmental Mechanism and Malformation. *Baltimore, MD: WILLIAMS & WILKINS; 1985.*

mesoderm, after which they are termed secondary villi. With the appearance of blood vessels in this mesodermal core, they are referred to as tertiary villi. The process of villus formation continues throughout gestation; as pregnancy advances, the mesodermal core and blood vessels become more and more highly developed. A completed chorionic villus consists of an epithelium (trophoblasts), a mesenchymal core, and allantoic (vitelline in some cases as mentioned earlier) blood vessels.

The distribution of the chorionic folds or villi is characteristic of various species. Furthermore, the type of maternofetal interdigitation that describes the geometrical pattern in which the maternal tissues and the fetal surfaces are spatially arranged to form a placenta differs among species.[9]

When the maternal and fetal tissues interdigitate over the entire surface of contact, the type of placenta is called a *diffuse placenta.* The diffuse placenta is present in the pig[12] and the horse.[13] In some species such as ruminants,[14] the apposition zone is confined to discrete specialized sites named placentomes; this type is referred to as a *cotyledonary placenta.* Whereas a maternofetal connection is broadly established over the general surface of a diffuse placenta, formation of villi in cotyledonary placentae occurs only in specific areas apposed to specialized preexisting uterine caruncles. These caruncles are fleshy protuberances in the uterine wall equipped with vascular adaptations that facilitate establishment of the intimate connection between mother and fetus.

In carnivores,[15] the interdigitation zone forms a strap or girdle around the chorionic sac: hence the name *zonary or girdle placenta.* A *discoidal placenta* is a feature of rodents and primates (including humans). The area of interdigitation in discoidal placenta is concentrated in a specific placental disc.[9,16] In some primates (including the rhesus monkey), the exchange tissue is concentrated on two placental discs, giving rise to the variant of a *bidiscoidal placenta.*[9] Despite these structural variations, maternofetal exchange is not restricted to the concentrated placental region in different models; it can also take place in the "smooth" areas.[9]

The *folded*-type placenta, present in the pig, is the simplest form that describes the geometrical pattern of the maternal and fetal tissues. In this type of placenta, the chorionallantois shows numerous small folds that interlock with corresponding endometrial folds.[12] Carnivores have a *lamellar*-type placenta, with a more extensive array of branched folds that increase adherence and facilitate exchange. In most ruminants, as in horses and humans, the chorion forms tree-like villi which interdigitate into corresponding maternal crypts.[6,17] This arrangement is called *villous*-type placenta (Fig. 3).

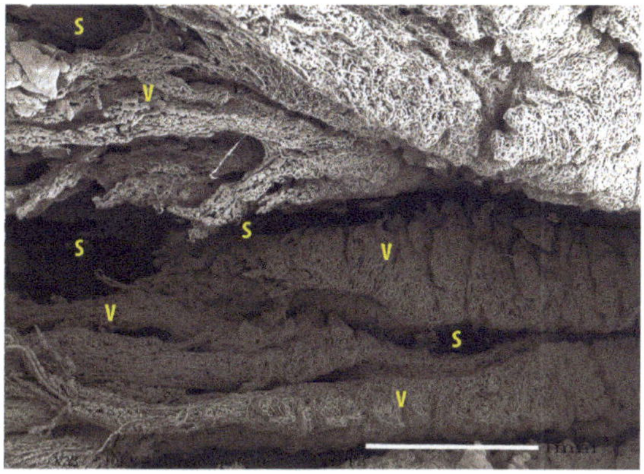

Fig. 3 Scanning electron micrograph of a microvascular corrosion cast of a sectioned cotyledon from a pregnant ewe showing an example of a villous-type arrangement. The maternal surface of the cotyledon is on the *left side* of the image. The entire image is filled with fetal villi (V) interspersed by spaces (S) for interdigitation of the maternal tissues. Scale bar = 1 mm. For preparation of microvascular corrosion casts, see Hafez *et al.*, 2007.[17]

A *labyrinthine* placenta is the most structurally elaborate type and is found in rodents.[18] In these species, the chorion is penetrated by a web-like arrangement of channels.[9]

2.3 Classification Based on the Histologic Nature of the Maternofetal Interface

Three main types are recognized, based on the cell layers comprising the interhemal area (Fig. 4): (1) *epitheliochorial* type (horses and pigs), (2) *endotheliochorial* type (carnivores), and (3) *hemochorial type* (primates and rodents).[19]

This classification was introduced by Grosser (1909)[20] (cited by Amoroso, 1952[10]; Wooding and Burton, 2008[5]). In all placental types, regardless of species, the three fetal layers of the chorioallantoic placenta are present. These are the endothelium lining of the allantoic blood vessels, the chorioallantoic mesodermal connective tissue, and the chorionic epithelial cell layer (trophoblast).[3] Histological classification of the placenta is based on which, if any, of the three *maternal* layers is retained: the maternal vascular endothelium, the uterine connective tissue, and the surface epithelium (endometrium).

The *epitheliochorial type* is the most superficial placenta, lacking significant invasion of the uterine tissues. In this type, which is found in horses and pigs,

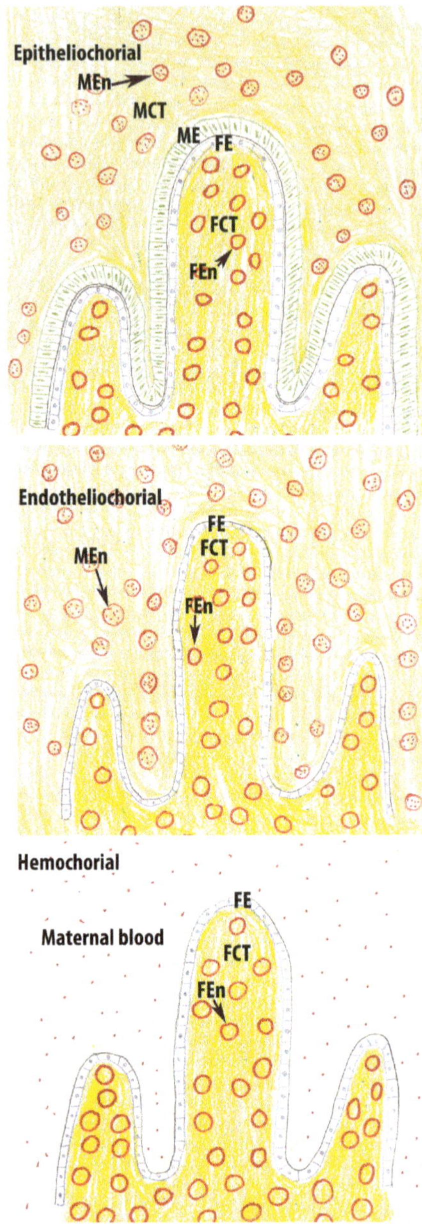

Fig. 4 Schematic representation of the types of placenta based on the layers comprising the interhemal interface. In an epitheliochorial type, all six layers exist: *FE*, fetal chorionic epithelium; *FCT*, fetal chorioallantoic mesodermal connective tissue; *FEn*, fetal endothelium; *ME*, maternal epithelium; *MCT*, maternal connective tissue; *MEn*, maternal endothelium. In an endotheliochorial type, the maternal epithelium and connective tissue have been eroded. In a hemochorial type, the fetal villi are bathed in maternal blood.

all six layers are present and there is no loss of endometrial tissue during attachment or throughout gestation. The fetal chorionic epithelium is in contact with the uterine epithelium.

The term *synepitheliochorial* is used precisely to describe the placenta of ruminants[14,21] because of the fusion of the binucleate trophoblasts with the uterine epithelium. The term "syndesmochorial" was used in older literature to describe the type of placenta where the endometrium epithelium is removed with implantation,[20] but findings from electron microscopic examination eliminated this type from the classification scheme.[5,14,21]

In *endotheliochorial* placentae, both the uterine epithelium and underlying connective tissue are lost during attachment, so that the maternal vascular endothelium comes into direct contact with the fetal chorionic epithelium. This placental type occurs in carnivores.[15]

The hemochorial placenta is the most invasive type. Here, the uterine vascular endothelium is lost along with the maternal epithelium and connective tissue. This produces a situation in which there is direct association between the chorionic epithelium and maternal blood; i.e., the surface of the chorion is bathed in maternal blood. There are hemomonochorial (primates), hemodichorial (rabbits), and hemotrichorial (rats and mice) placentas, with one, two, and three trophoblast layers, respectively.[6]

There are some limitations to classifying placenta types based on histology, even though this system is commonly used. This classification scheme is useful only for chorioallantoic placentas, such that the transitory choriovitelline placentae are not included. In addition, in some animals such as small ruminants, the number of retained maternal layers differs in various parts of the placenta and with stage of gestation.[3,5,14]

Older literature[20] posited the histological classification as a direct indicator of transplacental exchange efficiency, on the principle that the greater the number of layers between the maternal and fetal blood vessels, the lower the presumed transplacental efficacy. In this conceptualization, the efficiency of maternofetal exchange would be affected in part by the "interhemal distance" of the placenta. Therefore, epitheliochorial placentas might be considered as less "efficient" at the transfer than those with a shorter interhemal distance. However, this assumption was proven to be an unwarranted oversimplification.[5,22] Several other factors, notably the species-specific degree of permeability of the various layers making up the maternofetal barrier, the actual thickness of these layers, and the spatial relationship between fetal and maternal vasculatures, may be more important determinants of the efficiency of transplacental exchange.[23]

The efficiency of transplacental nutrient and gas exchange may be influenced by the relative arrangement of the capillary bed between the maternal and fetal vasculatures, and this arrangement differs significantly from one species to the next (Fig. 5).[5,9] In *concurrent* flow, both fetal and maternal blood streams would run in the same direction. This arrangement is not known to predominate in any mammal studied to date[9] and morphologically has not been demonstrated. It may, therefore, not exist as a functioning principle in nature.[24,25] In the *countercurrent* system, as represented by the Guinea Pig placenta, the fetal and maternal blood streams flow in opposite directions.[26] Cats and pigs possess a variant form of placental circulatory relationship termed a *crosscurrent* system,[24] which is intermediate with respect to the efficiency of exchange between an exclusively concurrent or exclusively countercurrent arrangement. In humans and goats, there

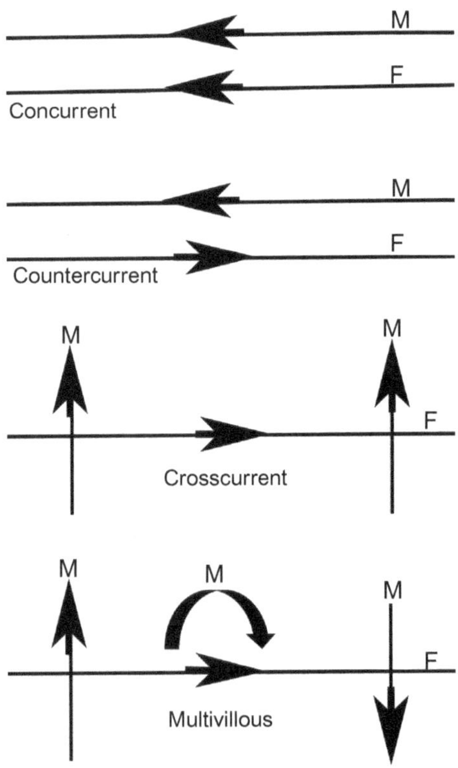

Fig. 5 Schematic arrangement of the maternal and fetal vasculature. *M*, maternal blood stream; *F*, fetal blood stream. The *arrows* indicate the direction of blood flow. *Redrawn after Leiser R, Kaufmann P. Placental structure: in a comparative aspect.* Exp Clin Endocrinol. *1994;102:122–134.*

is a combination of these three (counter, cross, and concurrent) systems, termed *multivillous* system.[17,24,27]

2.4 Classification Based on the Tissues Lost During Parturition

When apposition occurs without fusion, the maternal endometrium remains intact and is not sloughed off during parturition; this is the case in the *adeciduate* placenta.[3] When the trophoblast has fused with the endometrium or grown into it, tearing will result when the placenta separates at birth. This is the case of species with endotheliochorial or hemochorial placentae. These species are said to have *deciduate* placentas. Ruminants have a special "partially deciduate"-type placenta because only a portion of the endometrium sloughs off after parturition. This may be as a result of the fusion of binucleate trophoblasts with the uterine epithelium.[14]

The following section will deal with the anatomy of the placenta of selected species including humans, rodents, carnivores, ruminants, horses, and pigs. For details, see Ramzey, 1975; Noden and De Lahunta, 1985; Shanklin, 1986; Kaufmann *et al.*, 1988; Leiser *et al.*, 1997; Wooding and Burton, 2008; Verma and Verma, 2013 (human) [1,3,5,16,27,28,30]; Ramzey, 1975; Noden and De Lahunta, 1985; Wooding and Burton, 2008 (Rodents) [1,3,5]; Wynn and Corbett, 1969; Wooding and Burton, 2008 (Carnivores) [15,21]; Leiser *et al.*, 1997; Hafez *et al.*, 2007; Hafez *et al.*, 2010 (Ruminants) [17,23,28]; Samuel *et al.*, 1974; Noden and De Lahunta, 1985 (Horse) [3,13]; Friess *et al.*, 1980; Noden and De Lahunta, 1985; Leiser and Dantzer, 1988 (Pig).[3,12,29]

3. PLACENTA OF HUMANS

A fully discoid form of human placenta has been seen as early as 6 weeks postconceptional such observations are limited due to the rarity of intact specimens.[16] Morphologically, the human placenta is a discoidal, villous type. Histologically, the interhemal barrier is hemomonochorial. The maternal and fetal blood streams form a multivillous system. Humans have a deciduate-type placenta due to great invasion of the maternal tissues and subsequent loss during parturition. The maternofetal interdigitation areas are structurally chorioamniotic. This occurs as a result of the expansion of the amnion and the fusion of the amniotic and chorionic somatic mesodermal layers. Functionally, however, it is chorioallantoic based on the source of blood vessels, as the allantoic blood vessels invade the chorionic mesoderm.

Placenta formation begins with the attachment of the blastocyst to the endometrium, followed by nidation or interstitial implantation (embedding of the blastocyst in the uterine endometrium). At the time of attachment, the blastocyst is composed of a single layer of cytotrophoblasts. The cytotrophoblast is the primary epithelium. The trophoblasts located at the contact site with the uterine epithelium undergo massive proliferation and differentiate into syncytiotrophoblast and underlying cytotrophoblast.

Cytotrophoblasts are large, cuboidal, pale-staining cells with prominent, vesicular nuclei and few organelles. These mononucleated cells lie deep to the syncytial layer. The cytotrophoblast immediately surrounds the mesenchymal core of each villus. A periodic acid Schiff-positive basement membrane is laid down between these cells and the mesenchyme. Outside the cytotrophoblast is a syncytial layer of variable nucleation state.

The syncytiotrophoblast forms the outer covering of the villi. It is a multinucleated syncytial epithelial cell layer that contains a high concentration of organelles. There are microvilli over the entire surface, which provides a means for increasing the structure's surface area. Syncytiotrophoblast cells are generated from fusion of cytotrophoblasts. These syncytial cells are highly invasive, and wherever they contact the endometrium, the epithelium, underlying connective tissues and the maternal blood vessels' endothelium are destroyed. Cytotrophoblasts are the stem cells of the fetal epithelium, and the syncytiotrophoblasts are the most invasive cells that produce the hemochorial status. An intermediate form of trophoblast can also be found.[30]

This site of massive cell proliferation and profound differentiation develops into the chorion frondosum, in which the chorionic villi continue to grow with vigor and become highly branched. The outward side of the chorion frondosum, the site of contact with the endometrium, is referred to as the chorion laeve indicating that it is a relatively smooth area. Lacunar spaces develop at the site of the chorion frondosum. These become intervillous spaces that fill with maternal blood. As a result of the massive loss of endometrial tissues including the maternal vascular endothelium, the chorionic villi are bathed in maternal blood, producing the hemochorial interhemal interface.

The decidual reaction refers to the process of transformation of the maternal stromal cells around the blastocyst. These cells enlarge, and their cytoplasm becomes filled with glycogen and lipid droplets. Decidual cells separate the invasive trophoblast from the myometrium. The area of endometrium directly underlying the chorion frondosum is termed the decidua basalis. The decidua capsularis is adjacent to the chorion laeve. The decidua parietalis forms later as a result of the embryo growth and fusion of the decidua capsularis with the tissue on the opposite site.

As noted earlier, the area of human placentation is discoidal, consisting of the chorion frondosum and the decidua basalis. This structural organization gives the full-term placenta its gross appearance as "discoid." The mature placental disc is subdivided into cotyledons or placentones by septae; the cotyledons of the human placenta should not be confused with the cotyledons of the ruminant placenta. Each cotyledon is a collection of villi. The core of each septum is composed of endometrial tissues that were spared when the trophoblast invaded the implantation site. The septae are bordered by the trophoblast.

4. PLACENTA OF RODENTS

The mouse and rat placentae are nearly identical. Morphologically, these rodents' placentae are of discoidal and labyrinthine type. The interhemal interface is hemotrichorial, due to the presence of three types of trophoblasts in an ectoplacental cone or the so-called trager that overlies the embryonic disc. The three types are: (1) mononuclear giant phagocytic cells that are invasive. These cells play an important role by eroding the endometrial and decidual tissue (including the capillary endothelium) thus creating the hemochorial state, (2) Syncytiotrophoblasts, and (3) cytotrophoblasts. The ectoplacental cone is separated from the embryonic epiblasts by the ectoplacental cavity.

Rodents have an inverted yolk sac placenta. The yolk sac placenta is extensive in early stages, persists and coexists with the chorioallantoic placenta, and may form some mature placental structures. It is the principal nutritive placenta in these animals. It is inverted in the sense that the fetal endoderm lies between the maternal tissue and the mesoderm. This is in contrast to other species where the fetal mesoderm lies between the ectoderm and endoderm. The yolk sac is formed initially from hypoblast endoderm without being surrounded by splanchnic mesoderm; it then expands and becomes apposed to the trophoblast; together these form a bilaminar omphalopleure. Later continuous expansion of the endoderm results in a collapse of the yolk sac, which brings the opposite layer of endoderm (accompanied by splanchnic mesoderm) in contact with the bilaminar omphalopleure, forming complete inverted yolk sac. The amnion and chorion develop within the ectoplacental cavity when the extraembryonic somatopleure develops and then folds across this cavity. Later the ectoplacental tissue and allantoic blood vessels invade the endometrium forming a discoid placentone. Thus, a chorioallantoic placenta is also established. Despite the lack of the allantoic sac, the allantois still provides the vascular components of the chorioallantoic placenta. Mice and rats have a deciduate placenta due to the great invasion of the maternal tissues and subsequent loss during parturition.

5. PLACENTA OF CARNIVORES

In carnivores, the choriovitelline placenta is formed early in gestation and remains well developed up to 21–24 days. Later, this attachment between the yolk sac splanchnopleure and the chorionic somatopleure, which appears as a broad longitudinal band along the trophoblast on the abembryonic side, is broken down except at its extremities. The portions attached at the extremities remain functioning well after the establishment of the chorioallantoic placenta. The growth of the allantois and the vascularization of the chorion establish the definitive chorioallantoic placenta. The allantois also vascularizes the amnion in this species.

Morphologically, placenta of carnivores is zonary. The area of contact is confined to an equatorial band around the central third of the chorion. The rest of the wall of the chorionic sac around the zonary placenta is a thin vascularized chorion lacking chorionic villi. Paraplacental zones are features of the carnivore placenta. These are the peripheral poles of the fetal membranes. These poles are loosely apposed to the maternal epithelium and free of structural change. The villi are arranged in complex arrays, which gives the carnivores a labyrinthine placenta according to Anderson (1969),[31] or a lamellar type according to Leiser and Kaufmann (1994).[9]

Hematomas—extravasation of maternal blood—are characteristic of the carnivore placenta. These zones are specialized areas in which the chorion and maternal tissue are separated by a stagnant maternal blood. The dog and cat's marginal hematomas occur at the periphery of the zonary placenta (Fig. 3). The maternal epithelium degenerates, causing bleeding into the spaces surrounded by the labyrinth that results in the appearance of marginal hematomas.

The interhemal interface is endotheliochorial. Two distinct layers of syncytiotrophoblasts on the top of cytotrophoblasts are found, as is the case of the placenta of humans. This is probably due to the common need for endometrial invasion in the hemochorial human and endotheliochorial carnivore placentae. The cytotrophoblasts are found adjacent to the chorionic mesoderm, and the syncytiotrophoblast exists superficial to the cytotrophoblast and covers the maternal capillaries. The syncytiotrophoblasts cause loss of the maternal tissue during placentation and parturition, thus classifying the placenta of carnivores as a deciduate type. Transitional trophoblasts have also been reported.[15] A crosscurrent system of the fetal and maternal blood streams was reported in cats.[24]

Folding of the somatopleure results in amniogenesis, but fusion of the chorioamniotic folds dorsally is followed by complete separation of the amnion and chorion, so that the fetus is covered with amnion during gestation and at birth.

6. PLACENTA OF RUMINANTS

The goat will be used as a representative ruminant, keeping in mind that there are some differences in developmental timing and structure among ruminant species. According to the conventional placental classification schemes, the goat placenta is regarded as chorioallantoic. The yolk sac is functional for a short period of time. The mesamnion persists, so the fetus is born uncovered with the amnion in ruminants.

Morphologically, the placenta of the goat is cotyledonary and villous. The areas of villus formation are discrete specialized sites: formation of villi occurs only on areas in apposition to the preexisting uterine caruncles, which are normal features of the nonpregnant uterus. The villous processes extend into crypts that develop in proliferating caruncles; as the villi develop into a tree-like structure by secondary and tertiary branching, the adjacent endometrium on the caruncle undergoes hypertrophy and grows around the villi. The result is a more or less complete interdigitation of branched fetal villi within the walls of the maternal crypts. These specialized areas of fetal tissues are termed cotyledons; a cotyledon and its caruncle together form the functional unit of the ruminant placenta, the *placentome*. Some of the placentomes that form at the level of the dorsum of the fetus are structurally chorioamniotic, but still they are vascularized by the allantoic vessels. Placentomes have a convex surface in cattle and a concave surface in sheep and goats. The center of the concavity tends to be wider in goats than in sheep (Fig. 6).

The placenta of ruminants demonstrates some accessory placental structures. Sometimes calculi can be found floating in the amniotic or allantoic fluid. They are protein and calcium oxalate in nature,[3] consisting of a

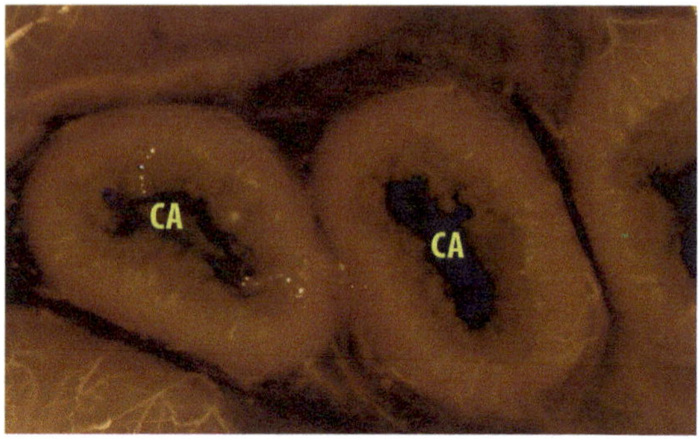

Fig. 6 A caruncle from a pregnant doe; note the wide center of its concave surface.

nucleus of cellular debris surrounded by deposits of mucoprotein, calcium, and phosphate. Ischemic zones or necrotic tips occur at the poles of fetal membranes.

Areolae are domes of trophoblasts covering the openings of the uterine glands. The allantochorion is not attached to the endometrial epithelium at the areolae. The columnar cytotrophoblasts forming the top layer of the domes are highly phagocytic and absorb the "uterine milk" secreted by the glands. Areolae are found in sheep, horse, and pig placentae.

Amniotic plaques are localized accumulations of amniotic epithelium that appear on the inner surface of the amnion and its reflection on the umbilical cord. They are rich in glycogen and found especially in ruminants and the horse.[1] The interhemal barrier is classified as *synepitheliochorial*[21] because of the fusion of the binucleate trophoblasts with the uterine epithelium. However, loss of the maternal epithelium has been observed toward the end of pregnancy in goats.[14]

The general arrangement of the maternofetal barrier can be seen in Fig. 7. This is most easily grasped at 7 weeks of pregnancy,[14] when the architecture of regions of the maternal side has not yet been distorted by the invasion of the fetal villi. The typical organization shows two populations of fetal epithelial cells: cytotrophoblasts and binucleate trophoblasts, the so-called giant cells. These overlie a core of mucous connective tissue (Wharton's jelly) and are served by fetal blood vessels within the connective tissue. Giant cells are characteristic large cells with two nuclei (Fig. 8), originating from cytotrophoblasts. Binucleated giant cells constitute about 20% of the fetal placenta, and they are the chief secretory source of placental lactogen. Also these cells secrete pregnancy-specific protein B, a protein of pregnancy unique to ruminants. They are also the sites for progesterone and estrogen synthesis.[32] Fig. 8 shows immunoreactivity of fetal and maternal tissues with

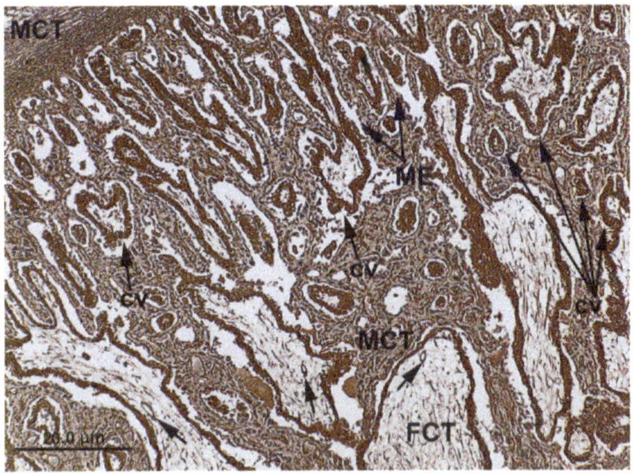

Fig. 7 Light microscopic sections of a placentome at 7 weeks of gestation showing the general organization of the maternofetal barrier. Maternal epithelial tissue (ME) can be seen adjacent to the chorionic villi (CV). The maternal connective tissue (MCT) can be seen in the *upper left corner* and surrounding the chorionic villi (CV). Maternal blood vessels could be found within the connective tissue (not shown in this figure. Chorionic villi (some are marked CV) are numerous in the image; they are covered with fetal epithelium, which is composed of cytotrophoblasts and binucleate trophoblasts. Fetal connective tissue (FCT) can be seen within the core of the villi. Fetal blood vessels can be seen within the connective tissue of the chorionic villi (*arrows*). Bar = 20 μm.

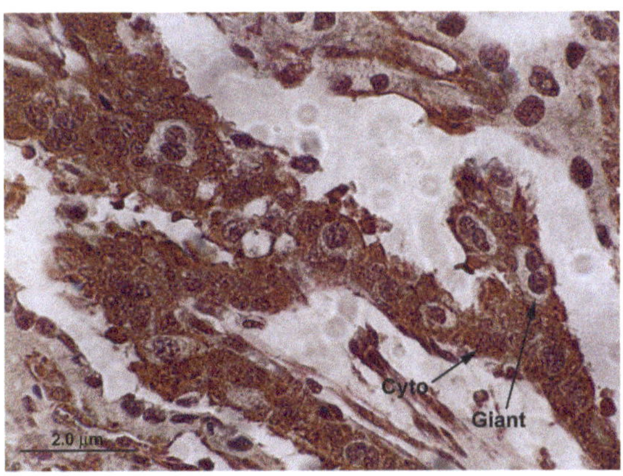

Fig. 8 High magnification image of a chorionic villus. The chorionic villus is covered with fetal epithelium, which is composed of cytotrophoblasts (Cyto) and binucleate trophoblasts (Giant). Bar = 2 μm.

antivascular endothelial growth factor antibody. Intense immunostaining can be seen in cytotrophoblasts of the fetal epithelium, but not in binucleate trophoblasts, facilitating visual distinction between these two populations of cells.

To resolve conflicting findings in a series of studies on the arrangement of the maternal and fetal blood vessels in small ruminants,[33–35] Hafez *et al.*, 2007[23] examined the system using scanning electron microscopy of both cut and intact caruncles and concluded that the goat possesses a multivillous-type placenta.[23] Goats develop a partially adeciduate status due to necrosis of maternal crypts.

7. PLACENTA OF THE HORSE

The blastocyst in the mare lies relatively unattached in the uterus until a comparatively late stage of development. The villi develop and attach to the uterine mucosa at about 10 weeks after fertilization.[36] The horse also has a villous-type placenta. The area of attachment and villous development is diffuse. The interhemal interface is epitheliochorial, and hence, this is an adeciduate placenta. The area of contact develops specialized branched villi, which fit into corresponding crypts on the endometrium. These clusters of fetal tissues, microcotyledons, are present over most of the placental surface, except in the areas of apposition with endometrial glands, a structure similar to the areolae of the ruminant and pig. The surfaces of these structures develop microvilli. The trophoblasts on the tips of the folds are relatively flat while those on the bases are columnar.

A choriovitelline placenta forms early in gestation and remains functioning for a longer period than in any other mammal. It remains functioning throughout the first quarter of gestation and is primary source of nutrients during that period. A large vessel, the sinus terminalis, forms circumferentially around the trilaminar omphalopleure. A band of elongated trophoblasts forms adjacent to the sinus terminalis at the boundary between the allantochorion and yolk sac; this is called the chorionic girdle. The band changes position as the allantochorionic attachment spreads and the yolk sac–chorion attachment recedes. Endometrial cups (source of eCG, Equine Chorionic Gonadotropin) are trophoblast-derived, gonadotropin-secreting cells, constituting hypertrophic endometrial glands in contact with the chorionic girdle. These cells exist only in the first half of gestation. They appear as gray tubular projections that are separate from each other.

The permanent placenta is chorioallantoic, forming and becoming functional after the transitory choriovitelline placenta. The allantoic blood vessels expand and invade the chorionic mesoderm, at first except for the area of the yolk sac attachment. Tsutsumi (1962)[36] demonstrated that the fetal blood stream runs in the opposite direction of the maternal blood stream. Amniotic plaques are present in the horse amnion. As in the case of carnivores, the foal is born covered with the amnion due to the loss of chorioamniotic raphe.

8. PLACENTA OF THE PIG

Attachment occurs at about 15 days after fertilization in the pig.[36] The choriovitelline placenta provides nourishment only early in pregnancy before establishment of the chorioallantoic placenta. The formation of the chorion and amnion is a result of bilateral folding of the extraembryonic somatopleure, as described earlier. The mesamnion persists, so the fetus is born uncovered by the amnion. The maternal and fetal tissues interdigitate over the entire surface of contact; hence, the pig possess a diffuse placenta.

Three zones develop on the chorionic surface, a placental zone in the center. In this region, transverse folds are formed which fit into similarly oriented grooves in the endometrium. Later, microvilli are developed on the surface of these folds, which further expand the surface area of contact. Also in this area, areolae are found. The peripheral paraplacental zones are smooth without folds or areolae. Distal areas exhibit ischemic zones or necrotic tips. The interhemal interface is epitheliochorial with no significant loss of maternal tissues, and hence, the pig placenta is adeciduate.

The maternal and fetal blood streams run in a mixture of one-way cross-current and countercurrent directions.[29] It is countercurrent because the maternal and fetal blood capillaries run opposite to each other, i.e., chorio-uterine vs utero-chorionic, respectively. This is a crosscurrent principle because some fetal blood flows in a one-way direction across the maternal flow.[29]

9. PLACENTAL VASCULATURE

Perhaps, one of the most important aspects of placental studies is the vasculature, since this component directly relates to the principal placental function (gas, nutrient, and waste exchange between the mother and fetus) and of course to the survival of the fetus to term.

9.1 Maternal Vasculature

9.1.1 Arterial Supply

In humans, the uterus is mainly supplied by the uterine artery (a. uterina), but a potential contribution from the ovarian artery may also be present.

The uterine artery is a branch of the internal iliac artery.[37] In ruminants, the uterus is supplied by three sources: (1) The uterine branch (r. uterinus) of the ovarian artery, which may arise in the form of several branches and anastomoses with cranial branches of the uterine artery; (2) The uterine artery (a. uterina), which arises with the umbilical artery as a common trunk off the internal iliac artery, at least in the goat. In the cow and ewe, it arises as the first vessel emerging from the umbilical artery. The latter arises from the internal iliac artery. The uterine artery runs through the broad ligament toward the uterus; upon reaching the mesometrium, it divides into several consecutive branches, which supply the uterine wall through a series of branches that run transversely over the dorsal and ventral aspects of the uterus. Cranial branches of the uterine artery anastomose with divisions of the uterine branch of the ovarian artery. Caudal branches of the uterine artery anastomose with the divisions of the uterine branch of the vaginal artery. There is a marked increase in size of the uterine artery during pregnancy. The fremitus of the uterine artery can be palpated rectally in the cow during gestation; 3. The uterine branch (r. uterinus) of the vaginal artery, supplies the horns through anastomoses with the uterine artery. Sometimes in the ewe, an additional uterine branch arises off the umbilical artery to supply the uterus. In the horse, the uterine artery originates from the external iliac artery. In the dog, the uterine branch of the vaginal artery is the designated uterine artery, which, together with the uterine branch of the ovarian artery, supplies blood to the uterus of the bitch. In all species many branches of the two uterine arteries anastomose on the dorsal/posterior surface of the uterus. For details about branching patterns in domestic animals, see schummer *et al.*, 1981.[38]

Our laboratory has provided a full description of the branching of the uterine arteries in the goat.[39] Primary and/or secondary branches of the caudal and cranial branches of the uterine arteries give rise to arcuate arteries, which form an arch that follows the contour of the lesser curvature of the uterus. Radial arteries arise (or radiate, hence the name) from arcuate arteries. These arteries are longer than the areas of the uterus through which they travel; therefore, they follow a helical course. As gestation advances and the size of the uterus increases, these arteries are drawn out straight. Each radial artery can supply more than one caruncle, and individual caruncles can be supplied by more than one radial artery. In contrast, spiral arteries in humans are branches of radial arteries and terminate in sinuses bringing blood into the intervillous space.[40]

Microvascular corrosion casting of the goat caruncle during pregnancy showed an increased complexity of vasculature with advancing gestation[17]; at 4 weeks, the surface showed a pattern of ridges separated by troughs. At later stages, branches of radial arteries penetrated the periphery forming an extensive mesh of capillaries on the concave surface. Capillary diameters increased significantly during pregnancy, especially after 4 weeks, when large flattened sinusoids formed (Fig. 9). Capillary sinusoids of irregular form and diameter were observed on the fetal surface of the caruncle at all stages in the goat and in sheep.[17,23] These sinusoids may reduce blood flow resistance and subsequently increase transplacental exchange capacity compensating for any negative consequences of the placental architecture.

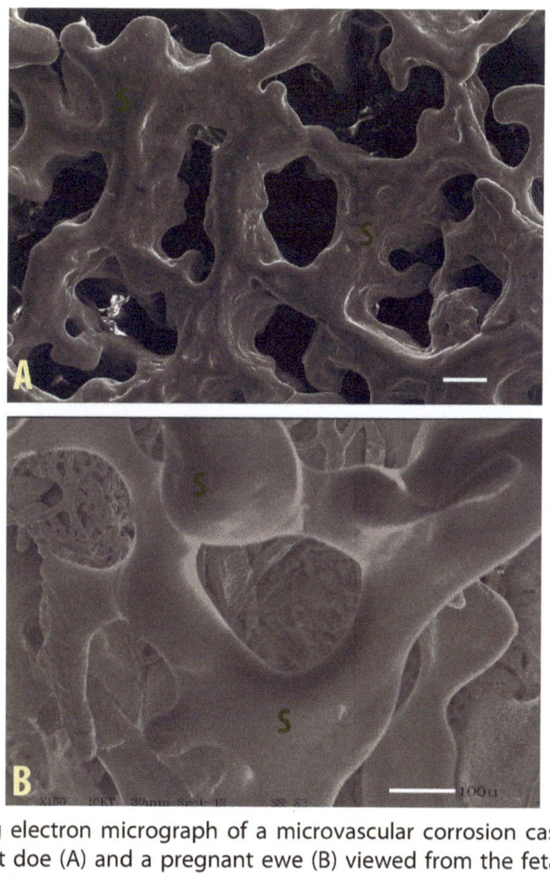

Fig. 9 Scanning electron micrograph of a microvascular corrosion cast of a caruncle from a pregnant doe (A) and a pregnant ewe (B) viewed from the fetal side showing capillary sinusoids (S) of large diameter. The sinusoids almost fill the entire image. Bar = 500 μm in (A) and 100 μm in (B).

9.1.2 Venous Drainage

In humans, the uterus is drained via the uterine vein. The uterine veins form a uterine venous plexus on each side of the cervix. Veins from the uterine plexus drain into the internal iliac veins.[37] In ruminants, blood drainage from the uterus occurs through: (1) The uterine branch of the ovarian vein (r. uterinus), which is its largest branch to the uterine horns, where it anastomoses with uterine vein. In the cow, the uterine vein which arises from internal iliac vein may or may not be present; however, in the ewe and doe, it is absent in most cases; (2) The uterine branch of the vaginal vein (r. uterinus). In the cow, in addition to the earlier mentioned routes, the uterus is drained by the accessory vaginal vein (v. vaginalis accessoria) which arises from internal iliac vein between the uterine and vaginal veins and caudal to the cranial gluteal vein.

The uterus of the mare is drained by the uterine branch of the ovarian vein and the uterine branch of vaginal vein. The uterine vein is the least important component, arising from the external iliac vein. In bitches, venous drainage occurs via the uterine branch of the ovarian vein and the uterine vein which is a branch of the vaginal vein.

9.2 Fetal Vasculature

In most species, two arteries and one vein pass through the umbilical cord. The vein bifurcates later in the amniotic portion of the umbilical cord. The umbilical arteries are ventral branches of the caudal dorsal aortae. The umbilical veins carry blood from the placenta to the caudal (inferior) vena cava via the ductus venosus. Several branches arise from the umbilical vessels and disperse radially on the allantoic side of the allantochorionic membrane. These branches branch further to be distributed to chorionic villi. Kaufmann *et al.* (1988)[27] discussed details of the fetal vascular system in human.

A microvascular corrosion cast of the cotyledon in sheep[23] revealed a cup-shaped structure with wide and narrow sides. Cotyledonary vessels enter and leave the cotyledon from the narrow side. A cotyledonary artery gives proximal collateral branches immediately after entering the cotyledon and then further branches to supply the remaining portion of the cotyledon. Vessel branches break into a mesh of capillaries forming the fetal vascular villi. Fetal villi that were nearest to the center of the cotyledon were the longest. Capillaries forming villi were in the form of a web-like mesh, were irregular in size, and had sinusoidal dilations (Fig. 10). The vascularity increase with advancing pregnancy and branching of the fetal villi become

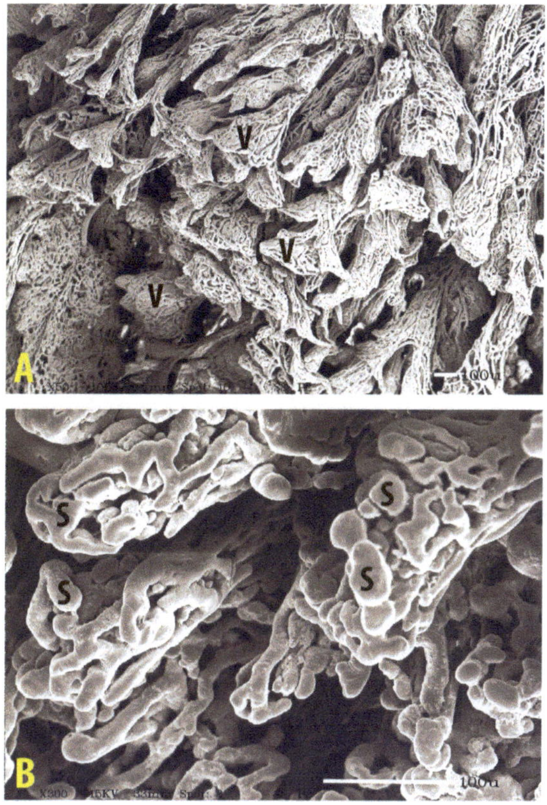

Fig. 10 Scanning electron micrograph of a microvascular corrosion cast of a cotyledon from a pregnant ewe viewed from the maternal side showing in (A) extensive branching of fetal villi. The entire image is filled with fetal villi; some are labelled "V." Panel (B) is a higher magnification image showing the sinusoidal dilations on the surface of the fetal villi; some are labelled "S." Scale bar = 100 μm.

more abundant. This extensive branching presumably allows a higher degree of invasion and surface contact to maternal tissues. Branching of fetal villi influences the elaboration of maternal crypts in all stages of gestation. However, correspondence between crypts and villi is restricted to distal portions of fetal villi.

10. SUMMARY

This chapter presents a merger of classic general classifications of the placenta and some species-specific anatomical features. The placenta has been the subject of extensive research in many species and will continue be an important research topic owing to its complex structure and function. Continuous development and remodeling of the placenta occurs simultaneously with the growth and remodeling of the fetus. This adds to the placenta's complexity and fascination.

The placenta has distinctive characteristics when compared across species, but no matter how different the morphology of the placenta is from one species to another, it still does what it is supposed to do. In all species, the placenta is equipped with a large surface area for exchange of myriad molecules, and its vascular system has structural refinements facilitating effective transfer of nutrients and gases. Elucidating the origin of placental structural defects is vital not only for understanding the pathophysiology of impaired fetal growth and the resulting weakened newborn in humans and animals but also to facilitate the choice of suitable animal models for placental research.

REFERENCES

[1] Ramsey E. *The Placenta of Laboratory Animals and Man*. New York: Holt, Rinehart and Winston, Inc.; 1975.
[2] Beck F. Comparative placental morphology and function. *Environ Health Perspect*. 1976;18:5–12.
[3] Noden D, De Lahunta A. *The Embryology of Domestic Animals Developmental Mechanism and Malformation*. Baltimore, MD: Williams & Wilkins; 1985.
[4] Lala PK, Chatterjee-Hasrouni S, Kearns M, Montgomery B, Colavincenzo V. Immunobiology of the feto-maternal interface. *Immunol Rev*. 1983;75:87–116.
[5] Wooding P, Burton GJ. *Comparative Placentation: Structures, Functions and Evolution*. Verlag Berlin Heidelberg: Springer; 2008.
[6] Steven D. Anatomy of the placental barrier. In: Steven D, ed. *Comparative Placentation. Essays in Structure and Function*. London, UK/New York: Academic Press; 1975. xiii + 315 p.
[7] Perry JS. The mammalian fetal membranes. *J Reprod Fertil*. 1981;62:321–335.
[8] Mossman HW. *Vertebrate Fetal Membranes: Comparative Ontogeny and Morphology; Evolution; Phylogenetic Significance; Basic Functions; Research Opportunities*. London: Rutgers University Press; 1987.
[9] Leiser R, Kaufmann P. Placental structure: in a comparative aspect. *Exp Clin Endocrinol*. 1994;102:122–134.
[10] Amoroso EC. Placentation. In: Parkes AS, ed. *Marshall's Physiology of Reproduction*. 3rd ed. London: Longmans Green; 1952:127–316.
[11] Kaufmann P, Burton G. Anatomy and genesis of the placenta. In: Knobil E, Neill JD, eds. *The Physiology of Reproduction*. 2nd ed. New York, NY: Raven Press Ltd.; 1994:441–484.
[12] Friess AE, Sinowatz F, Skolek-Winnisch R, Traautner W. The placenta of the pig. I. Finestructural changes of the placental barrier during pregnancy. *Anat Embryol (Berl)*. 1980;158:179–191.
[13] Samuel CA, Allen WR, Steven DH. Studies on the equine placenta. I. Development of the microcotyledons. *J Reprod Fertil*. 1974;41:441–445.
[14] Hafez SA. *Advanced Studies in Veterinary Anatomy: Angiogenesis in Caprine Reproductive Organs of Non-Pregnant and Pregnant Normal and Swainsonine-Treated Does*. Virginia:

Virginia Polytechnic Institute and State University; 2005. http://theses.lib.vt.edu/theses/available/etd-04212005092936/unrestricted/Dissertation.pdf.
[15] Wynn RM, Corbett JR. Ultrastructure of the canine placenta and amnion. *Am J Obstet Gynecol.* 1969;103:878–887.
[16] Shanklin DR. Anatomy of the placenta in human growth. In: Falkner F, Tanner JM, eds. *Human Growth A Comprehensive Treatise: Developmental Biology and Prenatal Growth.* New Yok, NY: Plenum Press; 1986:199–219.
[17] Hafez SA, Caceci T, Freeman LE, Panter KE. Angiogenesis in the caprine caruncles in non-pregnant and pregnant normal and swainsonine-treated does. *Anat Rec (Hoboken).* 2007;290:761–769.
[18] Kaufmann P, Davidoff M. The guinea-pig placenta. *Adv Anat Embryol Cell Biol.* 1977;53:5–91.
[19] Furukawa S, Kuroda Y, Sugiyama A. A comparison of the histological structure of the placenta in experimental animals. *J Toxicol Pathol.* 2014;27:11–18.
[20] Grosser O. *Vergleichende Anatomie und Entwicklungsgeschichte der Eihäute und der Placenta.* Vienna: W. Braumüller; 1909.
[21] Wooding FBP. Current topic: the synepitheliochorial placenta of ruminants: binucleated cell fusions and hormone production. *Placenta.* 1992;13:101–113.
[22] Faber JJ, Thornburg KL, Nancy DB. Physiology of placental transfer in mammals. *Am Zool.* 1992;32:343–354.
[23] Hafez SA, Borowicz P, Reynolds LP, Redmer DA. Maternal and fetal microvasculature in sheep placenta at several stages of gestation. *J Anat.* 2010;216:292–300.
[24] Dantzer V, Leiser R, Kaufmann P, Luckhardt M. Comparative morphological aspects of placental vascularization. In: Kaufmann P, Miller RK, Boston MA, eds. *Placental Vascularization and Blood Flow: Basic Research and Clinical Applications.* USA: Springer; 1988:235–260.
[25] Leiser R, Koob B. Structural and functional aspects of placental microvasculature studied from corrosion casts. In: Motta PM, Murakami T, Fujita H, eds. *Scanning Electron Microscopy of Vascular Casts: Methods and Applications.* Boston, MA: Springer; 1992:261–277.
[26] Bailey DJ. Proceedings: counter-current flow of maternal and foetal blood-streams in guinea-pig placenta. *J Physiol.* 1974;242:104–105.
[27] Kaufmann P, Luckhardt M, Leiser R. Three-dimensional representation of the fetal vessel system in the human placenta. In: Kaufmann P, Miller RK, eds. *Placental Vascularization and Blood Flow: Basic Research and Clinical Applications.* Boston, MA: Springer; 1988:113–137.
[28] Leiser R, Krebs C, Ebert B, Dantzer V. Placental vascular corrosion cast studies: a comparison between ruminants and humans. *Microsc Res Tech.* 1997;38:76–87.
[29] Leiser R, Dantzer V. Structural and functional aspects of porcine placental microvasculature. *Anat Embryol (Berl).* 1988;177:409–419.
[30] Verma U, Verma N. An overview of development, function and diseases of the placenta. In: Nicholson R, ed. *The Placenta: Development, Function and Diseases.* New York: Nova Science Publishers, Inc.; 2013:1–30.
[31] Anderson JW. Ultrastructure of the placenta and fetal membranes of the dog I. The placental labyrinth. *Anat Rec (Hoboken).* 1969;165:15–35.
[32] Senger PL. *Early embryogenesis and maternal recognition of pregnancy. Pathways to Pregnancy and Parturition.* 3rd ed. Washington: Current Conceptions, Inc.; 2015.
[33] Barcroft J, Barron DH. Observation upon the form and relations of the maternal and fetal vessels in the placenta of the sheep. *Anat Rec (Hoboken).* 1946;94:569–595.
[34] Makowski EL. Maternal and fetal vascular nets in placentas of sheep and goats. *Am J Obstet Gynecol.* 1968;100:283–288.
[35] Leiser R. Mikrovasckularisation der ziegenplazenta, dargestellt mit rasterelektronisch untersuchten gefassausgussen. *Schweiz Arch Tierheilkd.* 1987;129:59–74.

[36] Tsutsumi Y. The vascular pattern of the placenta in farm animals. *J Facul Agr Hokkaido Univ Sapporo*. 1962;52:372–482.
[37] Moore KL, Dalley AF. *Agur AMR. Pelvis and perineum. Moore Clinically Oriented Anatomy*. 7th ed. Philadelphia, PA: Wolters Kluwer/Lippincott Williams & Wilkins; 2014382–399.
[38] Schummer A, Wilkens H, Vollmerhaus B, Habermehl K. *The Circulatory System, the Skin, and the Cutaneous Organs of the Domestic Mammals*. Berlin & Hamburg: Verlag Paul Parey; 1981.
[39] Hafez SA, Freeman LE, Caceci T, Smith BJ. Study of the vasculature of the caprine reproductive organs using the tissue-clearing technique, with special reference to the angioarchitecture of the utero-ovarian vessels and the adaptation of the ovarian and/or vaginal arteries to multiple pregnancies. *Anat Rec (Hoboken)*. 2007;290:389–405.
[40] Ramsey EM, Donner MW. Placental vasculature and circulation. In: Kaufmann P, Miller RK, eds. *Placental Vascularization and Blood Flow: Basic Research and Clinical Applications*. Boston, MA: Springer; 1988:217–233.

CHAPTER TWO

Cell- and Tissue-Based Models for Study of Placental Development

W.R. Huckle[1]
Virginia-Maryland College of Veterinary Medicine, Virginia Polytechnic Institute & State University, Blacksburg, VA, United States
[1]Corresponding author e-mail address: wrhuckle@vt.edu

Contents

Abstract

Decades of research into the molecular mechanisms by which the placenta forms and functions have sought to improve prevention, diagnosis, and management of disorders of this vital tissue. This research has included development of experimental models intended to replicate behavior of the native placenta in both health and disease. Animal models devised in rodents, sheep, cattle, or other domestic animal species have the advantage of being biologically "complete," but all differ to some degree in developmental timing and anatomical details compared to the human, suggesting subtle differences in molecular mechanism. Consequently, investigators have resorted to simplified systems, characterizing the mechanisms of placental development by using explants of maternal and fetal tissue, primary cell cultures, and immortalized or choriocarcinoma-derived cell lines. Such studies have advanced our understanding of mechanisms by which trophoblasts and associated tissues invade the endometrium, produce chorionic gonadotropin, manage immune tolerance of the fetus, or elaborate proteins that may contribute to placental dysfunction. More recently, use of three-dimensional spheroid cultures, computational modeling of placental tissue dynamics and blood flow, and bioengineering of tissue constructs have been undertaken, aimed to recapitulate the types of interactions that occur among diverse uterine and placental cell types in utero. New technologies and biological paradigms, stemming in part from the ongoing Human Placenta Project, promise to expand the array of available tools, increasing the likelihood that the years ahead will see significant improvements in the ability to prevent, diagnose, and treat life-threatening disorders of placental formation and function.

Progress in Molecular Biology and Translational Science, Volume 145
ISSN 1877-1173
http://dx.doi.org/10.1016/bs.pmbts.2016.12.002

1. INTRODUCTION

The formation and function of the placenta, as is true for all aspects of the human reproductive process, are subjects of great fascination not only for the remarkable place they occupy in biology but also for their early and lasting impact on human health. This volume focuses on research into placental development that may translate to improved human therapies; accordingly, the discussion here is confined to those systems of study that may inform prevention, diagnosis, or management of disorders of this fleeting but vital tissue.[1]

A broad spectrum of experimental approaches—from the whole maternal/fetal pair down to the single cell—has been taken to investigate placental biology and disease. Each approach has yielded important findings. What would be the properties of an ideal experimental model for human placental dysfunction? To ask the question is a virtual admission that so such model exists, but, in order best to evaluate those at our disposal, a reminder of the desirable features is worthwhile. Topping the list for animal models would be considerations of similarities in etiology, pathophysiology, and predictive value for identifying effective therapies. Where disorders paralleling those in humans do not occur spontaneously at a significant frequency in animals, it is possible to devise models by surgical intervention, dietary restriction, environmental stress, pharmacologic manipulation, or, in the present era, creation of animals with precise gene modifications useful for testing the involvement of particular molecules and pathways in disease susceptibility or resistance.[2–4]

As noted in "Comparative Placental Anatomy: Divergent Structures Serving a Common Purpose" by Hafez in this volume, the placenta across species serves the same fundamental functions, but the tissues in different species reach their mature, anatomically distinct states via routes whose commonality of mechanism is often more assumed than confirmed. Nevertheless, models developed in large animal species such as sheep have facilitated investigation of pregnancy-induced hypertension,[5] intrauterine growth restriction,[6,7] and gestational diabetes.[8] In rodents, examples of creatively devised models of preeclampsia abound, including an inbred mouse strain that spontaneously develops pregnancy-associated hypertension,[9] elicitation of preeclampsia-like features in mice of specific strain crosses[10] or following immune stimulation,[11] and rats subjected to stressors that mimic overcrowding[12] or gene-modified to express human angiotensinogen and renin.[13] While views vary on whether preeclampsia occurs spontaneously in nonhuman primates, baboons have been used to explore potential immune-induced mechanisms of pregnancy-associated hypertension.[14]

Alongside issues of experimental animal model fidelity to human disease, there lies the practical consideration of costs—to purchase, house, and husband animals in numbers adequate to comprise a robust experimental design and to justify their use ethically, especially where choosing primate models may be attractive owing to biological comparability to humans. Study of the placenta in nonhuman, nonrodent models, including some of the domestic species described later, is further complicated by a relative paucity of thoroughly annotated genomes and limitations in availability of cross-reacting immunological reagents. Historically in this field, as in others, investigators have overcome some of these pragmatic obstacles to great effect by resorting to simplified systems, characterizing the molecular mechanisms of placental development by using explants of maternal and fetal tissue, primary cell cultures, and immortalized cell lines. Newer technologies are now enabling a swing back toward the complexity inherent in the intact organism, including use of three-dimensional spheroid cultures and bioengineered tissue models intended to recapitulate the wealth of interactions that occur among diverse uterine and placental cell types in utero.

2. SIMPLIFIED EX VIVO MODELS

2.1 Tissue Explants and Monolayer Cell Cultures

The use of tissue fragments recovered from term human placentas to investigate the structure and behavior of syncytiotrophoblasts was described in a series of studies by Carr *et al.* in the mid-1960s.[15–17] Since then, explant culture has been applied variously to study ultrastructure of placental tissue from preterm fetal deaths[18] and steroid hormone effects on decidual tissue[19] among numerous other phenomena related to placental development[20] or effects of toxic substances on the fetal-placental unit.[21] Such studies have advanced our understanding of mechanisms by which trophoblasts and associated tissues invade the endometrium,[22] produce chorionic gonadotropin,[23] or elaborate proteins such as sFlt-1 that may contribute to the onset of preeclampsia.[24]

Where the welfare of human patients is at stake, availability of tissue must be governed by overriding ethical considerations and with consideration of the invasive nature of tissue procurement; thus, findings using tissue specimens recovered at gestational term are most highly represented in the literature. Nevertheless, a number of studies have accomplished explant studies of trophoblast outgrowth and invasion using tissue obtained preterm.[25–27] Cryopreservation of placental villous explants offers a possible mitigation of the problem of securing tissue whose availability is limited and often unpredictable.[28]

The appeal of explant culture for the study of placental development and physiology stems from its potential to retain and exhibit ex vivo those properties characteristic of the dynamic relationship between tissue of fetal and maternal origin. However, as in all realms of biomedical research, there has been a simultaneous drive to develop homogeneous cell culture systems representing the major differentiated phenotypes that constitute the developing and mature placenta. Efforts in this direction have yielded primary cultures of trophoblastic cells from human,[29] rat,[30] mouse,[31] and bovine[32] placenta. To overcome limitations imposed by cell senescence or phenotypic drift upon propagation in culture, intentional immortalization of primary trophoblastic cells from a variety of species has been undertaken, for example, using expression of SV40 T antigen[33] or telomerase[34] as immortalizing agents. Finally, spontaneously transformed cells, notably, the BeWo,[29] Jar,[35] and JEG-3[36] lines, have been isolated from human gestational choriocarcinomas and characterized for retention of trophoblast-like properties. These cells, together with the more recently derived first trimester extravillous trophoblast cell line SGHPL-4,[37] have been employed subsequently in hundreds of studies described in the literature, greatly enhancing our appreciation of trophoblast migration, endometrial invasion and vascular remodeling, placental metabolite transport, endocrine function and responsiveness, and immune tolerance.[38]

2.2 Spheroids and Bioengineered Tissue Aggregates

Monolayer cell cultures, as relatively homogeneous systems whose environment and treatment conditions can be readily controlled, provide many advantages in the design of experiments. At the same time, their homogeneity presents a severe limitation for the study of placenta, where multiple cell genotypes and phenotypes coexist in close proximity and must interact in a highly coordinated fashion to support a healthy gestation. In an effort to restore a three-dimensional relationships for placental cells to experience in culture, investigators have prepared nonadherent "spheroid" cultures of cytotrophoblasts[39,40] or uterine endometrium[41] in order to model in vitro trophoblast invasion using tissue from normal or preeclamptic pregnancies.[42,43] Three-dimensional cultures of trophoblast cells have shown a capacity to develop spaces that resemble vasculature with an interior lining of trophoblast giant cells, mimicking to a degree events occurring during placental development in vivo.[44]

More recent efforts to create ex vivo experimental settings in which to reproduce more faithfully the biological complexity of the placenta have involved a combination of computational modeling and tissue engineering.[45–48] For example, Clark *et al.*[49] have devised mathematical models describing formation of chorionic and villous vessels as well as placental shape. (It should be noted that, although current work is enabled by advances in computational processing speed and modeling principles, their successful application to understanding placental development is built upon years of meticulous morphometric analysis.[50]) Information gained from such modeling exercises, together with long-standing knowledge of the gross and microanatomical cellular relationships, has led to the advent of microfluidic[51] or bioprinted multicellular assemblies[52] meant to better represent the dynamics of placental development and function.

3. CONCLUSIONS AND FUTURE PROSPECTS

Greater than 50 years of research into placental biology and pathophysiology have generated not only a wealth of knowledge that has advanced maternal and fetal/neonatal health, but also a wide array of experimental perspectives, reagents, animal models, and frameworks of data analysis ensuring that new knowledge will continue to accrue. Modern technologies and biological paradigms—some novel, some already known but likely to come of age in light of the priorities of the Human Placenta Project[53]—will markedly expand the pool of available tools. These include the ability to conduct transcriptional profiling in single cells recovered from among the diverse players at the maternal–fetal interface,[54] to investigate involvement of regulated DNA methylation in placental development,[55] and to thoroughly evaluate the roles of microRNAs in early trophoblast invasion[56] or in dysfunctional angiogenesis.[57] The great sensitivity of these analytical techniques suggests that, once new markers of important placenta-related disorders are validated preclinically and beyond, minimally invasive biopsy procedures may allow safe and highly targeted diagnosis. Moreover, refined application of modern imaging modalities, such as magnetic resonance[58] and ultrasound,[59] to monitor placental health represents powerful noninvasive diagnostic possibilities. Finally, the convergence of expanded knowledge of placental biology with advancing sophistication of semisynthetic tissue engineering technologies increases the likelihood that artificial placentas, envisioned to support the survival of extremely preterm infants, may be realized.[60] Thus, there is every reason to hope that the years ahead will see significant improvements in the ability to prevent, diagnose, and treat preeclampsia, gestational diabetes, intrauterine growth restriction, prematurity, and other life-threatening disorders of placental formation and function.

REFERENCES

1. Benirschke K. Needs for animal models of human diseases of the reproductive system. *Am J Pathol.* 1980;101(3 suppl):S229–S239.
2. Miano JM, Zhu QM, Lowenstein CJ. A CRISPR path to engineering new genetic mouse models for cardiovascular research. *Arterioscler Thromb Vasc Biol.* 2016;36(6):1058–1075.
3. Reynolds LP, Borowicz PP, Vonnahme KA, et al. Animal models of placental angiogenesis. *Placenta.* 2005;26(10):689–708.
4. Pereira FT, Oliveira LJ, Barreto Rda S, et al. Fetal-maternal interactions in the synepitheliochorial placenta using the eGFP cloned cattle model. *PLoS One.* 2013;8(5): e64399.
5. Thatcher CD, Keith Jr JC. Pregnancy-induced hypertension: development of a model in the pregnant sheep. *Am J Obstet Gynecol.* 1986;155(1):201–207.
6. Creasy RK, Barrett CT, de Swiet M, Kahanpaa KV, Rudolph AM. Experimental intrauterine growth retardation in the sheep. *Am J Obstet Gynecol.* 1972;112(4):566–573.
7. Cheung CY, Bogic L, Gagnon R, Harding R, Brace RA. Morphologic alterations in ovine placenta and fetal liver following induced severe placental insufficiency. *J Soc Gynecol Investig.* 2004;11(8):521–528.
8. Dickinson JE, Meyer BA, Chmielowiec S, Palmer SM. Streptozocin-induced diabetes mellitus in the pregnant ewe. *Am J Obstet Gynecol.* 1991;165(6 pt 1):1673–1677.
9. Davisson RL, Hoffmann DS, Butz GM, et al. Discovery of a spontaneous genetic mouse model of preeclampsia. *Hypertension.* 2002;39(2 pt 2):337–342.
10. Ahmed A, Singh J, Khan Y, Seshan SV, Girardi G. A new mouse model to explore therapies for preeclampsia. *PLoS One.* 2010;5(10): e13663.
11. Schmid M, Sollwedel A, Thuere C, et al. Murine pre-eclampsia induced by unspecific activation of the immune system correlates with alterations in the eNOS and AT1 receptor expression in the kidneys and placenta. *Placenta.* 2007;28(7):688–700.
12. Takiuti NH, Kahhale S, Zugaib M. Stress in pregnancy: a new Wistar rat model for human preeclampsia. *Am J Obstet Gynecol.* 2002;186(3):544–550.
13. Geusens N, Verlohren S, Luyten C, et al. Endovascular trophoblast invasion, spiral artery remodelling and uteroplacental haemodynamics in a transgenic rat model of pre-eclampsia. *Placenta.* 2008;29(7):614–623.
14. Sunderland N, Hennessy A, Makris A. Animal models of pre-eclampsia. *Am J Reprod Immunol.* 2011;65(6):533–541.
15. Carr MC. Human term placental villi in explant tissue culture. 1. Behavior. *Am J Obstet Gynecol.* 1964;88:584–591.
16. Carr MC, Preininger M. Human term placental villi in explant tissue culture. II. Dissociation of syncytiolytic and fibrinolytic activities. *Am J Obstet Gynecol.* 1965;93:259–265.

17. Carr MC, Preininger M. Human term placental villi in explant tissue culture. 3. Comparison of the effects of chicken plasma, sheep plasma, vitamin A, and hydrocarotisone in syncytial dissolution. *Am J Obstet Gynecol.* 1967;97(2):252–256.
18. Hustin J, Gaspard U. Comparison of histological changes seen in placental tissue cultures and in placentae obtained after fetal death. *Br J Obstet Gynaecol.* 1977;84(3):210–215.
19. Daly DC, Maslar IA, Riddick DH. Term decidua response to estradiol and progesterone. *Am J Obstet Gynecol.* 1983;145(6):679–683.
20. Steinberg ML, Robins JC. Cellular models of trophoblast differentiation. *Semin Reprod Med.* 2016;34(1):50–56.
21. Gohner C, Svensson-Arvelund J, Pfarrer C, et al. The placenta in toxicology. Part IV: battery of toxicological test systems based on human placenta. *Toxicol Pathol.* 2014;42(2):345–351.
22. Newby D, Marks L, Cousins F, Duffie E, Lyall F. Villous explant culture: characterization and evaluation of a model to study trophoblast invasion. *Hypertens Pregnancy.* 2005;24(1):75–91.
23. Ahmed NA, Murphy BE. The effects of various hormones on human chorionic gonadotropin production in early and late placental explant cultures. *Am J Obstet Gynecol.* 1988;159(5):1220–1227.
24. Rajakumar A, Powers RW, Hubel CA, et al. Novel soluble Flt-1 isoforms in plasma and cultured placental explants from normotensive pregnant and preeclamptic women. *Placenta.* 2009;30(1):25–34.
25. Bauer S, Pollheimer J, Hartmann J, Husslein P, Aplin JD, Knofler M. Tumor necrosis factor-alpha inhibits trophoblast migration through elevation of plasminogen activator inhibitor-1 in first-trimester villous explant cultures. *J Clin Endocrinol Metab.* 2004;89(2):812–822.
26. Hu Y, Tan R, MacCalman CD, et al. IFN-gamma-mediated extravillous trophoblast outgrowth inhibition in first trimester explant culture: a role for insulin-like growth factors. *Mol Hum Reprod.* 2008;14(5):281–289.
27. James JL, Stone PR, Chamley LW. The effects of oxygen concentration and gestational age on extravillous trophoblast outgrowth in a human first trimester villous explant model. *Hum Reprod.* 2006;21(10):2699–2705.
28. Huppertz B, Kivity V, Sammar M, et al. Cryogenic and low temperature preservation of human placental villous explants—a new way to explore drugs in pregnancy disorders. *Placenta.* 2011;32(suppl):S65–S76.
29. Pattillo RA, Gey GO. The establishment of a cell line of human hormone-synthesizing trophoblastic cells in vitro. *Cancer Res.* 1968;28(7):1231–1236.
30. Faria TN, Soares MJ. Trophoblast cell differentiation: establishment, characterization, and modulation of a rat trophoblast cell line expressing members of the placental prolactin family. *Endocrinology.* 1991;129(6):2895–2906.
31. Tanaka S, Kunath T, Hadjantonakis AK, Nagy A, Rossant J. Promotion of trophoblast stem cell proliferation by FGF4. *Science.* 1998;282(5396):2072–2075.
32. Stringfellow DA, Gray BW, Lauerman LH, Thomson MS, Rhodes PJ, Bird RC. Monolayer culture of cells originating from a preimplantation bovine embryo. *In Vitro Cell Dev Biol.* 1987;23(11):750–754.
33. Lei KJ, Gluzman Y, Pan CJ, Chou JY. Immortalization of virus-free human placental cells that express tissue-specific functions. *Mol Endocrinol.* 1992;6(5):703–712.
34. Wang YL, Qiu W, Feng HC, et al. Immortalization of normal human cytotrophoblast cells by reconstitution of telomeric reverse transcriptase activity. *Mol Hum Reprod.* 2006;12(7):451–460.
35. Pattillo RA, Ruckert A, Hussa R, Bernstein R, Delfs E. The jar cell line—continuous human multi hormone production and controls. *In Vitro.* 1971;6:398–399.

36. Kohler PO, Bridson WE, Hammond JM, Weintraub B, Kirschner MA, Van Thiel DH. Clonal lines of human choriocarcinoma cells in culture. *Acta Endocrinol Suppl (Copenh)*. 1971;153:137–153.
37. Choy MY, Manyonda IT. The phagocytic activity of human first trimester extravillous trophoblast. *Hum Reprod*. 1998;13(10):2941–2949.
38. Orendi K, Kivity V, Sammar M, et al. Placental and trophoblastic in vitro models to study preventive and therapeutic agents for preeclampsia. *Placenta*. 2011;32(suppl): S49–S54.
39. White TE, Saltzman RA, Di Sant'Agnese PA, Keng PC, Sutherland RM, Miller RK. Human choriocarcinoma (JAr) cells grown as multicellular spheroids. *Placenta*. 1988;9(6):583–598.
40. John NJ, Linke M, Denker HW. Retinoic acid decreases attachment of JAR choriocarcinoma spheroids to a human endometrial cell monolayer in vitro. *Placenta*. 1993;14(1):13–24.
41. Yamauchi N, Yamada O, Takahashi T, et al. A three-dimensional cell culture model for bovine endometrium: regeneration of a multicellular spheroid using ascorbate. *Placenta*. 2003;24(2–3):258–269.
42. LaMarca HL, Ott CM, Honer Zu Bentrup K, et al. Three-dimensional growth of extravillous cytotrophoblasts promotes differentiation and invasion. *Placenta*. 2005;26(10):709–720.
43. Korff T, Krauss T, Augustin HG. Three-dimensional spheroidal culture of cytotrophoblast cells mimics the phenotype and differentiation of cytotrophoblasts from normal and preeclamptic pregnancies. *Exp Cell Res*. 2004;297(2):415–423.
44. Rai A, Cross JC. Three-dimensional cultures of trophoblast stem cells autonomously develop vascular-like spaces lined by trophoblast giant cells. *Dev Biol*. 2015;398(1):110–119.
45. Haeger JD, Hambruch N, Pfarrer C. The bovine placenta in vivo and in vitro. *Theriogenology*. 2016;86(1):306–312.
46. Lin M, Mauroy B, James JL, Tawhai MH, Clark AR. A multiscale model of placental oxygen exchange: the effect of villous tree structure on exchange efficiency. *J Theor Biol*. 2016;408:1–12.
47. Blundell C, Tess ER, Schanzer AS, et al. A microphysiological model of the human placental barrier. *Lab Chip*. 2016;16(16):3065–3073.
48. Panitchob N, Widdows KL, Crocker IP, et al. Computational modelling of amino acid exchange and facilitated transport in placental membrane vesicles. *J Theor Biol*. 2015;365:352–364.
49. Clark AR, Lin M, Tawhai M, Saghian R, James JL. Multiscale modelling of the feto-placental vasculature. *Interface Focus*. 2015;5(2):20140078.
50. Teasdale F. Idiopathic intrauterine growth retardation: histomorphometry of the human placenta. *Placenta*. 1984;5(1):83–92.
51. Lee JS, Romero R, Han YM, et al. Placenta-on-a-chip: a novel platform to study the biology of the human placenta. *J Matern Fetal Neonatal Med*. 2016;29(7):1046–1054.
52. Kuo C-Y, Eranki A, Placone JK, et al. Development of a 3D printed, bioengineered placenta model to evaluate the role of trophoblast migration in preeclampsia. *ACS Biomater Sci Eng*. 2016;2(10):1817–1826.
53. Guttmacher AE, Maddox YT, Spong CY. The human placenta project: placental structure, development, and function in real time. *Placenta*. 2014;35(5):303–304.
54. Nelson AC, Mould AW, Bikoff EK, Robertson EJ. Single-cell RNA-seq reveals cell type-specific transcriptional signatures at the maternal-foetal interface during pregnancy. *Nat Commun*. 2016;7:11414.
55. Bianco-Miotto T, Mayne BT, Buckberry S, Breen J, Rodriguez Lopez CM, Roberts CT. Recent progress towards understanding the role of DNA methylation in human placental development. *Reproduction*. 2016;152(1):R23–R30.

56. Xie L, Sadovsky Y. The function of miR-519d in cell migration, invasion, and proliferation suggests a role in early placentation. *Placenta*. 2016;48:34–37.
57. Escudero CA, Herlitz K, Troncoso F, et al. Role of extracellular vesicles and microRNAs on dysfunctional angiogenesis during preeclamptic pregnancies. *Front Physiol*. 2016;7:98.
58. Siauve N, Chalouhi GE, Deloison B, et al. Functional imaging of the human placenta with magnetic resonance. *Am J Obstet Gynecol*. 2015;213(4 suppl):S103–S114.
59. Zaidi SF, Moshiri M, Osman S, et al. Comprehensive imaging review of abnormalities of the placenta. *Ultrasound Q*. 2016;32(1):25–42.
60. Bird SD. Artificial placenta: analysis of recent progress. *Eur J Obstet Gynecol Reprod Biol*. 2016;208:61–70.

CHAPTER THREE

Transcription Factors That Regulate Trophoblast Development and Function

K.J. Baines*, S.J. Renaud*,†,1
*The University of Western Ontario, London, ON, Canada
†Children's Health Research Institute, The University of Western Ontario, London, ON, Canada
[1]Corresponding author: e-mail address: srenaud4@uwo.ca

Contents

Abstract

The placenta is a transient organ that plays a critical role in sustaining pregnancy and supporting fetal growth and nutrition. The placental epithelium is comprised of trophoblast cells. Trophoblast cells are the first cell type to differentiate during embryogenesis and ultimately diversify into a heterogeneous population of cells specializing in distinct functions essential for placentation. The emergence of the trophoblast lineage and subsequent specialization into distinct trophoblast sublineages is tightly regulated by transcription factors. This chapter will provide an overview of transcription factors that regulate trophoblast development and function. The chapter is divided into three sections. In the first section, a generalized outline of trophoblast ontogeny and a functional description of different trophoblast sublineages will be provided. In the second section, transcription factors involved in emergence of the trophoblast lineage and maintenance of trophoblast stem cells will be discussed. In the third section, transcription factors implicated in the formation and function of villous and extravillous cytotrophoblast lineages will be described.

Progress in Molecular Biology and Translational Science, Volume 145
ISSN 1877-1173
http://dx.doi.org/10.1016/bs.pmbts.2016.12.003

1. INTRODUCTION

The placenta is a remarkable but temporary organ that develops during pregnancy in most mammals. It attaches to the uterine lining and performs a plethora of functions that help sustain fetal growth and development. Most placental functions are performed by trophoblast cells, which are the epithelial cells of the placenta. Trophoblast cells commence as multipotent progenitor cells but ultimately differentiate into several distinct trophoblast sublineages that perform unique functions. The emergence of the trophoblast lineage, and subsequent differentiation into specialized trophoblast sublineages, is a complex process involving the integration of environmental signals that leads to changes in the expression and activity of transcription factors. The purpose of this chapter is to provide an overview of transcription factors implicated as regulators of trophoblast development and function. When possible, emphasis will be placed on those factors deemed important for human trophoblast development. However, since a role for many of these factors in trophoblast development was first identified through experiments using mouse mutagenesis, we will also discuss relevant transcription factors in the context of mouse studies and other animals when appropriate. Before we summarize transcription factors involved in trophoblast development, the different trophoblast sublineages comprising the placenta will be introduced.

1.1 Early Trophoblast Development

The union of a sperm and ovum produces a zygote. The zygote undergoes a series of rapid mitotic divisions as it travels through the oviduct to the uterine cavity. When the zygote reaches 12–16 cells (in humans: 3–4 days postfertilization), it forms a sphere referred to as a morula, named due to its resemblance to a mulberry (*Lat. morus; mulberry*). At this stage, individual cells are referred to as blastomeres. Due to spatial localization, some blastomeres become confined to the periphery of the morula, whereas others remain inside. Blastomeres located at the periphery of the morula undergo significant morphological changes, including compaction (cells become tightly bound together until almost indistinguishable from one another) and polarization (microvilli and plasma membrane components become restricted to the free surface).[1] The tight adhesion of cells on the periphery of the morula causes fluid to accumulate within the center. The fluid ultimately coalesces to form a blastocoel and at this stage the morula becomes a blastocyst (approximately 32–64 cell stage; in humans, this occurs approximately 4–5 days postfertilization). At the blastocyst stage, two distinct populations of cells can be identified. On the inside of the blastocyst is a pluripotent inner cell mass, from which all germ layers of the embryo arise, as well as the extraembryonic mesoderm, yolk sac, amnion, and allantois. The cells confined to the outside constitute the trophectoderm, which exclusively gives rise to trophoblast cells (Fig. 1A). Thus, the emergence

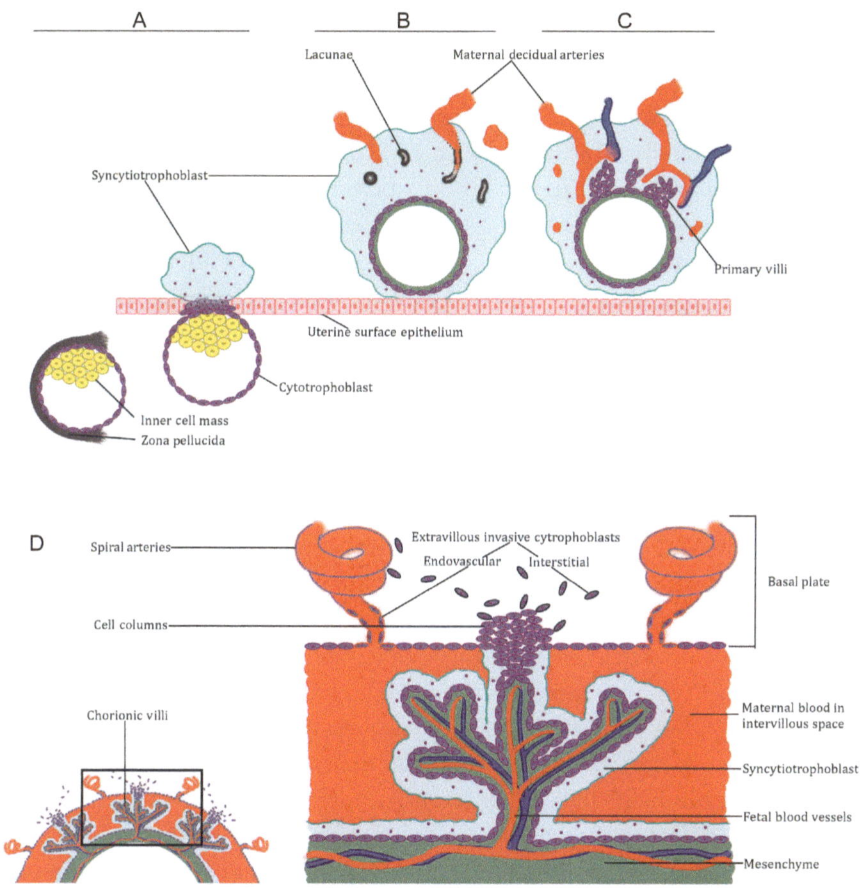

Fig. 1 Overview of human placental development. (A) Preimplantation and periimplantation blastocyst, showing both polar and mural trophectoderm, and the emergence of the primitive syncytiotrophoblast from the polar trophectoderm. (B) and (C) Postimplantation trophoblast development. Note the development of lacunae in the syncytiotrophoblast and the emergence of the primary villi. (D) Anatomy of the definitive placenta, showing both the villous trophoblast organized into chorionic villi, and the extravillous cytotrophoblast cells organized into trophoblast cell columns.

of the trophectoderm is the first lineage commitment event during embryogenesis. The exact stage at which the trophoblast lineage is fully committed varies among species and is still subject to debate.[2]

Once the developing blastocyst hatches from the zona pellucida (the glycoprotein layer initially surrounding the ovum), it adheres to the uterine surface epithelium and implants into the uterine endometrium (decidua). In humans, attachment of the blastocyst occurs around 6–8 days after fertilization and is mediated by the polar trophectoderm—the trophoblast cells directly contiguous with the inner cell mass.[3] The next phases of human postimplantation trophoblast development are not entirely clear, but the few images that are available appear to show an erosive trophoblastic syncytium (syncytiotrophoblast) emerging from the polar trophectoderm and spreading into the decidua basalis (Fig. 1A).[4] Syncytiotrophoblast is a mitotically inactive, multinucleated trophoblastic mass that is generated by fusion of mononuclear cytotrophoblast cells. Small chambers (lacunae) form within the syncytiotrophoblast, which connect together and fill with maternal blood percolating from eroded decidual blood vessels (Fig. 1B).[5] Subsequently, cytotrophoblast cells proliferate rapidly to form large finger-like villous projections that penetrate the entire depth of the syncytiotrophoblast (Fig. 1C). At first, these villi radiate out from the entire conceptus, but they gradually regress except adjacent to the decidua basalis, where the definitive placenta will develop. Ultimately, the villi become filled with mesenchyme originating from the extraembryonic somatic mesoderm. Blood vessels are formed within this mesenchyme, which then connect to the fetal circulation via the umbilical cord.[6] At this point, the fundamentals of the villous circulation are in place.

Early postimplantation trophoblast development in mice differs somewhat from that which occurs in humans (depicted in Fig. 2). In mice, attachment of the blastocyst is mediated by the mural trophectoderm (trophoblast cells on the opposite pole of the inner cell mass). Around this time, the first definitive trophoblast cells to differentiate are primary trophoblast giant cells.[7] Trophoblast giant cells are aptly named for their enormity, a feature caused by continuous endoreduplication (polyploidy) of their DNA despite cessation of cell division. Trophoblast giant cells envelop the entire conceptus except for the polar trophectoderm. The polar trophectoderm forms a cylindrical structure containing two separate regions: the extraembryonic ectoderm adjacent to the inner cell mass and the ectoplacental cone. Within these regions, trophoblast cells are diploid and progressively differentiate into the distinct regions of the definitive placenta (Fig. 2).

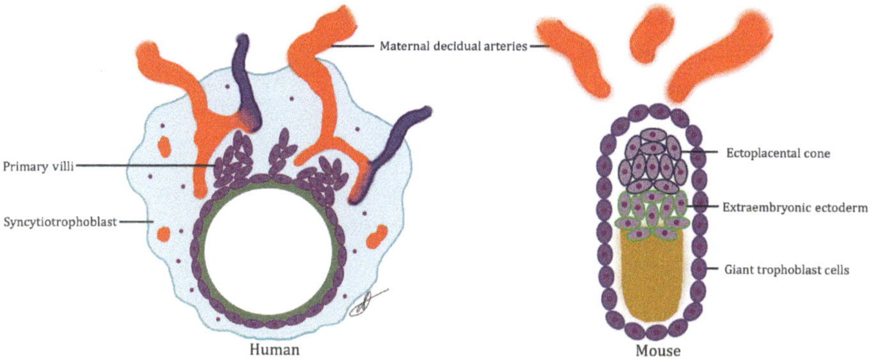

Fig. 2 Schematic of human vs mouse periimplantation development. *Left panel*: human early postimplantation placenta. Note the extensive syncytiotrophoblast development and primary villi. *Right panel*: mouse early postimplantation development. Note the arrangement into the extraembryonic ectoderm and ectoplacental cone.

In mice, trophoblast stem (TS) cells—multipotent cells that can give rise to all specialized trophoblast lineages but not embryonic or yolk sac lineages—can be readily derived from mouse embryos at the preimplantation blastocyst stage and from the postimplantation extraembryonic ectoderm.[8] When injected into blastocysts, mouse TS cells contribute to the formation of all trophoblast lineages, but not to the formation of the embryo.[9] Mouse TS cells can be maintained indefinitely in vitro, so long as fibroblast growth factor (FGF)-4/FGFR2 and nodal/activin/transforming growth factor beta signaling pathways remain stimulated.[8–10] Removal of these factors stimulates differentiation into specialized trophoblast sublineages.[8,9] TS cells have subsequently been derived from various other species, suggesting that the existence of TS cells is likely a conserved feature of early mammalian development. However, a multipotent TS cell population that gives rise to all trophoblast cell derivatives has yet to be definitively identified in humans. As discussed later in the chapter, the as-yet unsuccessful attempts to derive and culture established human TS cells may relate to the anatomical differences in postimplantation trophoblast development between mice and humans, as well as different signaling and transcription factor requirements needed to maintain TS cells in an undifferentiated state.

1.2 Trophoblast Cells in the Definitive Placenta

The definitive placenta can be subdivided into two anatomically distinct compartments, each providing a specialized function. In general, trophoblast cells residing near the embryo specialize in promoting the exchange of substances between maternal and fetal blood, whereas trophoblast cells developing adjacent to the decidua basalis interact with the decidual stroma and facilitate enhanced blood flow to the placenta.

1.2.1 Development of Trophoblast Cells Specializing in the Exchange of Nutrients Between Maternal and Fetal Blood

In humans, the maternal–fetal exchange surface is organized into complex tree-like structures called chorionic villi (Fig. 1D). The chorionic villi are comprised of an outer lining of trophoblast cells (collectively referred to as chorion in this region because of their close apposition to the extraembryonic membranes) and an inner core of vascularized mesoderm. Blood vessels form within this vascularized mesoderm, and these vessels connect with the fetal circulation via the umbilical vessels. Maternal blood directly bathes the chorionic villi, and nutrients in maternal blood are transported across the outer trophoblastic lining into the villous core, where they can then enter the fetal circulation. This type of placentation—where maternal blood directly contacts trophoblast cells—is referred to as hemochorial and is found in various other primates as well as many rodents, including common laboratory animals like mice and rats.[11]

The formation of the chorionic villi is dependent on the fusion of the chorion with the allantois—the site where vascularized mesoderm originates. Thus, the definitive placenta is referred to as chorioallantoic, since it comprises structures derived from the chorion (trophoblast) and allantois (mesoderm). Following chorioallantoic fusion, trophoblast cells are induced to undergo extensive branching morphogenesis via instructive signals by the allantoic mesoderm, culminating in tree-like villous formations with extensive surface areas. In their final form, the chorionic villi either attach to the decidua basalis as large anchoring villi or branch repeatedly into smaller terminal branches that float within the intervillous space. The intervillous space subsequently becomes filled with maternal blood, which bathes the terminal villous branches and is the site where exchange of nutrients and gases occurs.

The outer trophoblast layer that lines the chorionic villi and directly contacts maternal blood is syncytiotrophoblast, the phenotypic characteristics of which were introduced earlier. Syncytiotrophoblast specializes in both the regulation of nutrient exchange between maternal and fetal blood and the production of a variety of hormones. In essence, syncytiotrophoblast forms the functional epithelial barrier separating maternal and fetal blood. However, syncytiotrophoblast has a limited lifespan and is continuously sloughed off into the maternal circulation throughout pregnancy.[12] Replenishment of the syncytiotrophoblast layer is accomplished by differentiation and fusion of cytotrophoblast cells, which form a continuous monolayer directly beneath the syncytiotrophoblast layer during early pregnancy and act as the progenitor cells of the placental epithelium. Through continuous proliferation, expansion, and differentiation, cytotrophoblast cells are responsible for the rapid placental growth during the first half of gestation.[13] However, the relative number of cytotrophoblast cells decreases as gestation advances, and by term they exist only as scattered clusters of cells.

Cytotrophoblast differentiation into syncytiotrophoblast is a complex process characterized by three interdependent steps. First, differentiation requires repression of genes promoting cytotrophoblast self-renewal and junctional proteins such as E-cadherin. The second step involves induction of genes associated with syncytiotrophoblast function referred to as "biochemical differentiation." Biochemical differentiation includes induction of hormones and hormone biosynthetic machinery associated with the syncytiotrophoblast's endocrine function, including human chorionic gonadotropin subunits (encoded by the *CGA* and *CGB* genes), human placental lactogen (encoded by the *CSH1* gene), aromatase (encoded by the *CYP19A1* gene), corticotropin releasing hormone (encoded by the *CRH* gene), placental growth factor (encoded by the *PGF* gene), pregnancy-specific glycoproteins (PSGs), and hydroxysteroids (HSD)3B1 and HSD11B2. Third, cytotrophoblast cells undergo "morphological differentiation," which consists of fusion into the syncytiotrophoblast. This process involves induction of the *syncytin* genes, *ERVW-1* and *ERVFRD-1*, which are expressed specifically in differentiating trophoblast cells and encode proteins that facilitate cell–cell fusion.[14,15]

The site of maternal–fetal exchange in mice is functionally equivalent and anatomically similar to the chorionic villi.[16] This site is referred to as the labyrinth zone because of its maze-like appearance in cross section. Like the chorionic villi, the most notable trophoblast cell population that comprises the labyrinth zone is syncytiotrophoblast. The villous core consists of vascularized mesenchyme, which is similar in appearance in mice as in humans. Like the chorionic villi, the formation of the labyrinth zone is dependent on chorioallantoic fusion and extensive branching morphogenesis. However, some notable differences between the structure and composition of the labyrinth zone and chorionic villi exist. For instance, the trophoblast portion of the labyrinth zone is formed from precursors located in the extraembryonic ectoderm, and trophoblast syncytialization is not apparent until after chorioallantoic fusion. This is in contrast to humans, where trophoblast syncytialization and primary villous formation precede chorioallantoic fusion. Another key difference is that the labyrinth zone in mice is trilaminar, consisting of a single, outer layer of mononuclear trophoblast cells adjacent to maternal blood, and two layers of syncytiotrophoblast. Conversely, the chorionic villi consist of a single layer of syncytiotrophoblast. A third key difference is the absence of an underlying, proliferative cytotrophoblast layer in mouse placentae. The perpetuation of a progenitor population of cytotrophoblast cells may relate to the much longer duration of human pregnancies in comparison to mice.

1.2.2 Development of Trophoblast Cells That Interact With the Decidual Stroma and Enhance Maternal Blood Flow to the Placenta

Trophoblast cells situated near the decidua basalis at the tips of the anchoring villi do not undergo branching morphogenesis like their counterparts lining the chorionic villi, nor do they fuse to form a multinucleated syncytiotrophoblast. Instead, these cells proliferate and stratify, forming highly compact cell columns breached only by channels carrying maternal blood toward and away from the placenta (Fig. 1D). The trophoblast cells within this structure are collectively referred to as extravillous cytotrophoblast cells, appropriately named based on their external location relative to the chorionic villi.

Extravillous cytotrophoblast cells constitute a heterogeneous population of cells. In the compact cell columns, extravillous cytotrophoblast cells are highly proliferative. In the distal aspects of the cell columns, close to where the cells contact the decidua basalis, extravillous cytotrophoblast cells stop proliferating and develop invasive properties. The transition of extravillous cytotrophoblast cells from proliferating to invasive phenotypes involves changes in the expression of genes that promote cessation of proliferation and modification in the expression of cell adhesion molecules, growth factor receptors, and proteases that facilitate cell motility.[17–20] Invasive, postmitotic extravillous cytotrophoblast cells depart the stratified extravillous cell layer where they originate and migrate deeply into the decidua basalis, where they transform the uterine vasculature in order to supply the placenta with sufficient maternal blood—a critical step in establishing uteroplacental circulation.[21] Invasive extravillous cytotrophoblast cells migrate via two routes: interstitially through the uterine stroma and endovascularly via the lumen of the spiral arterioles (which are the primary feeder vessels of the placenta). At first, extravillous cytotrophoblast cells clog the spiral arterioles, allowing blood plasma to seep into the intervillous space but hindering the flow of blood cells, thereby causing the oxygen levels in the placenta to remain low.[22] Low oxygen levels are critical for cytotrophoblast proliferation and invasion, and are also thought to reduce the risk of DNA damage by reactive oxygen species during critical phases of organogenesis.[23] The endovascular trophoblast plugs disappear by the end of the first trimester, facilitating enhanced oxygen delivery to the placental exchange surface to support further placental development and fetal growth.

In addition to regulating oxygen delivery during the first trimester of pregnancy, another key function performed by invasive trophoblast cells is the transformation of the spiral arterioles into high-capacitance conduits, thereby ensuring a consistent supply of maternal blood to the exchange surface of the placenta during the second half of pregnancy. Transformation entails trophoblast cells supplanting the endothelium, and producing a fibrous extracellular matrix that replaces the elastic laminae and vascular smooth muscle.[24] The trophoblast cells situated within the vessel lumen subsequently adopt a pseudoendothelial phenotype, which helps promote vascular integrity and deters coagulation and immune reactivity.[25] A large number of pregnancy complications, including preeclampsia, fetal growth restriction, and placental accreta/increta, are associated with defective spiral artery remodeling by invasive extravillous cytotrophoblast cells, emphasizing the importance of this process for human placental development. A flowchart illustrating the trophoblast multilineage differentiation pathway during human placental development is shown in Fig. 3.

A compact cellular layer comprised of trophoblast cells also develops at the decidual–trophoblast interface in mouse placentas. This zone is referred to as the junctional zone and is comprised of cells derived from the ectoplacental cone. In the junctional zone, three trophoblast subtypes can be identified: secondary trophoblast giant cells, spongiotrophoblast cells, and glycogen cells.[26] Secondary trophoblast giant cells originate at the periphery of the ectoplacental cone and directly border the decidua basalis during early pregnancy, where they are thought to perform similar functions as distal extravillous cytotrophoblast cells adjacent to the decidua.[27] Spongiotrophoblast cells form a stratified cell layer sandwiched between the secondary trophoblast giant cells and the labyrinth zone. They comprise

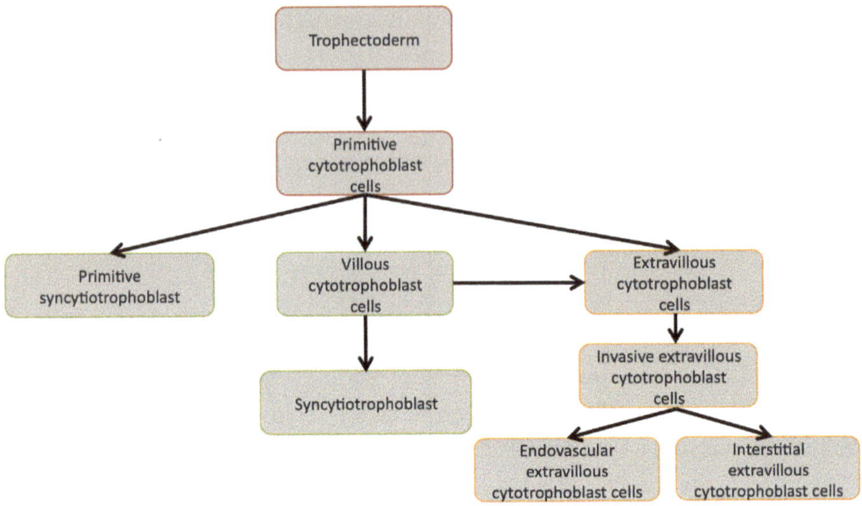

Fig. 3 Flowchart showing human trophoblast multilineage differentiation pathway.

the bulk of the junctional zone, but the functional significance of these cells is not fully understood. Glycogen cells—aptly named due to their accumulation of glycogen—appear during the last half of pregnancy and appear to be the cellular source of invasive trophoblast cells. Invasive trophoblast cells migrate into the decidua through both interstitial and endovascular routes. It should be noted that trophoblast invasion commences at midgestation in mice and other rodents, and continues until near term while in humans, trophoblast invasion occurs during the first half of pregnancy. However, the ultimate goal of spiral arteriole transformation is comparable in humans as in rodents, particularly in larger rodents with more extensive trophoblast invasion such as rats and guinea pigs.[28–31] In mice, spiral arteriole remodeling may be less dependent on trophoblast cells and more reliant on innate immune cell populations residing in the decidua basalis.[32] A flowchart depicting the trophoblast multilineage differentiation pathway in mouse placental development is shown in Fig. 4.

2. TRANSCRIPTION FACTORS IMPLICATED IN TROPHOBLAST DEVELOPMENT

2.1 Early Trophoblast Development

Coordinated expression of a precise set of transcription factors at specific times during development is necessary to facilitate emergence of the

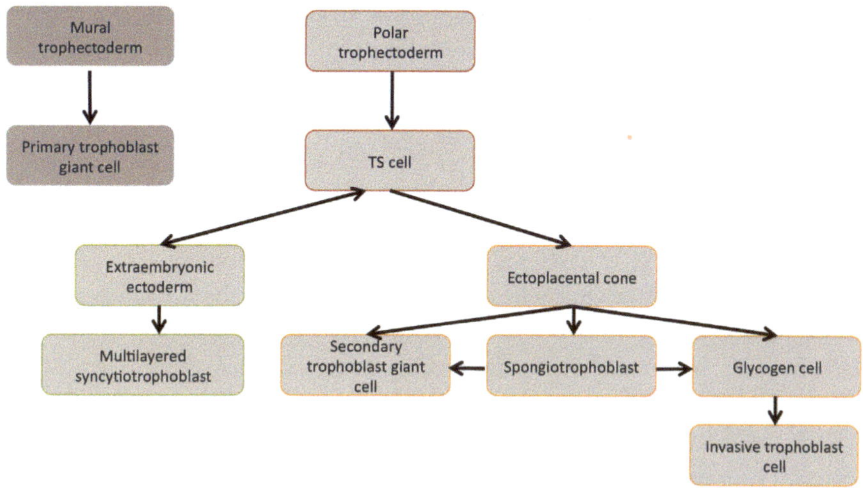

Fig. 4 Flowchart showing mouse trophoblast multilineage differentiation pathway.

trophoblast lineage, to maintain trophoblast cells in a state of multipotentiality, or to promote differentiation programs that facilitate the various functions executed by trophoblast cells. Most of our knowledge of factors deemed important for trophoblast development, particularly in the earliest stages of trophoblast lineage emergence, has been discovered using mice. Two reasons account for why most discoveries emanate from mouse studies. The first reason is due to the relative ease of engineering the mouse genome and the availability of a wide variety of mutant mouse strains. Consequently, many factors important for trophoblast development in mice were discovered serendipitously by observing defects in implantation or placentation following genetic disruption of specific transcription factors. The second reason is that the first successful derivation, maintenance, and characterization of TS cells originated from mice. Identification of transcription factors regulating TS cell fate and differentiation often originates from analysis of mouse TS cells. At this point in the chapter, we will review some of the most relevant transcription factors for trophoblast specification and TS cell maintenance in mice. We will then discuss progress, differences, and limitations in TS cell derivation from humans. Of note, there are many transcription factors that, when mutated in mice, cause placental abnormalities. For a more detailed review of transcription factors, chromatin remodelers, and epigenetic factors implicated in mouse placental development, the reader is referred to several excellent recent reviews.[33,34]

2.1.1 *Transcription Factors Implicated in the Emergence of the Trophoblast Lineage and Development of TS Cells in Mice*

2.1.1.1 Position Sensing, YAP/WWTR1, and TEAD4

Blastomeres of a mouse embryo are morphologically indistinguishable from each other up until the 8-cell stage. These cells have the capacity to contribute to all embryonic and extraembryonic lineages. Subsequent cell divisions cause positional asymmetries, with some blastomeres confined to the inside of the morula, and other blastomeres forced to the outside. By the late 32-cell stage, a distinct epithelial trophectoderm is evident at the periphery of the blastocyst. At this stage, cells within the trophectoderm exclusively differentiate toward the trophoblast lineage and are incapable of contributing to other embryonic lineages. Cells within the inside of the morula contribute solely to derivatives of the inner cell mass. What is the difference between blastomeres at the 8-cell stage and those at the 32-cell stage? From an architectural perspective, cells within the inside of the morula form interactions with adjacent cells on all borders (apolar), whereas cells facing the outside of the morula have a free edge adjacent to the zona pellucida (polar). A prevailing hypothesis is that position-sensing differences between inside cells and outside cells of a morula, attuned to the degree of cell contact, dictate whether cells will assume a trophectoderm or inner cell mass fate.[35–37]

How positional asymmetries modulate transcription factor expression patterns that define cell fates is still somewhat uncertain, but the Hippo-signaling pathway has been implicated. The Hippo pathway is a network of kinases that relay position-sensing information to influence cell proliferation, apoptosis, and gene expression. In outside cells of a morula, the Hippo-signaling pathway is inactivated. This enables nuclear shuttling of transcriptional cofactors such as *yes-associated protein* (YAP) and WW *domain containing transcription regulator 1* (WWTR1, also known as TAZ), where they act as coactivators for *TEA domain family member* (TEAD) transcription factors.[38] TEAD1, 2, and 4 are all expressed in preimplantation mouse embryos.[39] In particular, an essential role for the transcription factor TEAD4 has been shown for mouse trophectoderm development. TEAD4 is expressed in all cells of a blastocyst, but it is transcriptionally active only in outside cells presumably due to the nuclear localization of TEAD4 cofactors such as YAP and WWTR1 in these cells,[38] or due directly to nuclear translocation of TEAD4 itself.[40] *Tead4*-deficient mouse embryos are incapable of forming a blastocyst, do not implant, and do not express any downstream factors consistent with trophoblast cells.[41] Instead, the outside cells in *Tead4*$^{-/-}$ embryos express transcriptional signatures consistent with those of cells derived from inner cell mass. Ectopic expression of *Tead4* in mouse embryonic stem (ES) cells is sufficient to induce gene signatures consistent with trophoblast cells. Thus, TEAD4 is the earliest known transcription factor vital for trophoblast development and has been proposed to be a master orchestrator of the trophectoderm-specific transcriptional program. Of note, mice deficient in either *Wwtr1* or *Yap1* do not exhibit preimplantation defects.[42,43] However, deletion of both *Wwtr1* and *Yap1* causes preimplantation development to arrest at the morula stage prior to lineage specification,[38] indicating that there is likely some functional redundancy between these cofactors in early trophoblast development.

2.1.1.2 CDX2

Activation of TEAD4 is a prerequisite for upregulation of the homeobox transcription factor *caudal-type homeobox 2* (CDX2). *Cdx2* is a member of the caudal-type family of CDX homeobox genes and is the most well-known transcription factor involved in development and maintenance of trophectoderm. Expression of CDX2 is diffusely evident in all cells of an 8–16 cell embryo, but expression becomes progressively more intense in nuclei of outer cells during the blastocyst stage. CDX2 expression remains high in polar trophectoderm of a periimplantation blastocyst and in the proximal few rows of extraembryonic ectoderm up to gestational day 6.5.[44,45] Thus, CDX2 expression is associated with cells that form the trophectoderm layer and later, in multipotential trophoblast cells that have the capacity to self-renew. There is ample evidence to support a role for CDX2 as an early initiator of trophectoderm formation, but exactly when it acts during embryonic development is not entirely clear. Mouse embryos lacking *Cdx2* do not progress past the blastocyst stage.[46] These *Cdx2*$^{-/-}$ embryos initiate trophectoderm formation, including an upregulation of several genes associated with trophoblast cells, including *Fgfr2, Gata3,* and *Eomes* (albeit at reduced levels compared to wild-type cells), and show limited development of a blastocoelic cavity.[44,47] However, these embryos fail to maintain epithelial integrity, and do not implant.[44] In line with the essential role of CDX2 in early trophectoderm development, derivation of mouse TS cells is not attainable from *Cdx2*$^{-/-}$ embryos.[48] Other reports have noted that *Cdx2* expression from the zygote is not the only source of CDX2. Maternal *Cdx2* mRNA is also detectable in early zygotes. Although maternal *Cdx2* is not required for embryonic development per se,[49] depletion of both maternal and zygotic *Cdx2* using RNA-interference (RNAi) results in embryo arrest prior to blastocyst cavitation, increased blastomere cell cycle lengths, enhanced blastomere cell death, and alterations in cell polarity and compaction.[50,51] Thus, CDX2 is one of the very few transcription factors that, when mutated in mice, causes impairment of the trophectoderm leading to preimplantation lethality.

CDX2 is thought to function by both promoting the expression of genes implicated in trophectoderm formation and repressing transcription factors vital for maintaining pluripotency in the inner cell mass, namely *Pou5f1* and *Nanog*.[48,52] In the early morula stage, CDX2 and POU5F1 are expressed throughout and are locked in a state of mutual repression, with CDX2 attempting to activate trophectoderm gene signatures and POU5F1 striving to maintain cells in a state of pluripotency.[45,48] Upon further development, activation of TEAD4 in outside cells stimulates additional *Cdx2* expression, overwhelming POU5F1 and directing cells to adopt a trophectoderm fate. CDX2-deficient embryos fail to repress *Pou5f1* and *Nanog* in outside cells of a blastocyst, whereas ectopic expression of *Cdx2* is sufficient to drive trophectoderm development in murine ES cells.[44,48,53] Importantly, repression of *Pou5f1* is sufficient to drive mouse ES cells toward trophectoderm, even in the absence of *Cdx2* or its upstream regulator *Tead4*.[48,54] Thus, repression of *Pou5f1* by CDX2 is a likely mechanism for promoting initial trophectoderm formation in mice. CDX2 expression is a requirement for TS cell self-renewal and is downregulated as TS cells differentiate into more committed trophoblast lineages.[44] Thus, CDX2 must have additional functions vital for maintaining TS cells in a stem state besides repression of *Pou5f1* and/or *Nanog*. Collectively, evidence implicates CDX2 as a central transcription factor in early trophoblast development.

2.1.1.3 GATA3

An additional transcriptional target of TEAD4 is *Gata3*. GATA3 belongs to the GATA family of transcription factors. GATA3 is selectively expressed in the trophectoderm lineage of the blastocyst and in undifferentiated TS cells,[55] and is later expressed in the ectoplacental cone and differentiated trophoblast lineages, namely trophoblast giant cells.[56] GATA3 is capable of inducing genes associated with a trophoblast cell phenotype in mouse ES cells,[56,57] including *Cdx2*,[55] but it does not maintain TS cells in a stem state. Instead, GATA3 promotes trophoblast differentiation. RNAi-mediated depletion of *Gata3* causes early developmental delays in mouse embryos, but *Gata3*-null embryos survive until gestational day 10.5, whereupon they exhibit defects in placental development and a variety of embryonic tissues. However, it should be noted that a more severe phenotype in early trophoblast formation may be apparent in *Gata3*-null embryos if not for a compensatory upregulation of GATA2, which shares similar transcriptional targets as GATA3.[58,59]

2.1.1.4 EOMESODERMIN

EOMES is a T-box transcription factor encoded by the gene *Eomes*. EOMES is expressed in the trophectoderm of the blastocyst and, following implantation, is localized in the proximal few rows of extraembryonic ectoderm.[60] Consistent with this expression pattern in vivo, *Eomes* is expressed in mouse TS cells maintained in the stem state and is downregulated during differentiation. Mouse embryos deficient in *Eomes* form an initial trophectoderm and implant into the uterus, but die shortly thereafter due to a severe defect in TS cell proliferation.[44,60] Likewise, RNAi-mediated depletion of *Eomes* in mouse TS cells interferes with self-renewal and promotes differentiation.[61] EOMES binds to DNA sequences in conjunction with several other transcriptional regulators promoting the TS cell stem state, including TFAP2C (discussed later), and the chromatin remodeling factor SMARCA4, indicating that these transcription factors may act in concert to maintain TS cells in an undifferentiated state.

2.1.1.5 TFAP2C

TFAP2C, a member of the *activating enhancer-binding protein 2* (AP-2) family of transcription factors, is expressed in the trophectoderm of the early blastocyst, and expression is maintained in all trophoblast derivatives except syncytiotrophoblast.[62,63] Mouse TS cells express high levels of *Tfap2c* in both the stem state and following differentiation. *Tfap2c*-deficient mouse embryos form a trophectoderm and implant into the uterus, but die around E7.5 with severe defects in trophoblast development.[63] Of note, TFAP2A is also expressed in trophectoderm and combined deficiency of both TFAP2A and TFAP2C causes earlier lethality than either gene alone.[64] Thus, it is possible that these genes may have redundant functions in trophectoderm formation. Additionally, TFAP2 family members may have an earlier importance in trophectoderm formation. For example, ectopic expression of *Tfap2c* stimulates genes associated with trophoblast formation, including *Cdx2,* and represses the ES cell pluripotency marker *Nanog*.[62] Moreover, RNAi-mediated depletion of both maternal and zygotic copies of *Tfap2c* in early mouse embryos results in impaired expression of genes associated with cell polarity and tight junction formation, deregulation of Hippo-signaling during positional sensing, and ultimately, reduced CDX2 expression.[65,66] Thus, in addition to being important for trophoblast development during later stages, TFAP2C may also have important regulatory roles in trophectoderm lineage specification.

2.1.1.6 ESRRB, SOX2, and FOXD3

Estrogen-related receptor-beta (ESRRB), *sex-determining region Y—box 2* (SOX2), and *forkhead box D3* (FOXD3) are unique among the transcription factors mentioned thus far, since they are well known as pivotal regulators of pluripotency in mouse ES cells.[67–70] However, all factors are also highly expressed in mouse TS cells, and mice deficient in any one of these factors exhibit severe deficits in placental development.

Esrrb$^{-/-}$ embryos die before gestational day 10.5 with an overabundance of trophoblast giant cells and defective development of the labyrinth zone. Tetraploid chimeras consisting of wild-type trophoblast and *Esrrb*-mutant embryo are viable, confirming that the lethal phenotype is defective trophoblast development.[71] ESRRB is highly expressed in mouse TS cells and declines as cells differentiate. TS cells cannot be derived from mouse embryos lacking *Esrrb*.[72] *Esrrb* expression is dependent on FGF signaling, which is required for mouse TS cell derivation and maintenance, and binds within or adjacent to genes encoding various transcription factors implicated in maintaining TS cells in a stem state, including *Cdx2, Eomes, Tfap2c, Sox2*, and *Elf5*.[73] Collectively, this evidence implicates *Esrrb* in the maintenance of the TS cell stem state.

Sox2 is a critical gene for the induction and regulation of pluripotency. It is highly expressed in TS cells and is downregulated during differentiation. In TS cells, SOX2 forms a complex with TFAP2C to regulate target genes involved in maintaining TS cell stemness.[74] Like ESRRB, SOX2 is a critical downstream effector of FGF signaling during trophoblast development, and TS cells cannot be derived from *Sox2*$^{-/-}$ embryos. Interestingly, ectopic expression of SOX2 together with ESRRB or TFAP2C can replace the dependency of TS cells on FGF signaling.[74] SOX2-deficient embryos implant normally, but die shortly thereafter with severe defects in both embryonic and trophoblast lineage development, although trophoblast giant cells are detectable.[70] Tetraploid embryos harboring SOX2 deficiency specifically in trophoblast cells also die shortly after implantation, but slightly later than whole embryo knockouts (approximately gestational day 7.5).[70] Depletion of both maternal and zygotic *Sox2* via RNAi results in embryo arrest around the morula to blastocyst transition, with reduced expression of a variety of trophectoderm markers including *Cdx2*, *Gata3*, and *Eomes*.[75]

Thus, like its binding partner TFAP2C, SOX2 may have additional regulatory roles in early trophectoderm development.

Foxd3$^{-/-}$ mice die around gestational day 6.5 with defects in epiblast formation.[68] However, similar to ESRRB and SOX2 noted earlier, derivation of TS cells is not possible from FOXD3-null mice. Chimeras consisting of *Foxd3*-deficient trophoblast cells and wild-type inner cell mass exhibit delayed embryonic lethality until gestational day 9.5, with severe deficits in the development of the labyrinth zone and spongiotrophoblast layers and an overabundance of trophoblast giant cells.[76] Thus, like SOX2 and ESRRB, FOXD3 may have a general role in maintaining cells in a stem state and preventing their precocious differentiation.

2.1.1.7 The ID Family

The *inhibitor of differentiation/inhibitor of DNA-binding* (ID) family of transcriptional regulators is a subfamily of basic helix-loop-helix (bHLH) transcription factors that lack a basic DNA-binding domain. Although essentially not transcription factors per se, these factors regulate many genes by heterodimerizing with other bHLH proteins, including those required for growth and differentiation, and prevent their transcriptional activation by blocking chromatin binding.[77] Four mammalian ID isoforms, ID1–4, have been identified. *Id1*, *Id2*, and *Id3* are highly expressed in mouse and rat TS cells and are robustly downregulated during in vitro differentiation.[10,78] Mice deficient in any of the *Id* genes are viable, although this may relate to their overlapping expression patterns and/or functional redundancy. Interestingly, a study using single-cell RNA-sequencing identified *Id2* as the first transcriptional regulator evident in nascent trophectoderm cells of a mouse early embryo.[79] However, the functional significance of ID2 in the early embryo has yet to be elucidated.

2.1.1.8 ARID3A

AT-rich interactive domain 3A (*ARID3A*), the founding member of the ARID family of transcription factors, is expressed modestly in self-renewing ES cells, but is upregulated following induction of differentiation.[80] ARID3A is also highly expressed in trophoblast cells. ARID3A-deficient mice exhibit defective embryonic and placental defects and die between gestational days 8.5 and 11.5.[81,82] ARID3A-overexpressing mouse ES cells adopt a TS cell-like morphology and gene signature and, when injected into blastocysts, contribute to the formation of the trophectoderm.[81] ARID3A binds to highly similar target loci as POU5F1. Thus, ARID3A may act in concert with CDX2 to decrease POU5F1-dependent activation of pluripotency-associated genes, thereby facilitating formation of trophectoderm.

2.1.1.9 ETS2 and ELF5

ETS2 and *E74-like factor 5* (ELF5) belong to the ETS superfamily of transcription factors, characterized by a winged helix-turn-helix structure. Both ETS2 and ELF5 function nonredundantly to maintain the TS cell stem state. Mice exhibiting a targeted deletion in the *Ets2* gene die around gestational day 8.5 with severe defects in trophoblast development, including an absence of chorion, failure to form an extraembryonic ectoderm, and disruption in embryonic anteroposterior patterning.[83,84] Embryonic lethality is reversed in tetraploid chimeras, which have wild-type trophectoderm and *Ets2*-mutant embryos.[83] This finding was confirmed in mice harboring an Ets2 deletion specifically in lineages derived from epiblast.[85] Thus, the cause of embryonic lethality is defective trophoblast development. TS cells cannot be derived from *Ets2*-deficient mouse embryos. Depletion of *Ets2* in TS cells reduces proliferation and inhibits the expression of several genes implicated in maintaining the TS cell stem state, including *Cdx2, Eomes,* and *Esrrb*.[85] Thus, ETS2 contributes to the maintenance of the TS cell stem state.

ELF5 is also implicated in the maintenance of mouse TS cell stemness. Similar to *Ets2*-deficient mice, mice deficient in *Elf5* implant and form an ectoplacental cone, but exhibit a complete absence of extraembryonic ectoderm. No TS cells can be derived from *Elf5*-null embryos.[86] Induced ELF5 expression in mouse ES cells stimulates genes associated with trophectoderm in a similar manner as ectopic expression of TEAD4, CDX2, EOMES, GATA3, or TFAP2C. Mouse ES cells exhibit extensive methylation of the *Elf5* gene (associated with gene repression), whereas in mouse TS cells, the *Elf5* gene is hypomethylated (gene is activated). Interestingly, mouse ES cells harboring a defect in DNA methylation express ELF5 and display trophectoderm markers. Thus, the methylation status of ELF5 may be a key determining characteristic distinguishing the transcriptional circuitry of ES cells from that of TS cells.[87,88] ELF5 has been proposed to participate in a self-promoting transcription factor circuit involving CDX2, EOMES, and TFAP2C to activate genes that maintain TS cells in a stem state, including the promotion of their own expression and other genes such as SOX2.[61,89–91] Thus, despite the fact that ELF5 is likely dispensable for preimplantation trophectoderm formation, this transcription factor assists in reinforcing the stem state and preventing precocious trophoblast differentiation.

2.1.2 Transcription Factors Involved in Trophoblast Specification and Development of TS Cells in Humans

As described earlier, studies using mice have revealed a variety of transcription factors implicated in the emergence and early development of trophoblast cells. It stands to reason that knowledge gained from mice can be extrapolated to better understand early trophoblast development in humans. Studies using human material are limited due to obvious ethical constraints and small sample sizes, and thus, it is not feasible to confirm mechanistically whether transcription factors that promote emergence and development of trophoblast cells in mice are the same as those which drive trophoblast development in humans. Some conserved features are evident when comparing mouse and human trophoblast development, but there are also significant differences (reviewed in Refs. 92–94) that need to be considered.

The first key difference when comparing early trophoblast development in mice and humans is the timing and anatomy of implantation. Mouse trophoblast commitment and establishment of TS cells occurs in the preimplantation blastocyst stage. Implantation occurs around gestational day 4.5 and is characterized by adhesion of the mural trophectoderm to the uterine surface epithelium and encapsulation of the conceptus. Trophoblast cells continue to proliferate shortly following implantation, at which juncture they form the extraembryonic ectoderm—a rich source of TS cells. This is in stark contrast to preimplantation embryonic development in humans, which includes at least one additional round of cell division prior to implantation around gestational days 6.5–8.0.[1,92] Implantation is accomplished by a highly invasive trophectoderm at the polar regions of the blastocyst that immediately differentiates into an erosive syncytium.[95] Progenitor mononuclear cytotrophoblast cells, which may act like a TS cell population to replenish the chorionic villi, are not evident until later in gestation. Consequently, a "classical" TS cell population that exhibits characteristics similar to mouse TS cells and that forms a structure comparable to mouse extraembryonic ectoderm has not yet been identified in humans.

A second major difference when comparing mouse and human early trophoblast development is that there appears to be different signaling requirements to facilitate lineage derivation from early mouse embryos compared to early human embryos. For example, mouse ES cells are reliant on leukemia inhibitory factor (LIF) for their self-renewal, whereas LIF has no apparent effect on ES cell derivation and self-renewal in primates and humans.[96] TS cells have never been successfully derived from human embryos, so there is no consensus on the signaling requirements for their self-renewal. However, since their original derivation in mice, TS cells have subsequently been identified and cultured in a variety of other species, including rat,[78,97] common vole,[98] and rhesus monkey.[99] There appears to be significant heterogeneity in the capacity to derive TS cells (or TS-like cells) from different species. Sustained FGF signaling is only critical for TS cell maintenance in mice.[8,9] In rats, several TS cell lines have been derived, some of which are dependent on FGF signaling and others which are not. Neither common vole nor rhesus monkey TS cells are dependent on FGF signaling. Trophoblast cells with stem-like characteristics have also been derived from porcine and bovine embryos, neither of which are reliant on exogenous FGF.[100,101] Thus, FGF signaling does not appear to be a universal requirement for TS cell establishment and self-renewal. Transcript levels of *FGFR*s are not detectable in human trophectoderm and FGFs do not improve human blastocyst outgrowth quality, suggesting that putative human "TS" cells do not require FGF to maintain their self-renewal.[102] Not surprisingly, the few attempts at deriving TS cells from human embryos using protocols commonly applied for mouse TS cell derivation were unsuccessful.[7,102] However, it is certainly reasonable to predict that, if the signaling components necessary for TS cell self-renewal in human embryos were discovered, human TS cells could eventually be derived and cultured.

Due to the manifest anatomical and probable molecular differences in trophectoderm establishment and maintenance of TS cells between mice and humans, there are likely to be differences in the transcription factor components required for the emergence and early development of trophoblast cells between these species. It remains to be determined whether some of the transcription factors that are induced in response to FGF signaling in mice (which facilitate self-renewal in mouse TS cells) are also relevant for human trophoblast development. When assessing transcription factor expression in human trophectoderm there appears to be some overlap with mouse embryos. For example, like mouse embryos, human embryos express POU5F1 and NANOG in all cells until the blastocyst stage, when POU5F1 becomes restricted to the epiblast.[103–105] TEAD4 has also been identified in the nuclei of both human and rhesus monkey trophectoderm of the early blastocyst and excluded from the inner cell mass, indicating that TEAD4 nuclear localization may be a conserved feature of trophectoderm specification.[40] CDX2 is also readily detectable in the trophectoderm of human blastocysts.[103,105] However, key differences in the timing of expression of lineage-associated factors have also been noted between mouse and human embryos. For example, in the mouse, CDX2 is expressed at the morula stage, whereas CDX2 expression is not detected until the full blastocyst stage in human embryos, and expression overlaps with POU5F1 until just prior to implantation.[103,106] A possible explanation for the overlap in CDX2 and POU5F1 observed in human (and other species such as cattle, pig, primate, and rabbit) trophectoderm in the blastocyst stage but not in mice relates to a difference in the regulatory elements of *Pou5f1*, which contains CDX2- and TFAP2C-binding sites capable of directly repressing *Pou5f1* in mice (and possibly rats) but not in other species.[107] Thus, the rapid repression of POU5F1 that facilitates trophectoderm commitment observed in preimplantation mouse embryo development may not be representative of the initial stages of trophoblast development in humans and many other species. In line with the apparent delay in trophoblast commitment in human embryos, trophectoderm cells reaggregated at the full blastocyst stage can reform blastocysts and contribute to NANOG-expressing inner cell mass.[108] Thus, unlike the situation in mice, human outer cell mass cells do not appear to be fully committed to the trophoblast lineage at the early blastocyst stage.

Microarray analyses from dissected blastocysts, as well as single-cell RNA-sequencing of human embryos at different stages, reveal some similarities with mouse embryos, as well as some compelling differences in the timing and transcription factor circuitry associated with each lineage.[103,105,109–113] For example, *CDX2, GATA2*, and *GATA3* are readily detectable in both human and mouse trophectoderm, underscoring the importance of these factors in regulating trophectoderm development. However, unlike mouse trophectoderm, *EOMES* and *ELF5* are undetectable in human trophectoderm.[109,110] Also unlike mouse blastocysts, *ID2* was most readily detectable in the primitive endoderm layer of human blastocysts and not in the trophectoderm layer. In human embryos, *TFAP2C* and *ESRRB* are not expressed until the 8-cell stage, whereas in mice, the orthologs *Esrrb* and *Tfap2c* are detectable throughout preimplantation development. In addition, TFAP2C expression in human embryos is not exclusive to the trophectoderm like it is in mice, but is rather diffusely evident in all lineages of a human blastocyst.[109] Thus, significant differences exist between mice and humans with respect to the transcription factor circuitry evident in preimplantation trophoblast development. It should be noted that the expression of transcription factors immediately following implantation into the uterus has not been examined in humans due to ethical constraints. It would be instructive (but not ethically feasible) to determine whether peri- or postimplantation human trophoblast cells contain a transcription factor expression profile that is more consistent with transcription factor expression patterns in mouse trophectoderm.

Although identification and derivation of definitive human TS cells has been elusive, human ES cells differentiate into trophoblast-like cells, particularly after treatment with bone morphogenetic protein-4 (BMP4).[114–116] This is another feature distinguishing human preimplantation development from mouse preimplantation development, since mouse ES cells do not readily differentiate into extraembryonic lineages.[117] Human ES cell-derived trophoblast cells express many genes characteristic of differentiated human trophoblast. As of yet, culture conditions have not been established to maximize self-renewal of these human ES cell-derived trophoblast cells. However, over the time course of BMP4-mediated trophoblast differentiation, a variety of transcription factors associated with mouse TS cell development are at least transiently induced in this model, including *CDX2, TFAP2C, ID2, ELF5*, and *GATA3*.[114,115,117–119] Once committed to the trophoblast lineage, the cells differentiate. Thus, BMP4-treated human ES cells may provide insights into the origin of putative human TS cells. It should be noted that trophoblast-like cells derived from ES cells are not universally accepted as a *bona fide* model of trophoblast cell differentiation.[120,121] Thus, data obtained using these cells must be interpreted with caution.

Since TS cells have also been successfully derived from mouse conceptuses several days after implantation, investigators have sought to identify TS cell populations in human placenta during the first trimester using the same transcription factor criteria used to identify mouse TS cells. As described earlier, this may be problematic since it is not likely that all transcription factors expressed in mouse TS cells will be similarly expressed in a putative human TS cell population. Villous cytotrophoblast cells preside in the first-trimester placenta, are highly proliferative and act as progenitor cells that contribute to differentiated villous and extravillous trophoblast populations. Although there is little evidence that these cells are equivalent to mouse TS cells, Hemberger *et al.* identified subpopulations of ELF5 and CDX2-positive cells in villous cytotrophoblast cells, with higher transcript expression levels of these factors in first-trimester tissue compared to term.[122] *EOMES* was also detected, although expression was not evident until later in the first trimester, and there was no difference in expression between early and late gestation. Exactly which cells expressed *EOMES* was also not assessed.[122]

CDX2, ELF5, and *EOMES* were also detected in human trophoblast-like cell lines, and have been used, in conjunction with other transcription factors (MYC and KLF4) to generate a putative "induced" TS cell line from human fetal fibroblasts.[123] Another study identified CDX2^{-}, POU5F1^{+}, EOMES^{+}, and GATA4^{+} mesenchymal cells in the chorion as a potential niche of TS-like cells.[124,125] With respect to other transcription factors implicated in the early stages of mouse TS cell development, GATA3 is expressed in both villous and extravillous cytotrophoblast cells.[126] TFAP2A and TFAP2C are expressed in all trophoblast derivatives of the human placenta, which are similar to their broad expression pattern in the fully developed mouse placenta, and may have a role in the differentiation and function of villous and extravillous cytotrophoblast cells.[127]

2.2 Transcription Factors Implicated in Villous and Extravillous Cytotrophoblast Formation and Function

Once the trophoblast lineage has become established and the placenta has formed, trophoblast cells can be anatomically divided into those that comprise the chorionic villi (villous) and those located outside the villi (extravillous).

Trophoblast cells comprising the chorionic villi include villous cytotrophoblast cells and syncytiotrophoblast. Although there is still much to be discovered, studies have revealed a number of transcription factors that play important roles in the differentiation and function of trophoblast cells within the chorionic villi. These transcription factors can broadly be divided into (1) those that promote self-renewal and proliferation of villous cytotrophoblast cells, (2) those that promote morphological differentiation, and (3) those that facilitate biochemical differentiation. Although the latter two are independent events, they are not always easy to distinguish and may have significant overlap.[128]

Extravillous cytotrophoblast cells can be subdivided into proliferative extravillous cytotrophoblast cells that comprise the stratified cell columns and invasive extravillous cytotrophoblast cells. Invasive extravillous cytotrophoblast cells can be further subdivided into those that line the spiral arteries (invasive endovascular extravillous cytotrophoblast cells) and those that migrate through the uterine stroma (invasive interstitial extravillous cytotrophoblast cells). Although much is still unknown, a few transcription factors have been identified that regulate extravillous cytotrophoblast proliferation and invasion. Transcription factors can broadly be divided into (1) those that promote extravillous cytotrophoblast proliferation and (2) those that promote extravillous cytotrophoblast growth arrest and stimulate invasive properties. Interestingly, although villous and extravillous cytotrophoblast cells develop and function differently, there are many transcription factors that play important/similar roles in both cell types. Additional transcription factors may have roles in further development of the invasive extravillous cytotrophoblast lineage, such as those that promote cessation of invasion during later pregnancy or those stimulating pseudoendothelial gene signatures in endovascular cells; to date, these transcription factors have not been well defined.

This section will summarize transcription factors that have a role in villous and/or extravillous cytotrophoblast (1) self-renewal and (2) differentiation and function. We will also indicate parallels with mouse (and other species) trophoblast development when appropriate.

2.2.1 Transcription Factors Implicated in Villous and Extravillous Cytotrophoblast Cell Self-Renewal

2.2.1.1 HIF

The ability of cells to sense and respond to oxygen levels is mediated primarily through the *hypoxia-inducible factor* (HIF) transcription factor complex.[129] HIF is a basic helix-loop-helix/Per-Arnt-Sim transcription factor comprised of two subunits: an alpha subunit (with three isoforms: HIF-1alpha, HIF-2alpha, and HIF-3alpha) and a beta subunit (HIF-1beta, also called *aryl hydrocarbon receptor nuclear translocator,* or ARNT). HIF controls the expression of many genes involved in cell growth, differentiation, and motility by binding to hypoxia responsive elements proximate to target genes.

Since early placental development takes place in an environment characterized by low oxygen levels, HIF is implicated as a key regulator of placental development. Mice deficient in *Hif1b, or Hif1a* and *Hif2a,* display severe defects in placental development leading to embryonic lethality around gestational day 10.5.[130,131] In these placentas, there is variable chorioallantoic fusion leading to deficient generation of the labyrinth zone and greatly reduced spongiotrophoblast formation. In particular, hypoxia-HIF signaling appears to play an important role in the development of the junctional zone and the invasive trophoblast lineage. Exposure of mouse or rat TS cells to hypoxia increases the expression of genes associated with spongiotrophoblast lineages.[130–132] Conversely, TS cells lacking *Hif1b,* or *Hif1a/Hif2a,* have a reduced capacity to generate junctional zone lineages, are less invasive, and exhibit enhanced evidence of syncytiotrophoblast generation.[130,131,133] In human placenta, both HIF1A and HIF2A are highly expressed in trophoblast cells during early pregnancy, but their expressions decline as gestation advances and they are weakly expressed or absent by the end of the first trimester.[134] Culturing human villous cytotrophoblast cells under low oxygen conditions prevents syncytiotrophoblast formation in vitro, although whether this is a direct result of HIF transcriptional activity is not yet known.[135–137] In extravillous cytotrophoblast cells, a variety of studies support the notion that hypoxia-HIF signaling promotes proliferation of extravillous cytotrophoblast cells and inhibits their differentiation into invasive cytotrophoblast cells.[138–141] In this context, hypoxia-HIF signaling would create a large supply of extravillous cytotrophoblast cells to support placental growth. However, other reports suggest the opposite effect—that hypoxia-HIF signaling directly stimulates trophoblast invasion.[142–144] These discrepancies may relate to differences related to in vitro experimentation. It is possible that hypoxia-HIF signaling stimulates both extravillous cytotrophoblast proliferation and invasion. In support of this notion, pregnant mice and rats placed into hypoxic chambers during early pregnancy exhibit robust expansion of the junctional zone, indicative of increased proliferation, as well as precocious trophoblast invasion into the decidua.[145] In sum, HIF activation may provide the placenta with a means to adapt to reduced oxygen levels by stimulating development of extravillous cytotrophoblast cells, which ultimately helps the placenta to obtain a greater supply of blood.

2.2.1.2 ASCL2

Achaete-scute family bHLH transcription factor 2 (ASCL2) is a bHLH transcription factor. ASCL2 has the distinction of being the first transcription factor that was discovered in mice to have an essential role specifically for trophoblast development.[146] In mice, *Ascl2* is a maternally expressed imprinted gene that encodes a protein detected in proliferative trophoblast cells in the placenta during early development and subsequently in the intestine, where it controls self-renewal of cells that regenerate the intestinal crypts.[147] Mice deficient in *Ascl2* die at midgestation due to defective spongiotrophoblast development and expansion of the trophoblast giant cell layer.[146,148] Hypomorphic expression of *Ascl2* impairs both spongiotrophoblast and labyrinth development and causes reduced placental and neonatal weights.[149] In human placenta, ASCL2 is expressed in proliferative villous and extravillous cytotrophoblast cells.[150,151] ASCL2 expression is highly expressed in freshly isolated primary villous cytotrophoblast cells. Its expression declines as cells differentiate, but it is maintained in villous cytotrophoblast cells cultured at low oxygen levels. Ectopic expression of

ASCL2 prevents villous cytotrophoblast cell differentiation into syncytiotrophoblast.[137] At least part of the mechanism for how ASCL2 prevents human villous cytotrophoblast differentiation is facilitating the nuclear accumulation of other bHLH factors, such as upstream stimulatory factors (USF) 1 and 2, which negatively regulate genes associated with villous cytotrophoblast differentiation.[152,153] Thus, a key mode of action for ASCL2 is to prevent cytotrophoblast differentiation and promote proliferation, thereby ensuring a consistent pool of proliferating villous cytotrophoblast cells. Although the function of ASCL2 in extravillous cytotrophoblast cells has not yet been defined, the expression of ASCL2 in these cells and its absence in postmitotic invasive extravillous cytotrophoblast cells suggest it may play a similar role in promoting proliferation and preventing differentiation.[150,151]

2.2.1.3 TP63

Tumor protein p63 (TP63) is a member of the p53 family of transcription factors. Via the use of alternative promoters, *TP63* generates transcripts encoding two major protein isoforms: TAp63 and ΔNp63.[154] ΔNp63 is highly expressed in the basal layer of stratified epithelia of a variety of tissues, including skin, breast, and prostate, where it maintains self-renewal and prevents differentiation.[155,156] In human placenta, ΔNp63 is expressed in proliferating villous cytotrophoblast cells and at the base of the cytotrophoblast cell columns, and expression is absent in syncytiotrophoblast and extravillous cytotrophoblast cells.[157,158] Ectopic expression of ΔNp63 promotes villous cytotrophoblast proliferation, prevents syncytiotrophoblast formation in primary cytotrophoblast cells and BMP4-treated human ES cells, and inhibits extravillous cytotrophoblast differentiation and migration.[118,159] Thus, TP63 appears to reinforce the villous cytotrophoblast cell progenitor state and inhibit differentiation.

2.2.1.4 The ID Family

The ID family (ID1–ID4) of transcriptional regulators function primarily as dominant-negative inhibitors of bHLH transcription factors and are usually implicated in the promotion of proliferation and inhibition of differentiation.[77] All four ID proteins are expressed in villous cytotrophoblast cells and proliferative extravillous cytotrophoblast cells and are absent in syncytiotrophoblast, with higher expression in diseases characterized by unregulated trophoblast proliferation.[160] ID1, ID2, and ID3 are also detected in primary villous cytotrophoblast cells, and their expression progressively declines during syncytiotrophoblast formation. Interestingly, ID1 expression is maintained in villous cytotrophoblast cells cultured in low oxygen conditions, although whether ID1 expression is a cause or consequence of the reduced cytotrophoblast differentiation in low oxygen culture conditions is not known.[137] In extravillous cytotrophoblast cells, increased expression of ID2 is sufficient to reduce differentiation and invasion in vitro.[161] Collectively, ID proteins likely have a role in maintaining the progenitor state of human cytotrophoblast cells.

2.2.1.5 MYC

MYC encodes an oncogenic bHLH–leucine zipper transcription factor that regulates a diverse set of genes and is highly implicated in the regulation of cell self-renewal.[162] MYC is expressed in cytotrophoblast cells and the proliferative cell columns of the placenta.[163,164] Placental *MYC* expression progressively declines from early pregnancy to term and during primary villous cytotrophoblast differentiation in vitro.[163] Interestingly, MYC indirectly represses cytotrophoblast differentiation via transcriptional activation of the microRNA-17–92 and microRNA-106a–363 clusters. These microRNAs, in turn, directly inhibit transcripts encoding proteins associated with cytotrophoblast differentiation, such as *CYP19A1* and *glial cells missing-1* (*GCM1;* discussed later).[165] Thus, MYC may maintain cytotrophoblast cells in a state of self-renewal by indirectly inhibiting the expression of factors that promote differentiation.

2.2.1.6 FOS

FOS is the founding member of the FOS family of leucine zipper transcription factors, which heterodimerize with JUN family members to form the activator protein-1 (AP-1) transcription factor complex. In the mouse, FOS is expressed in placenta, and mice-lacking *Fos* have reduced placental and fetal weights.[166] In human placenta, members of the FOS family are expressed in different trophoblast derivatives, suggesting that they may have roles in trophoblast development.[167] FOS is expressed in proliferating villous and extravillous cytotrophoblast cells, but is not expressed in more differentiated lineages. In immortalized human extravillous cytotrophoblast cell lines, knockdown of *FOS* causes cells to stop proliferating and gain invasive capabilities.[17] These data suggest that FOS is essential for maintaining undifferentiated, proliferative phenotypes in extravillous cytotrophoblast cells.

2.2.2 Transcription Factors Associated With Villous and Extravillous Cytotrophoblast Differentiation and Function

2.2.2.1 GCM1

In mice and humans, *GCM1* encodes a transcription factor that is expressed mostly, if not exclusively, in trophoblast cells. In mice, GCM1 is initially expressed in the basal layer of the chorion, and subsequently in the syncytiotrophoblast layer of the labyrinth.[168,169] GCM1 regulates branching morphogenesis during formation of the labyrinth, activates *syncytin* genes to promote syncytiotrophoblast generation, and transitions cells away from a proliferative state in preparation for cell fusion.[168,170,171] Mice-lacking *Gcm1* die at E10.5 with severe deficits in the formation of the placental labyrinth zone.[168] *Gcm1*-heterozygote mice are viable, but exhibit deformities in the syncytiotrophoblast layer.[172] In humans, *GCM1* transcripts are prevalent throughout villous and extravillous cytotrophoblast cells, whereas GCM1 protein is expressed only in subsets of these cells.[173] The differing RNA and protein expression patterns in human placenta likely relate to posttranslational acetylation, desumolyation, and dephosphorylation of GCM1, which stabilize the protein.[174,175] GCM1 is also inhibited under conditions favoring proliferation at the expense of differentiation, such as during culture under low oxygen conditions.[176] The nuclear factors MYC, GATA3, nuclear factor E2, and histone deacetylase-5 all negatively regulate GCM1 transcription and/or activity; thus, these transcription factors may play a role in preventing or fine-tuning GCM1 activity.[165,174,177,178] GCM1 appears to promote syncytiotrophoblast formation in human placenta in a similar manner as mice, through transcriptional activation of *syncytin* genes that stimulate cell–cell fusion, as well as transitioning cells away from a proliferative state.[179,180] GCM1 is also vital for activation of various genes encoding syncytiotrophoblast-specific hormones, such as *PGF* and *CYP19A1*.[181,182] Due to the relative exclusivity of GCM1 expression in trophoblast cells and the conserved importance of this transcription factor for development of the maternal–fetal exchange surface of both mice and humans, GCM1 is considered to be a master orchestrator of syncytiotrophoblast generation.

In addition to its essential role in syncytiotrophoblast formation, GCM1 is implicated in the regulation of extravillous cytotrophoblast differentiation. GCM1 is expressed in extravillous cytotrophoblast cells and directly induces genes involved in the promotion of invasion.[183] Ectopic expression of *GCM1* stimulates invasion of human choriocarcinoma cells, whereas knockdown of *Gcm1* in first-trimester explants cultured on Matrigel causes an accumulation of proliferating cytotrophoblast cells and a substantial reduction of invasion.[179] Thus, GCM1 may facilitate the transition, in both villous and extravillous cytotrophoblast cells, from a proliferative to a differentiated phenotype.

2.2.2.2 CREB

The cyclic adenosine monophosphate (cAMP)-activated protein kinase (PKA) pathway is a primary regulator of syncytiotrophoblast fusion.[184] One of the major transcription factors activated by PKA is *cAMP* response element binding protein (CREB, also called *activating transcription factor 1*, ATF1). The CREB/ATF family is a large family of basic region leucine zipper transcription factors that either homodimerize or heterodimerize with each other or with members of the FOS and JUN family. CREB mediates the transcriptional activation of a variety of genes implicated in cytotrophoblast differentiation, including *CGA, ERVFRD-1*, and the gene encoding the proprotein convertase, Furin.[185–187] CREB heterodimerization with another member of the CREB family, *old astrocyte specifically induced substance* (OASIS), transcriptionally regulates the expression of GCM1.[188] CREB/CREB-binding protein (CBP) also mediates the acetylation of GCM1, which confers stability to the protein.[189] Thus, activation of CREB is a key event in the initiation of GCM1 expression and cytotrophoblast differentiation.

2.2.2.3 OVOL1

OVO-like 1 (OVOL1) encodes a transcription repressor and is a member of the C_2H_2 zinc finger transcription factor family. This family of transcription factors regulates epithelial differentiation in several tissues.[190,191] *OVOL1* transcript is expressed in specific subsets of cytotrophoblast cells in human placenta. Depletion of OVOL1 in human cytotrophoblast cells, choriocarcinoma cells, or BMP4-treated human ES cells inhibits both biochemical and morphological differentiation of cytotrophoblast cells.[192] OVOL1 binds upstream of genes implicated in maintaining cytotrophoblast cell self-renewal, such as *MYC, ID1, TP63*, and *ASCL2* (discussed earlier). Thus, OVOL1 may have a key role in repressing the cytotrophoblast progenitor state in order to facilitate differentiation.

2.2.2.4 PPARG

Peroxisome proliferator-activated receptors (PPARs) are ligand-inducible members of the nuclear steroid receptor superfamily best known for their roles in cell growth, differentiation, and metabolism. There are three known members of this family: PPARalpha (PPARA), PPARbeta/delta (PPARD), and PPARgamma (PPARG).[193] Each subtype has a unique tissue distribution pattern, but all three PPARs are expressed in murine and human placenta.[194] In particular, PPARG is implicated as a key regulator of trophoblast differentiation. *Pparg*$^{-/-}$ mice die around gestational day 10 with defective labyrinthine precursor cell differentiation, leading to severe deformities in the labyrinth zone.[195] Conversely, administering the PPARG agonist rosiglitazone to pregnant mice from mid to late gestation alters placental morphology, reduces placental and fetal weight and the size of the spongiotrophoblast layer, and enhances trophoblast lipid uptake.[196] In line with the importance of PPARG in mouse trophoblast differentiation, *Pparg* is induced during TS cell differentiation in vitro, and it controls the expression of a variety of genes associated with trophoblast function, such as *Muc1*.[197] *Pparg*-null TS cells do not form differentiated labyrinthine trophoblast cells, whereas administration of rosiglitazone induces GCM1 expression and impairs trophoblast giant cell formation.[198] Collectively, these results implicate PPARG as a key inducer of mouse labyrinth trophoblast formation.

In human placenta, PPARG is localized to villous and extravillous cytotrophoblast cells, and syncytiotrophoblast. Activity of PPARG is enhanced during in vitro villous cytotrophoblast differentiation.[199] Administration of PPARG agonists augments villous cytotrophoblast differentiation, whereas inhibition of PPARG activity promotes proliferation and prevents differentiation.[200–202] At least part of the mechanism for how PPARG stimulates cytotrophoblast differentiation and syncytiotrophoblast function is due to direct PPARG binding to the *cis*-regulatory regions of *GCM1*, *CGB*, and *CSH1*, and by altering trophoblast metabolism, such as inducing free fatty acid uptake.[201–203] Interestingly, in extravillous cytotrophoblast cells, PPARG appears to have the reverse effect on differentiation: administration of PPARG agonists inhibits invasion.[204–206] Thus, PPARG has differential effects on trophoblast differentiation in humans and mice.

2.2.2.5 RXRA

Retinoic acid is a metabolite capable of stimulating differentiation of both mouse TS cells and human villous cytotrophoblast cells.[207,208] Retinoic acid mediates its effects on cells by acting as ligands for the retinoic acid receptors (RARs) and retinoid X receptors (RXRs)—members of the nuclear hormone receptor superfamily. There are three types of RARs: RARalpha

(RARA), RARbeta (RARB), and RARgamma (RARG); and three types of RXRs: RXRalpha (RXRA), RXRbeta (RXRB), and RXRgamma (RXRG). RARA and RXRA are expressed at high levels in human villous and extravillous cytotrophoblast cells, as well as in mouse placenta.[209,210] *Rxra*-deficient mice die at gestational day 10 with severe deficits in the development of the labyrinth zone.[211,212] The placental defects evident in *Rxra*-deficient mice are reminiscent of the defects in labyrinth zone formation observed in *Pparg*-deficient mice. This finding is logical, since DNA binding of PPARG requires heterodimerization with RXRs. RXRA–PPARG heterodimers mediate retinoic acid-induced villous cytotrophoblast differentiation, activate the expression of *CSH1* and *CGB*, and inhibit extravillous cytotrophoblast invasion.[202,203,205] Thus, RXRA and PPARG cooperate to transactivate a variety of genes associated with villous cytotrophoblast differentiation into syncytiotrophoblast and inhibit extravillous cytotrophoblast cell differentiation.

2.2.2.6 TFAP2A

The transcription factor TFAP2 was introduced in Section 2.1.1.5 as a putative regulator of trophectoderm development. Both TFAP2A and TFAP2C are expressed in human villous cytotrophoblast cells. During in vitro villous cytotrophoblast differentiation, TFAP2C expression is repressed, whereas TFAP2A expression is induced.[213] The induction of TFAP2A is at least partly dependent on transcription factors such as RXRA, nuclear receptor subfamily 2 group F member 2 (NR2F2), and v-ets avian erythroblastosis virus E26 oncogene homolog 1 (ETS1, which is discussed later).[214,215] TFAP2A does not regulate morphological differentiation of villous cytotrophoblast cells but is required for the induction of a variety of genes associated with syncytiotrophoblast endocrine function, including *CSH1*, *CGA*, *CGB*, *CRH*, *CYP19A1*, *PSG1*, and leucine aminopeptidase.[213,216] TFAP2A acts both by direct DNA binding adjacent to target genes and by potentiating CREB-mediated transcriptional activation. Thus, CREB and TFAP2A may work together to facilitate activation of gene expression programs required for biochemical differentiation of villous cytotrophoblast cells.

A variety of TFAP2 family members (TFAP2A, TFAP2B, and TFAP2C) are also expressed in extravillous cytotrophoblast cells, but their effects on development and function of the invasive extravillous cytotrophoblast lineage are not clearly defined. In some studies, TFAP2A and TFAP2C were found to inhibit the migration and invasion of cytotrophoblast cell lines, whereas in another study, TFAP2A promoted invasion in response to epidermal growth factor.[127,217,218] Further studies are, therefore, warranted to elucidate the specific role of TFAP2 family members on extravillous cytotrophoblast function.

2.2.2.7 DLX3

Distalless 3 (DLX3) is a homeodomain containing transcription factor. *Dlx3* is initially expressed in the ectoplacental cone and chorionic plate, and subsequently, in labyrinthine trophoblast cells. *Dlx3*$^{-/-}$ mice exhibit embryonic lethality around gestational day 10, with severe deformities of the spongiotrophoblast and labyrinth trophoblast layers.[219] In human placenta, DLX3 is expressed in villous cytotrophoblast cells, syncytiotrophoblast as well as in extravillous cytotrophoblast cells in the proximal regions of the cell columns.[220] DLX3-binding sites have been identified upstream of a variety of genes associated with syncytiotrophoblast function, including *CSH1*, *CGA*, and *HSD3B1*, indicating that this transcription factor may have a role in the biochemical differentiation of villous cytotrophoblast cells.[221–223]

2.2.2.8 TEAD3

TEAD3 is a transcription factor that is highly expressed in mouse and human placenta. In mice, *Tead3* is expressed in trophoblast giant cells and labyrinthine trophoblast cells, whereas in humans *TEAD3* is expressed in the syncytiotrophoblast layer.[224] *TEAD3* is induced during human villous cytotrophoblast cell differentiation into syncytiotrophoblast, where it plays a key role in transactivation of the *CSH1* and *HSD3B1* genes.[221,224,225] TEAD3 may thus play a role in biochemical differentiation of villous cytotrophoblast cells.

2.2.2.9 ETS1

ETS1 belongs to the ETS family of transcription factors, which was introduced in Section 2.1.1.9. Unlike *Ets2*$^{-/-}$ *mice*, *Ets1*$^{-/-}$ mice are viable and fertile, suggesting that ETS1 is not required for a successful pregnancy in mice.[226] However, ETS1 is expressed in human placenta, with the highest expression in first-trimester villous and extravillous cytotrophoblast cells and decreased expression by term.[227] Knockdown of *ETS1* inhibits both biochemical and morphological differentiation of villous cytotrophoblast cells, whereas ectopic expression of *ETS1* is sufficient to transactivate the promoters of *ERVW-1*, *TFAP2A*, *CSH1*, and *CRH*.[215] Thus, ETS1 activation is an essential transcriptional regulator of villous cytotrophoblast differentiation.

2.2.2.10 IKZF1

Ikaros family zinc finger protein 1, IKZF1 (also called *ikaros*), is a zinc finger DNA-binding protein. *IKZF1* is expressed at low levels in syncytiotrophoblast, but is expressed in extravillous cytotrophoblast cells. Expression of a dominant-negative *IKZF1* isoform in an immortalized extravillous cytotrophoblast cell line abrogated migration and invasion of these cells.[228] *IKZF1* is also able to potentiate TFAP2A transactivation of the leucine aminopeptidase gene, indicating it may have a role in TFAP2A-mediated biochemical differentiation of villous cytotrophoblast cells.[229] These data suggest that *Ikaros* is involved in promoting villous and extravillous cytotrophoblast differentiation.

2.2.2.11 KLF6

The *Krüppel-like factor* family of zinc finger transcription factors is comprised of 17 members, and many of these members are highly expressed in human and mouse placenta.[230] In particular, KLF6 is expressed throughout the in vitro differentiation of human villous cytotrophoblast cells.[231] Knockdown of *KLF6* in villous cytotrophoblast cells reduces morphological differentiation concomitant with reduced expression of *ERVW-1* and the cell cycle regulator *CDKN1A*.[232] Thus, KLF6 may be an important contributor to syncytiotrophoblast generation.

2.2.2.12 β-Catenin/BCL9L/TCF4

β-Catenin is a transcriptional coactivator that regulates major genes involved in cellular proliferation, self-renewal, and differentiation. In "unstimulated" cells, β-catenin resides in the cytoplasm and is constantly targeted for phosphorylation and degradation, thereby preventing β-catenin from translocating to the nucleus to activate target gene expression. Upon stimulation of specific signaling pathways (most notably signaling induced by Wnt ligands binding to *Frizzled* receptors), β-catenin protein is stabilized and shuttles to the nucleus, where it associates with coactivators such as T-cell factor (TCF)/lymphoid enhancer factor (LEF) family members to activate gene transcription.[233] A variety of β-catenin signaling components have been implicated in regulating placental development in mice. For example, mice-lacking β-catenin upstream activators *Wnt2*, *Wnt7b*, and *R-spondin3*, and the Wnt receptor *Frizzled 5* (*Fzd5*) all exhibit severe defects in labyrinth formation.[234–237] Similarly, *Tcf1*/*Lef1* double-deficient mice display severe defects in chorioallantoic fusion and placenta formation,[238] and mice harboring a genetic deficiency of B cell-like 9L (*Bcl9L*), a β-catenin coactivator, exhibit reduced *Gcm1* expression and deficient syncytiotrophoblast formation.[239] In humans, various Wnt ligands and *Frizzled* receptors are expressed in trophoblast cells.[240] β-Catenin expression is high in freshly isolated mononuclear cytotrophoblast cells, and expression dwindles as cells form syncytia in culture.[241] β-Catenin directly activates *GCM1* transcription by binding with cofactors such as TCF4 and BCL9L, suggesting that β-catenin/BCL9L/TCF4-mediated activation of GCM1 may be a conserved necessity for syncytiotrophoblast fusion.[239] β-Catenin activation may also play a role in the regulation of extravillous cytotrophoblast differentiation and function, since the transition from proliferative to postmitotic, invasive extravillous cytotrophoblast cells is associated with increased expression of TCF4 and nuclear accumulation of β-catenin.[242] β-Catenin activation is associated with increased cytotrophoblast invasion in response to a variety of cellular stimuli.[243] However, whether β-catenin activation directly causes enhanced invasiveness of cytotrophoblast cells or is a consequence of differentiation to a more invasive lineage is not yet known. Collectively, the data support a role for β-catenin as a key promoter of villous and extravillous cytotrophoblast differentiation.

2.2.2.13 FOSL1

FOS-like 1 (FOSL1; also called *FOS-related antigen 1* or FRA1) is another member of the FOS family of leucine zipper transcription factors that contributes to the formation of the AP-1 transcription factor complex. FOSL1-deficient mice exhibit placental and extraembryonic abnormalities leading to early embryonic death.[244] Furthermore, knockdown of *Fosl1* or its binding partner, *Junb*, in rat TS cells reduces differentiation and invasion of these cells, whereas in vivo trophoblast-specific disruption of FOSL1 in rats by using lentiviral delivery of *Fosl1* short hairpin RNAs inhibits the depth of endovascular trophoblast cell invasion.[245,246] A similar essential function for FOSL1 in the control of cell motility and invasiveness has been identified in human extravillous cytotrophoblast cells,[17] suggesting that the actions of FOSL1 on the invasive trophoblast lineage are conserved among different species exhibiting robust trophoblast invasion.

2.2.2.14 STAT3

Signal transducer and activator of transcription 3 (*STAT3*) is a transcription factor that mediates cellular responses to a variety of cytokines and growth factors. STAT3 activation is mediated through phosphorylation and either homodimerization or heterodimerization with another member of the

STAT family. Dimerized STAT3 translocates to the nucleus, where it activates a variety of genes associated with cell growth, survival, and motility.

STAT3-deficient mice exhibit early embryo lethality around E7.5, due most likely to deficient visceral endoderm formation.[247] Whether STAT3 activation is required for later stages of mouse placental development is not yet known; however, phosphorylated STAT3 is evident in trophoblast giant cells and spongiotrophoblast cells at midgestation, suggesting that it may have a role in development of the placenta.[248] Interestingly, a key inhibitor of STAT3 activation, *suppressor of cytokine signaling-3* (SOCS3), is also evident in mouse placenta, primarily in trophoblast giant cells.[248] SOCS3-deficient mice die at midgestation due to severe defects in spongiotrophoblast and labyrinthine trophoblast development, and an increase in the size and number of trophoblast giant cells.[249,250] Embryonic lethality in SOCS3-mutant embryos can be rescued by aggregating wild-type trophoblast cells with SOCS3-mutant embryos; or by injecting wild-type TS cells into mutant blastocysts, indicating that the cause of death in these embryos is due to compromised placental development.[250,251] Thus, intricate regulation of STAT3-SOCS3 signaling is required for proper development of the mouse placenta.

STAT3 has also been implicated in human trophoblast development. STAT3 is activated in human extravillous cytotrophoblast cells and cell lines in response to a variety of extracellular stimuli, and activation of STAT3 is associated with increased proliferation and invasiveness of these cells.[252–254] Moreover, depletion of STAT3 by RNA-interference reduces the invasive capacity of human cytotrophoblast cell lines.[254] Activation of STAT3 is also associated with promotion of cytotrophoblast fusion to form syncytiotrophoblast. Thus, STAT3 activation may have a critical role in regulating both villous and extravillous cytotrophoblast cell differentiation and function.

3. CONCLUSIONS

Trophoblast development is a complex process involving dynamic changes in the expression and activity of transcription factors. It is clear that transcription factors have critical roles in regulating all stages of trophoblast development (Table 1). However, transcription factors rarely work alone; rather, they operate in concert with chromatin remodeling factors, histone-modifying enzymes, and other signaling and epigenetic regulators. A deeper mechanistic understanding of how diverse subsets of transcription

Table 1 List of Transcription Factors Implicated in Trophoblast Development and Function

General Function	Name	Description
Early trophoblast development	TEAD4	TEA domain
	CDX2	Caudal-related homeobox
	GATA3	GATA-binding factors
	EOMES	T-box factor
	TFAP2C	Activator protein-2 family
	ESRRB	Nuclear receptor
	SOX2	SRY-related HMG-box
	FOXD3	Forkhead family
	ID (1–4)	Inhibitor of bHLH DNA binding
	ARID3A	ARID family
	ETS2	ETS family
	ELF5	ETS family
Villous and extravillous cytotrophoblast proliferation	HIF	bHLH
	ASCL2	bHLH
	TP63	p53 family
	MYC	bHLH–leucine zipper
	FOS	Leucine zipper, forms AP-1
Villous and extravillous cytotrophoblast differentiation	GCM1	GCM motif family
	CREB	Leucine zipper
	OVOL1	C2H2 zinc finger
	PPARG	Ligand-activated nuclear receptor
	RXRA	Ligand-activated nuclear receptor
	TFAP2A	Activator protein-2 family
	DLX3	Distalless homeobox
	TEAD3	TEA domain
	ETS1	ETS family

Table 1 List of Transcription Factors Implicated in Trophoblast Development and Function—cont'd

General Function	Name	Description
	IKZF1	Ikaros family
	KLF6	Krüppel-like factor
	β-Catenin/ LEF/TCF	High mobility group box-related family
	FOSL1	Leucine zipper, forms AP-1
	STAT3	STAT family

factors and epigenetic regulators integrate together to coordinate gene expression during trophoblast development will yield key insight into the fundamentals of placental biology.

ACKNOWLEDGMENTS

We would like to thank Alexandra Surugiu for illustrative assistance.

REFERENCES

1. Niakan KK, Han J, Pedersen RA, Simon C, Pera RA. Human pre-implantation embryo development. *Development*. 2012;139(5):829–841.
2. Rossant J, Tam PP. Blastocyst lineage formation, early embryonic asymmetries and axis patterning in the mouse. *Development*. 2009;136(5):701–713.
3. Genbacev OD, Prakobphol A, Foulk RA, et al. Trophoblast L-selectin-mediated adhesion at the maternal-fetal interface. *Science*. 2003;299(5605):405–408.
4. Benirschke K, Burton GJ, Baergen RN. *Pathology of the Human Placenta*. 6th ed. New York: Springer; 2012.
5. Enders AC. Trophoblast differentiation during the transition from trophoblastic plate to lacunar stage of implantation in the rhesus monkey and human. *Am J Anat*. 1989;186(1):85–98.
6. Enders AC, King BF. Formation and differentiation of extraembryonic mesoderm in the rhesus monkey. *Am J Anat*. 1988;181(4):327–340.
7. Rossant J, Cross JC. Placental development: lessons from mouse mutants. *Nat Rev Genet*. 2001;2(7):538–548.
8. Quinn J, Kunath T, Rossant J. Mouse trophoblast stem cells. *Methods Mol Med*. 2006;121:125–148.
9. Tanaka S, Kunath T, Hadjantonakis AK, Nagy A, Rossant J. Promotion of trophoblast stem cell proliferation by FGF4. *Science*. 1998;282(5396):2072–2075.
10. Erlebacher A, Price KA, Glimcher LH. Maintenance of mouse trophoblast stem cell proliferation by TGF-beta/activin. *Dev Biol*. 2004;275(1):158–169.
11. Carter AM, Enders AC. Comparative aspects of trophoblast development and placentation. *Reprod Biol Endocrinol*. 2004;2:46.
12. Knight M, Redman CW, Linton EA, Sargent IL. Shedding of syncytiotrophoblast microvilli into the maternal circulation in pre-eclamptic pregnancies. *Br J Obstet Gynaecol*. 1998;105(6):632–640.

13. Simpson RA, Mayhew TM, Barnes PR. From 13 weeks to term, the trophoblast of human placenta grows by the continuous recruitment of new proliferative units: a study of nuclear number using the disector. *Placenta*. 1992;13(5):501–512.
14. Mi S, Lee X, Li X, et al. Syncytin is a captive retroviral envelope protein involved in human placental morphogenesis. *Nature*. 2000;403(6771):785–789.
15. Blaise S, de Parseval N, Benit L, Heidmann T. Genomewide screening for fusogenic human endogenous retrovirus envelopes identifies syncytin 2, a gene conserved on primate evolution. *Proc Natl Acad Sci USA*. 2003;100(22):13013–13018.
16. Malassine A, Frendo JL, Evain-Brion D. A comparison of placental development and endocrine functions between the human and mouse model. *Hum Reprod Update*. 2003;9(6):531–539.
17. Renaud SJ, Kubota K, Rumi MA, Soares MJ. The FOS transcription factor family differentially controls trophoblast migration and invasion. *J Biol Chem*. 2014;289(8):5025–5039.
18. Wright JK, Dunk CE, Amsalem H, Maxwell C, Keating S, Lye SJ. HER1 signaling mediates extravillous trophoblast differentiation in humans. *Biol Reprod*. 2010;83(6):1036–1045.
19. Damsky CH, Fitzgerald ML, Fisher SJ. Distribution patterns of extracellular matrix components and adhesion receptors are intricately modulated during first trimester cytotrophoblast differentiation along the invasive pathway, in vivo. *J Clin Invest*. 1992;89(1):210–222.
20. Jokhi PP, King A, Loke YW. Reciprocal expression of epidermal growth factor receptor (EGF-R) and c-erbB2 by non-invasive and invasive human trophoblast populations. *Cytokine*. 1994;6(4):433–442.
21. Pijnenborg R, Vercruysse L, Hanssens M. The uterine spiral arteries in human pregnancy: facts and controversies. *Placenta*. 2006;27(9–10):939–958.
22. Rodesch F, Simon P, Donner C, Jauniaux E. Oxygen measurements in endometrial and trophoblastic tissues during early pregnancy. *Obstet Gynecol*. 1992;80(2):283–285.
23. Jauniaux E, Watson AL, Hempstock J, Bao YP, Skepper JN, Burton GJ. Onset of maternal arterial blood flow and placental oxidative stress. A possible factor in human early pregnancy failure. *Am J Pathol*. 2000;157(6):2111–2122.
24. Pijnenborg R, Robertson WB, Brosens I, Dixon G. Review article: trophoblast invasion and the establishment of haemochorial placentation in man and laboratory animals. *Placenta*. 1981;2(1):71–91.
25. Damsky CH, Fisher SJ. Trophoblast pseudo-vasculogenesis: faking it with endothelial adhesion receptors. *Curr Opin Cell Biol*. 1998;10(5):660–666.
26. Simmons DG, Cross JC. Determinants of trophoblast lineage and cell subtype specification in the mouse placenta. *Dev Biol*. 2005;284(1):12–24.
27. Simmons DG, Fortier AL, Cross JC. Diverse subtypes and developmental origins of trophoblast giant cells in the mouse placenta. *Dev Biol*. 2007;304(2):567–578.
28. Ain R, Canham LN, Soares MJ. Gestation stage-dependent intrauterine trophoblast cell invasion in the rat and mouse: novel endocrine phenotype and regulation. *Dev Biol*. 2003;260(1):176–190.
29. Vercruysse L, Caluwaerts S, Luyten C, Pijnenborg R. Interstitial trophoblast invasion in the decidua and mesometrial triangle during the last third of pregnancy in the rat. *Placenta*. 2006;27(1):22–33.
30. Caluwaerts S, Vercruysse L, Luyten C, Pijnenborg R. Endovascular trophoblast invasion and associated structural changes in uterine spiral arteries of the pregnant rat. *Placenta*. 2005;26(7):574–584.
31. Soares MJ, Chakraborty D, Karim Rumi MA, Konno T, Renaud SJ. Rat placentation: an experimental model for investigating the hemochorial maternal-fetal interface. *Placenta*. 2012;33(4):233–243.

32. Hofmann AP, Gerber SA, Croy BA. Uterine natural killer cells pace early development of mouse decidua basalis. *Mol Hum Reprod.* 2014;20(1):66–76.
33. Knott JG, Paul S. Transcriptional regulators of the trophoblast lineage in mammals with hemochorial placentation. *Reproduction.* 2014;148(6):R121–R136.
34. Maltepe E, Bakardjiev AI, Fisher SJ. The placenta: transcriptional, epigenetic, and physiological integration during development. *J Clin Invest.* 2010;120(4):1016–1025.
35. Suwinska A, Czolowska R, Ozdzenski W, Tarkowski AK. Blastomeres of the mouse embryo lose totipotency after the fifth cleavage division: expression of Cdx2 and Oct4 and developmental potential of inner and outer blastomeres of 16- and 32-cell embryos. *Dev Biol.* 2008;322(1):133–144.
36. Hillman N, Sherman MI, Graham C. The effect of spatial arrangement on cell determination during mouse development. *J Embryol Exp Morphol.* 1972;28(2):263–278.
37. Lorthongpanich C, Doris TP, Limviphuvadh V, Knowles BB, Solter D. Developmental fate and lineage commitment of singled mouse blastomeres. *Development.* 2012;139(20):3722–3731.
38. Nishioka N, Inoue K, Adachi K, et al. The Hippo signaling pathway components Lats and Yap pattern Tead4 activity to distinguish mouse trophectoderm from inner cell mass. *Dev Cell.* 2009;16(3):398–410.
39. Nishioka N, Yamamoto S, Kiyonari H, et al. Tead4 is required for specification of trophectoderm in pre-implantation mouse embryos. *Mech Dev.* 2008;125(3–4): 270–283.
40. Home P, Saha B, Ray S, et al. Altered subcellular localization of transcription factor TEAD4 regulates first mammalian cell lineage commitment. *Proc Natl Acad Sci USA.* 2012;109(19):7362–7367.
41. Yagi R, Kohn MJ, Karavanova I, et al. Transcription factor TEAD4 specifies the trophectoderm lineage at the beginning of mammalian development. *Development.* 2007;134(21):3827–3836.
42. Hossain Z, Ali SM, Ko HL, et al. Glomerulocystic kidney disease in mice with a targeted inactivation of Wwtr1. *Proc Natl Acad Sci USA.* 2007;104(5):1631–1636.
43. Morin-Kensicki EM, Boone BN, Howell M, et al. Defects in yolk sac vasculogenesis, chorioallantoic fusion, and embryonic axis elongation in mice with targeted disruption of Yap65. *Mol Cell Biol.* 2006;26(1):77–87.
44. Strumpf D, Mao CA, Yamanaka Y, et al. Cdx2 is required for correct cell fate specification and differentiation of trophectoderm in the mouse blastocyst. *Development.* 2005;132(9):2093–2102.
45. Ralston A, Rossant J. Cdx2 acts downstream of cell polarization to cell-autonomously promote trophectoderm fate in the early mouse embryo. *Dev Biol.* 2008;313(2): 614–629.
46. Chawengsaksophak K, James R, Hammond VE, Kontgen F, Beck F. Homeosis and intestinal tumours in Cdx2 mutant mice. *Nature.* 1997;386(6620):84–87.
47. Wu G, Gentile L, Fuchikami T, et al. Initiation of trophectoderm lineage specification in mouse embryos is independent of Cdx2. *Development.* 2010;137(24):4159–4169.
48. Niwa H, Toyooka Y, Shimosato D, et al. Interaction between Oct3/4 and Cdx2 determines trophectoderm differentiation. *Cell.* 2005;123(5):917–929.
49. Blij S, Frum T, Akyol A, Fearon E, Ralston A. Maternal Cdx2 is dispensable for mouse development. *Development.* 2012;139(21):3969–3972.
50. Jedrusik A, Cox A, Wicher KB, Glover DM, Zernicka-Goetz M. Maternal-zygotic knockout reveals a critical role of Cdx2 in the morula to blastocyst transition. *Dev Biol.* 2015;398(2):147–152.
51. Jedrusik A, Bruce AW, Tan MH, et al. Maternally and zygotically provided Cdx2 have novel and critical roles for early development of the mouse embryo. *Dev Biol.* 2010;344(1):66–78.

52. Dietrich JE, Hiiragi T. Stochastic patterning in the mouse pre-implantation embryo. *Development.* 2007;134(23):4219–4231.
53. Tolkunova E, Cavaleri F, Eckardt S, et al. The caudal-related protein cdx2 promotes trophoblast differentiation of mouse embryonic stem cells. *Stem Cells.* 2006;24(1):139–144.
54. Niwa H, Miyazaki J, Smith AG. Quantitative expression of Oct-3/4 defines differentiation, dedifferentiation or self-renewal of ES cells. *Nat Genet.* 2000;24(4): 372–376.
55. Home P, Ray S, Dutta D, Bronshteyn I, Larson M, Paul S. GATA3 is selectively expressed in the trophectoderm of peri-implantation embryo and directly regulates Cdx2 gene expression. *J Biol Chem.* 2009;284(42):28729–28737.
56. Ralston A, Cox BJ, Nishioka N, et al. Gata3 regulates trophoblast development downstream of Tead4 and in parallel to Cdx2. *Development.* 2010;137(3):395–403.
57. Nishiyama A, Xin L, Sharov AA, et al. Uncovering early response of gene regulatory networks in ESCs by systematic induction of transcription factors. *Cell Stem Cell.* 2009;5(4):420–433.
58. Ma GT, Roth ME, Groskopf JC, et al. GATA-2 and GATA-3 regulate trophoblast-specific gene expression in vivo. *Development.* 1997;124(4):907–914.
59. Ray S, Dutta D, Rumi MA, Kent LN, Soares MJ, Paul S. Context-dependent function of regulatory elements and a switch in chromatin occupancy between GATA3 and GATA2 regulate Gata2 transcription during trophoblast differentiation. *J Biol Chem.* 2009;284(8):4978–4988.
60. Russ AP, Wattler S, Colledge WH, et al. Eomesodermin is required for mouse trophoblast development and mesoderm formation. *Nature.* 2000;404(6773):95–99.
61. Kidder BL, Palmer S. Examination of transcriptional networks reveals an important role for TCFAP2C, SMARCA4, and EOMES in trophoblast stem cell maintenance. *Genome Res.* 2010;20(4):458–472.
62. Kuckenberg P, Buhl S, Woynecki T, et al. The transcription factor TCFAP2C/AP-2gamma cooperates with CDX2 to maintain trophectoderm formation. *Mol Cell Biol.* 2010;30(13):3310–3320.
63. Auman HJ, Nottoli T, Lakiza O, Winger Q, Donaldson S, Williams T. Transcription factor AP-2gamma is essential in the extra-embryonic lineages for early postimplantation development. *Development.* 2002;129(11):2733–2747.
64. Winger Q, Huang J, Auman HJ, Lewandoski M, Williams T. Analysis of transcription factor AP-2 expression and function during mouse preimplantation development. *Biol Reprod.* 2006;75(3):324–333.
65. Choi I, Carey TS, Wilson CA, Knott JG. Transcription factor AP-2gamma is a core regulator of tight junction biogenesis and cavity formation during mouse early embryogenesis. *Development.* 2012;139(24):4623–4632.
66. Cao Z, Carey TS, Ganguly A, Wilson CA, Paul S, Knott JG. Transcription factor AP-2gamma induces early Cdx2 expression and represses HIPPO signaling to specify the trophectoderm lineage. *Development.* 2015;142(9):1606–1615.
67. Martello G, Sugimoto T, Diamanti E, et al. Esrrb is a pivotal target of the Gsk3/Tcf3 axis regulating embryonic stem cell self-renewal. *Cell Stem Cell.* 2012;11(4):491–504.
68. Hanna LA, Foreman RK, Tarasenko IA, Kessler DS, Labosky PA. Requirement for Foxd3 in maintaining pluripotent cells of the early mouse embryo. *Genes Dev.* 2002;16(20):2650–2661.
69. Savarese F, Davila A, Nechanitzky R, et al. Satb1 and Satb2 regulate embryonic stem cell differentiation and Nanog expression. *Genes Dev.* 2009;23(22):2625–2638.
70. Avilion AA, Nicolis SK, Pevny LH, Perez L, Vivian N, Lovell-Badge R. Multipotent cell lineages in early mouse development depend on SOX2 function. *Genes Dev.* 2003;17(1):126–140.

71. Luo J, Sladek R, Bader JA, Matthyssen A, Rossant J, Giguere V. Placental abnormalities in mouse embryos lacking the orphan nuclear receptor ERR-beta. *Nature*. 1997;388(6644):778–782.
72. Tremblay GB, Kunath T, Bergeron D, et al. Diethylstilbestrol regulates trophoblast stem cell differentiation as a ligand of orphan nuclear receptor ERR beta. *Genes Dev*. 2001;15(7):833–838.
73. Latos PA, Goncalves A, Oxley D, Mohammed H, Turro E, Hemberger M. Fgf and Esrrb integrate epigenetic and transcriptional networks that regulate self-renewal of trophoblast stem cells. *Nat Commun*. 2015;6:7776.
74. Adachi K, Nikaido I, Ohta H, et al. Context-dependent wiring of Sox2 regulatory networks for self-renewal of embryonic and trophoblast stem cells. *Mol Cell*. 2013;52(3):380–392.
75. Keramari M, Razavi J, Ingman KA, et al. Sox2 is essential for formation of trophectoderm in the preimplantation embryo. *PLoS One*. 2010;5(11), e13952.
76. Tompers DM, Foreman RK, Wang Q, Kumanova M, Labosky PA. Foxd3 is required in the trophoblast progenitor cell lineage of the mouse embryo. *Dev Biol*. 2005;285(1):126–137.
77. Lasorella A, Uo T, Iavarone A. Id proteins at the cross-road of development and cancer. *Oncogene*. 2001;20(58):8326–8333.
78. Asanoma K, Rumi MA, Kent LN, et al. FGF4-dependent stem cells derived from rat blastocysts differentiate along the trophoblast lineage. *Dev Biol*. 2011;351(1):110–119.
79. Guo G, Huss M, Tong GQ, et al. Resolution of cell fate decisions revealed by single-cell gene expression analysis from zygote to blastocyst. *Dev Cell*. 2010;18(4):675–685.
80. Wang J, Rao S, Chu J, et al. A protein interaction network for pluripotency of embryonic stem cells. *Nature*. 2006;444(7117):364–368.
81. Rhee C, Lee BK, Beck S, et al. Arid3a is essential to execution of the first cell fate decision via direct embryonic and extraembryonic transcriptional regulation. *Genes Dev*. 2014;28(20):2219–2232.
82. Webb CF, Bryant J, Popowski M, et al. The ARID family transcription factor bright is required for both hematopoietic stem cell and B lineage development. *Mol Cell Biol*. 2011;31(5):1041–1053.
83. Yamamoto H, Flannery ML, Kupriyanov S, et al. Defective trophoblast function in mice with a targeted mutation of Ets2. *Genes Dev*. 1998;12(9):1315–1326.
84. Georgiades P, Rossant J. Ets2 is necessary in trophoblast for normal embryonic anteroposterior axis development. *Development*. 2006;133(6):1059–1068.
85. Wen F, Tynan JA, Cecena G, et al. Ets2 is required for trophoblast stem cell self-renewal. *Dev Biol*. 2007;312(1):284–299.
86. Donnison M, Beaton A, Davey HW, Broadhurst R, L'Huillier P, Pfeffer PL. Loss of the extraembryonic ectoderm in Elf5 mutants leads to defects in embryonic patterning. *Development*. 2005;132(10):2299–2308.
87. Latos PA, Helliwell C, Mosaku O, et al. NuRD-dependent DNA methylation prevents ES cells from accessing a trophectoderm fate. *Biol Open*. 2012;1(4):341–352.
88. Ng RK, Dean W, Dawson C, et al. Epigenetic restriction of embryonic cell lineage fate by methylation of Elf5. *Nat Cell Biol*. 2008;10(11):1280–1290.
89. Donnison M, Broadhurst R, Pfeffer PL. Elf5 and Ets2 maintain the mouse extraembryonic ectoderm in a dosage dependent synergistic manner. *Dev Biol*. 2015;397(1):77–88.
90. Latos PA, Sienerth AR, Murray A, et al. Elf5-centered transcription factor hub controls trophoblast stem cell self-renewal and differentiation through stoichiometry-sensitive shifts in target gene networks. *Genes Dev*. 2015;29(23):2435–2448.
91. Pearton DJ, Smith CS, Redgate E, van Leeuwen J, Donnison M, Pfeffer PL. Elf5 counteracts precocious trophoblast differentiation by maintaining Sox2 and 3 and inhibiting Hand1 expression. *Dev Biol*. 2014;392(2):344–357.

92. Cockburn K, Rossant J. Making the blastocyst: lessons from the mouse. *J Clin Invest.* 2010;120(4):995–1003.
93. Rossant J. Mouse and human blastocyst-derived stem cells: vive les differences. *Development.* 2015;142(1):9–12.
94. Douglas GC, VandeVoort CA, Kumar P, Chang TC, Golos TG. Trophoblast stem cells: models for investigating trophectoderm differentiation and placental development. *Endocr Rev.* 2009;30(3):228–240.
95. De Paepe C, Krivega M, Cauffman G, Geens M, Van de Velde H. Totipotency and lineage segregation in the human embryo. *Mol Hum Reprod.* 2014;20(7):599–618.
96. Ginis I, Luo Y, Miura T, et al. Differences between human and mouse embryonic stem cells. *Dev Biol.* 2004;269(2):360–380.
97. Chuykin I, Lapidus I, Popova E, et al. Characterization of trophoblast and extraembryonic endoderm cell lineages derived from rat preimplantation embryos. *PLoS One.* 2010;5(3), e9794.
98. Grigor'eva EV, Shevchenko AI, Mazurok NA, et al. FGF4 independent derivation of trophoblast stem cells from the common vole. *PLoS One.* 2009;4(9), e7161.
99. Vandevoort CA, Thirkill TL, Douglas GC. Blastocyst-derived trophoblast stem cells from the rhesus monkey. *Stem Cells Dev.* 2007;16(5):779–788.
100. Hashizume K, Shimada A, Nakano H, Takahashi T. Bovine trophoblast cell culture systems: a technique to culture bovine trophoblast cells without feeder cells. *Methods Mol Med.* 2006;121:179–188.
101. Flechon JE, Laurie S, Notarianni E. Isolation and characterization of a feeder-dependent, porcine trophectoderm cell line obtained from a 9-day blastocyst. *Placenta.* 1995;16(7):643–658.
102. Kunath T, Yamanaka Y, Detmar J, et al. Developmental differences in the expression of FGF receptors between human and mouse embryos. *Placenta.* 2014;35(12):1079–1088.
103. Niakan KK, Eggan K. Analysis of human embryos from zygote to blastocyst reveals distinct gene expression patterns relative to the mouse. *Dev Biol.* 2013;375(1):54–64.
104. Hansis C, Grifo JA, Krey LC. Oct-4 expression in inner cell mass and trophectoderm of human blastocysts. *Mol Hum Reprod.* 2000;6(11):999–1004.
105. Adjaye J, Huntriss J, Herwig R, et al. Primary differentiation in the human blastocyst: comparative molecular portraits of inner cell mass and trophectoderm cells. *Stem Cells.* 2005;23(10):1514–1525.
106. Cauffman G, Van de Velde H, Liebaers I, Van Steirteghem A. Oct-4 mRNA and protein expression during human preimplantation development. *Mol Hum Reprod.* 2005;11(3):173–181.
107. Berg DK, Smith CS, Pearton DJ, et al. Trophectoderm lineage determination in cattle. *Dev Cell.* 2011;20(2):244–255.
108. De Paepe C, Cauffman G, Verloes A, et al. Human trophectoderm cells are not yet committed. *Hum Reprod.* 2013;28(3):740–749.
109. Blakeley P, Fogarty NM, del Valle I, et al. Defining the three cell lineages of the human blastocyst by single-cell RNA-seq. *Development.* 2015;142(18):3151–3165.
110. Petropoulos S, Edsgard D, Reinius B, et al. Single-cell RNA-Seq reveals lineage and X chromosome dynamics in human preimplantation embryos. *Cell.* 2016;165(4):1012–1026.
111. Piras V, Tomita M, Selvarajoo K. Transcriptome-wide variability in single embryonic development cells. *Sci Rep.* 2014;4:7137.
112. Yan L, Yang M, Guo H, et al. Single-cell RNA-Seq profiling of human preimplantation embryos and embryonic stem cells. *Nat Struct Mol Biol.* 2013;20(9):1131–1139.
113. Deng Q, Ramskold D, Reinius B, Sandberg R. Single-cell RNA-seq reveals dynamic, random monoallelic gene expression in mammalian cells. *Science.* 2014;343(6167):193–196.

114. Xu RH, Chen X, Li DS, et al. BMP4 initiates human embryonic stem cell differentiation to trophoblast. *Nat Biotechnol.* 2002;20(12):1261–1264.
115. Amita M, Adachi K, Alexenko AP, et al. Complete and unidirectional conversion of human embryonic stem cells to trophoblast by BMP4. *Proc Natl Acad Sci USA.* 2013;110(13):E1212–E1221.
116. Zdravkovic T, Nazor KL, Larocque N, et al. Human stem cells from single blastomeres reveal pathways of embryonic or trophoblast fate specification. *Development.* 2015;142(23):4010–4025.
117. Rossant J. Stem cells in the mammalian blastocyst. *Harvey Lect.* 2001;97:17–40.
118. Li Y, Moretto-Zita M, Soncin F, et al. BMP4-directed trophoblast differentiation of human embryonic stem cells is mediated through a DeltaNp63+ cytotrophoblast stem cell state. *Development.* 2013;140(19):3965–3976.
119. Marchand M, Horcajadas JA, Esteban FJ, McElroy SL, Fisher SJ, Giudice LC. Transcriptomic signature of trophoblast differentiation in a human embryonic stem cell model. *Biol Reprod.* 2011;84(6):1258–1271.
120. Roberts RM, Loh KM, Amita M, et al. Differentiation of trophoblast cells from human embryonic stem cells: to be or not to be? *Reproduction.* 2014;147(5):D1–D12.
121. Lee CQ, Gardner L, Turco M, et al. What is trophoblast? A combination of criteria define human first-trimester trophoblast. *Stem Cell Reports.* 2016;6(2):257–272.
122. Hemberger M, Udayashankar R, Tesar P, Moore H, Burton GJ. ELF5-enforced transcriptional networks define an epigenetically regulated trophoblast stem cell compartment in the human placenta. *Hum Mol Genet.* 2010;19(12):2456–2467.
123. Chen Y, Wang K, Gong YG, Khoo SK, Leach R. Roles of CDX2 and EOMES in human induced trophoblast progenitor cells. *Biochem Biophys Res Commun.* 2013;431(2):197–202.
124. Genbacev O, Larocque N, Ona K, et al. Integrin alpha4-positive human trophoblast progenitors: functional characterization and transcriptional regulation. *Hum Reprod.* 2016;31(6):1300–1314.
125. Genbacev O, Donne M, Kapidzic M, et al. Establishment of human trophoblast progenitor cell lines from the chorion. *Stem Cells.* 2011;29(9):1427–1436.
126. Banet N, Gown AM, Shih Ie M, et al. GATA-3 expression in trophoblastic tissues: an immunohistochemical study of 445 cases, including diagnostic utility. *Am J Surg Pathol.* 2015;39(1):101–108.
127. Biadasiewicz K, Sonderegger S, Haslinger P, et al. Transcription factor AP-2alpha promotes EGF-dependent invasion of human trophoblast. *Endocrinology.* 2011;152(4): 1458–1469.
128. Orendi K, Gauster M, Moser G, Meiri H, Huppertz B. The choriocarcinoma cell line BeWo: syncytial fusion and expression of syncytium-specific proteins. *Reproduction.* 2010;140(5):759–766.
129. Semenza GL. Regulation of oxygen homeostasis by hypoxia-inducible factor 1. *Physiology (Bethesda).* 2009;24:97–106.
130. Adelman DM, Gertsenstein M, Nagy A, Simon MC, Maltepe E. Placental cell fates are regulated in vivo by HIF-mediated hypoxia responses. *Genes Dev.* 2000;14(24): 3191–3203.
131. Cowden Dahl KD, Fryer BH, Mack FA, et al. Hypoxia-inducible factors 1alpha and 2alpha regulate trophoblast differentiation. *Mol Cell Biol.* 2005;25(23):10479–10491.
132. Chakraborty D, Rumi MA, Konno T, Soares MJ. Natural killer cells direct hemochorial placentation by regulating hypoxia-inducible factor dependent trophoblast lineage decisions. *Proc Natl Acad Sci USA.* 2011;108(39):16295–16300.
133. Maltepe E, Krampitz GW, Okazaki KM, et al. Hypoxia-inducible factor-dependent histone deacetylase activity determines stem cell fate in the placenta. *Development.* 2005;132(15):3393–3403.

134. Rajakumar A, Conrad KP. Expression, ontogeny, and regulation of hypoxia-inducible transcription factors in the human placenta. *Biol Reprod.* 2000;63(2):559–569.
135. Genbacev O, Zhou Y, Ludlow JW, Fisher SJ. Regulation of human placental development by oxygen tension. *Science.* 1997;277(5332):1669–1672.
136. Alsat E, Wyplosz P, Malassine A, et al. Hypoxia impairs cell fusion and differentiation process in human cytotrophoblast, in vitro. *J Cell Physiol.* 1996;168(2):346–353.
137. Jiang B, Kamat A, Mendelson CR. Hypoxia prevents induction of aromatase expression in human trophoblast cells in culture: potential inhibitory role of the hypoxia-inducible transcription factor Mash-2 (mammalian Achaete-Scute homologous protein-2). *Mol Endocrinol.* 2000;14(10):1661–1673.
138. Caniggia I, Mostachfi H, Winter J, et al. Hypoxia-inducible factor-1 mediates the biological effects of oxygen on human trophoblast differentiation through TGFbeta(3). *J Clin Invest.* 2000;105(5):577–587.
139. Kilburn BA, Wang J, Duniec-Dmuchowski ZM, Leach RE, Romero R, Armant DR. Extracellular matrix composition and hypoxia regulate the expression of HLA-G and integrins in a human trophoblast cell line. *Biol Reprod.* 2000;62(3):739–747.
140. Zhou Y, Genbacev O, Damsky CH, Fisher SJ. Oxygen regulates human cytotrophoblast differentiation and invasion: implications for endovascular invasion in normal pregnancy and in pre-eclampsia. *J Reprod Immunol.* 1998;39(1–2):197–213.
141. James JL, Stone PR, Chamley LW. The effects of oxygen concentration and gestational age on extravillous trophoblast outgrowth in a human first trimester villous explant model. *Hum Reprod.* 2006;21(10):2699–2705.
142. Graham CH, Postovit LM, Park H, Canning MT, Fitzpatrick TE. Adriana and Luisa Castellucci award lecture 1999: role of oxygen in the regulation of trophoblast gene expression and invasion. *Placenta.* 2000;21(5–6):443–450.
143. Robins JC, Heizer A, Hardiman A, Hubert M, Handwerger S. Oxygen tension directs the differentiation pathway of human cytotrophoblast cells. *Placenta.* 2007;28(11–12): 1141–1146.
144. Hayashi M, Sakata M, Takeda T, et al. Up-regulation of c-met protooncogene product expression through hypoxia-inducible factor-1alpha is involved in trophoblast invasion under low-oxygen tension. *Endocrinology.* 2005;146(11):4682–4689.
145. Rosario GX, Konno T, Soares MJ. Maternal hypoxia activates endovascular trophoblast cell invasion. *Dev Biol.* 2008;314(2):362–375.
146. Guillemot F, Nagy A, Auerbach A, Rossant J, Joyner AL. Essential role of Mash-2 in extraembryonic development. *Nature.* 1994;371(6495):333–336.
147. Schuijers J, Junker JP, Mokry M, et al. Ascl2 acts as an R-spondin/Wnt-responsive switch to control stemness in intestinal crypts. *Cell Stem Cell.* 2015;16(2):158–170.
148. Tanaka M, Gertsenstein M, Rossant J, Nagy A. Mash2 acts cell autonomously in mouse spongiotrophoblast development. *Dev Biol.* 1997;190(1):55–65.
149. Oh-McGinnis R, Bogutz AB, Lefebvre L. Partial loss of Ascl2 function affects all three layers of the mature placenta and causes intrauterine growth restriction. *Dev Biol.* 2011;351(2):277–286.
150. Alders M, Hodges M, Hadjantonakis AK, et al. The human Achaete-Scute homologue 2 (ASCL2, HASH2) maps to chromosome 11p15.5, close to IGF2 and is expressed in extravillus trophoblasts. *Hum Mol Genet.* 1997;6(6):859–867.
151. Janatpour MJ, Utset MF, Cross JC, et al. A repertoire of differentially expressed transcription factors that offers insight into mechanisms of human cytotrophoblast differentiation. *Dev Genet.* 1999;25(2):146–157.
152. Jiang B, Mendelson CR. O_2 enhancement of human trophoblast differentiation and hCYP19 (aromatase) gene expression are mediated by proteasomal degradation of USF1 and USF2. *Mol Cell Biol.* 2005;25(20):8824–8833.

153. Jiang B, Mendelson CR. USF1 and USF2 mediate inhibition of human trophoblast differentiation and CYP19 gene expression by Mash-2 and hypoxia. *Mol Cell Biol.* 2003;23(17):6117–6128.
154. King KE, Weinberg WC. p63: defining roles in morphogenesis, homeostasis, and neoplasia of the epidermis. *Mol Carcinog.* 2007;46(8):716–724.
155. Romano RA, Smalley K, Magraw C, et al. DeltaNp63 knockout mice reveal its indispensable role as a master regulator of epithelial development and differentiation. *Development.* 2012;139(4):772–782.
156. Senoo M, Pinto F, Crum CP, McKeon F. p63 Is essential for the proliferative potential of stem cells in stratified epithelia. *Cell.* 2007;129(3):523–536.
157. Lee Y, Kim KR, McKeon F, et al. A unifying concept of trophoblastic differentiation and malignancy defined by biomarker expression. *Hum Pathol.* 2007;38(7):1003–1013.
158. Shih IM, Kurman RJ. p63 expression is useful in the distinction of epithelioid trophoblastic and placental site trophoblastic tumors by profiling trophoblastic subpopulations. *Am J Surg Pathol.* 2004;28(9):1177–1183.
159. Li Y, Moretto-Zita M, Leon-Garcia S, Parast MM. p63 inhibits extravillous trophoblast migration and maintains cells in a cytotrophoblast stem cell-like state. *Am J Pathol.* 2014;184(12):3332–3343.
160. Xue WC, Feng HC, Chan KY, et al. Id helix-loop-helix proteins are differentially expressed in gestational trophoblastic disease. *Histopathology.* 2005;47(3):303–309.
161. Janatpour MJ, McMaster MT, Genbacev O, et al. Id-2 regulates critical aspects of human cytotrophoblast differentiation, invasion and migration. *Development.* 2000;127(3):549–558.
162. Eilers M, Eisenman RN. Myc's broad reach. *Genes Dev.* 2008;22(20):2755–2766.
163. Roncalli M, Bulfamante G, Viale G, et al. C-myc and tumour suppressor gene product expression in developing and term human trophoblast. *Placenta.* 1994;15(4):399–409.
164. Diebold J, Arnholdt H, Lai MD, Lohrs U. C-myc expression in early human placenta—a critical evaluation of its localization. *Virchows Arch B Cell Pathol Incl Mol Pathol.* 1991;61(1):65–73.
165. Kumar P, Luo Y, Tudela C, Alexander JM, Mendelson CR. The c-Myc-regulated microRNA-17∼92 (miR-17∼92) and miR-106a∼363 clusters target hCYP19A1 and hGCM1 to inhibit human trophoblast differentiation. *Mol Cell Biol.* 2013;33(9):1782–1796.
166. Johnson RS, Spiegelman BM, Papaioannou V. Pleiotropic effects of a null mutation in the c-fos proto-oncogene. *Cell.* 1992;71(4):577–586.
167. Bamberger AM, Bamberger CM, Aupers S, Milde-Langosch K, Loning T, Makrigiannakis A. Expression pattern of the activating protein-1 family of transcription factors in the human placenta. *Mol Hum Reprod.* 2004;10(4):223–228.
168. Anson-Cartwright L, Dawson K, Holmyard D, Fisher SJ, Lazzarini RA, Cross JC. The glial cells missing-1 protein is essential for branching morphogenesis in the chorioallantoic placenta. *Nat Genet.* 2000;25(3):311–314.
169. Stecca B, Nait-Oumesmar B, Kelley KA, Voss AK, Thomas T, Lazzarini RA. Gcm1 expression defines three stages of chorio-allantoic interaction during placental development. *Mech Dev.* 2002;115(1–2):27–34.
170. Hughes M, Dobric N, Scott IC, et al. The Hand1, Stra13 and Gcm1 transcription factors override FGF signaling to promote terminal differentiation of trophoblast stem cells. *Dev Biol.* 2004;271(1):26–37.
171. Schubert SW, Lamoureux N, Kilian K, Klein-Hitpass L, Hashemolhosseini S. Identification of integrin-alpha4, Rb1, and syncytin a as murine placental target genes of the transcription factor GCMa/Gcm1. *J Biol Chem.* 2008;283(9):5460–5465.

172. Bainbridge SA, Minhas A, Whiteley KJ, et al. Effects of reduced Gcm1 expression on trophoblast morphology, fetoplacental vascularity, and pregnancy outcomes in mice. *Hypertension*. 2012;59(3):732–739.
173. Baczyk D, Satkunaratnam A, Nait-Oumesmar B, Huppertz B, Cross JC, Kingdom JC. Complex patterns of GCM1 mRNA and protein in villous and extravillous trophoblast cells of the human placenta. *Placenta*. 2004;25(6):553–559.
174. Chang CW, Cheong ML, Chang GD, Tsai MS, Chen H. Involvement of Epac1/Rap1/CaMKI/HDAC5 signaling cascade in the regulation of placental cell fusion. *Mol Hum Reprod*. 2013;19(11):745–755.
175. Lin FY, Chang CW, Cheong ML, et al. Dual-specificity phosphatase 23 mediates GCM1 dephosphorylation and activation. *Nucleic Acids Res*. 2011;39(3):848–861.
176. Chiang MH, Liang FY, Chen CP, et al. Mechanism of hypoxia-induced GCM1 degradation: implications for the pathogenesis of preeclampsia. *J Biol Chem*. 2009;284(26):17411–17419.
177. Kashif M, Hellwig A, Kolleker A, et al. p45NF-E2 represses Gcm1 in trophoblast cells to regulate syncytium formation, placental vascularization and embryonic growth. *Development*. 2011;138(11):2235–2247.
178. Chiu YH, Chen H. GATA3 inhibits GCM1 activity and trophoblast cell invasion. *Sci Rep*. 2016;6:21630.
179. Baczyk D, Drewlo S, Proctor L, Dunk C, Lye S, Kingdom J. Glial cell missing-1 transcription factor is required for the differentiation of the human trophoblast. *Cell Death Differ*. 2009;16(5):719–727.
180. Yu C, Shen K, Lin M, et al. GCMa regulates the syncytin-mediated trophoblastic fusion. *J Biol Chem*. 2002;277(51):50062–50068.
181. Yamada K, Ogawa H, Honda S, Harada N, Okazaki T. A GCM motif protein is involved in placenta-specific expression of human aromatase gene. *J Biol Chem*. 1999;274(45):32279–32286.
182. Chang M, Mukherjea D, Gobble RM, Groesch KA, Torry RJ, Torry DS. Glial cell missing 1 regulates placental growth factor (PGF) gene transcription in human trophoblast. *Biol Reprod*. 2008;78(5):841–851.
183. Wang LJ, Cheong ML, Lee YS, Lee MT, Chen H. High-temperature requirement protein A4 (HtrA4) suppresses the fusogenic activity of syncytin-1 and promotes trophoblast invasion. *Mol Cell Biol*. 2012;32(18):3707–3717.
184. Delidaki M, Gu M, Hein A, Vatish M, Grammatopoulos DK. Interplay of cAMP and MAPK pathways in hCG secretion and fusogenic gene expression in a trophoblast cell line. *Mol Cell Endocrinol*. 2011;332(1–2):213–220.
185. Ghosh D, Sachdev S, Hannink M, Roberts RM. Coordinate regulation of basal and cyclic 5'-adenosine monophosphate (cAMP)-activated expression of human chorionic gonadotropin-alpha by Ets-2 and cAMP-responsive element binding protein. *Mol Endocrinol*. 2005;19(4):1049–1066.
186. Zhou Z, Wang R, Yang X, et al. The cAMP-responsive element binding protein (CREB) transcription factor regulates furin expression during human trophoblast syncytialization. *Placenta*. 2014;35(11):907–918.
187. Toufaily C, Lokossou AG, Vargas A, Rassart E, Barbeau B. A CRE/AP-1-like motif is essential for induced syncytin-2 expression and fusion in human trophoblast-like model. *PLoS One*. 2015;10(3), e0121468.
188. Schubert SW, Abendroth A, Kilian K, et al. bZIP-Type transcription factors CREB and OASIS bind and stimulate the promoter of the mammalian transcription factor GCMa/Gcm1 in trophoblast cells. *Nucleic Acids Res*. 2008;36(11):3834–3846.
189. Chang CW, Chuang HC, Yu C, Yao TP, Chen H. Stimulation of GCMa transcriptional activity by cyclic AMP/protein kinase A signaling is attributed to CBP-mediated acetylation of GCMa. *Mol Cell Biol*. 2005;25(19):8401–8414.

190. Nair M, Teng A, Bilanchone V, Agrawal A, Li B, Dai X. Ovol1 regulates the growth arrest of embryonic epidermal progenitor cells and represses c-myc transcription. *J Cell Biol.* 2006;173(2):253–264.
191. Payre F, Vincent A, Carreno S. ovo/svb integrates Wingless and DER pathways to control epidermis differentiation. *Nature.* 1999;400(6741):271–275.
192. Renaud SJ, Chakraborty D, Mason CW, Rumi MA, Vivian JL, Soares MJ. OVO-like 1 regulates progenitor cell fate in human trophoblast development. *Proc Natl Acad Sci USA.* 2015;112:E6175–E6184.
193. Berger J, Moller DE. The mechanisms of action of PPARs. *Annu Rev Med.* 2002;53:409–435.
194. Wang Q, Fujii H, Knipp GT. Expression of PPAR and RXR isoforms in the developing rat and human term placentas. *Placenta.* 2002;23(8–9):661–671.
195. Barak Y, Nelson MC, Ong ES, et al. PPAR gamma is required for placental, cardiac, and adipose tissue development. *Mol Cell.* 1999;4(4):585–595.
196. Schaiff WT, Knapp Jr FF, Barak Y, Biron-Shental T, Nelson DM, Sadovsky Y. Ligand-activated peroxisome proliferator activated receptor gamma alters placental morphology and placental fatty acid uptake in mice. *Endocrinology.* 2007;148(8):3625–3634.
197. Shalom-Barak T, Nicholas JM, Wang Y, et al. Peroxisome proliferator-activated receptor gamma controls Muc1 transcription in trophoblasts. *Mol Cell Biol.* 2004;24(24):10661–10669.
198. Parast MM, Yu H, Ciric A, Salata MW, Davis V, Milstone DS. PPARgamma regulates trophoblast proliferation and promotes labyrinthine trilineage differentiation. *PLoS One.* 2009;4(11), e8055.
199. Capparuccia L, Marzioni D, Giordano A, et al. PPARgamma expression in normal human placenta, hydatidiform mole and choriocarcinoma. *Mol Hum Reprod.* 2002;8(6):574–579.
200. Schaiff WT, Carlson MG, Smith SD, Levy R, Nelson DM, Sadovsky Y. Peroxisome proliferator-activated receptor-gamma modulates differentiation of human trophoblast in a ligand-specific manner. *J Clin Endocrinol Metab.* 2000;85(10):3874–3881.
201. Levytska K, Drewlo S, Baczyk D, Kingdom J. PPAR-gamma regulates trophoblast differentiation in the BeWo cell model. *PPAR Res.* 2014;2014:637251.
202. Tarrade A, Schoonjans K, Guibourdenche J, et al. PPAR gamma/RXR alpha heterodimers are involved in human CG beta synthesis and human trophoblast differentiation. *Endocrinology.* 2001;142(10):4504–4514.
203. Schaiff WT, Bildirici I, Cheong M, Chern PL, Nelson DM, Sadovsky Y. Peroxisome proliferator-activated receptor-gamma and retinoid X receptor signaling regulate fatty acid uptake by primary human placental trophoblasts. *J Clin Endocrinol Metab.* 2005;90(7):4267–4275.
204. Fournier T, Pavan L, Tarrade A, et al. The role of PPAR-gamma/RXR-alpha heterodimers in the regulation of human trophoblast invasion. *Ann NY Acad Sci.* 2002;973:26–30.
205. Tarrade A, Schoonjans K, Pavan L, et al. PPARgamma/RXRalpha heterodimers control human trophoblast invasion. *J Clin Endocrinol Metab.* 2001;86(10):5017–5024.
206. Fournier T, Handschuh K, Tsatsaris V, Evain-Brion D. Involvement of PPARgamma in human trophoblast invasion. *Placenta.* 2007;28(suppl A):S76–S81.
207. Stephanou A, Handwerger S. Retinoic acid and thyroid hormone regulate placental lactogen expression in human trophoblast cells. *Endocrinology.* 1995;136(3):933–938.
208. Lu XJ, Deb S, Soares MJ. Spontaneous differentiation of trophoblast cells along the spongiotrophoblast cell pathway: expression of members of the placental prolactin gene family and modulation by retinoic acid. *Dev Biol.* 1994;163(1):86–97.

209. Tarrade A, Rochette-Egly C, Guibourdenche J, Evain-Brion D. The expression of nuclear retinoid receptors in human implantation. *Placenta*. 2000;21(7):703–710.
210. Sapin V, Ward SJ, Bronner S, Chambon P, Dolle P. Differential expression of transcripts encoding retinoid binding proteins and retinoic acid receptors during placentation of the mouse. *Dev Dyn*. 1997;208(2):199–210.
211. Sapin V, Dolle P, Hindelang C, Kastner P, Chambon P. Defects of the chorioallantoic placenta in mouse RXRalpha null fetuses. *Dev Biol*. 1997;191(1):29–41.
212. Wendling O, Chambon P, Mark M. Retinoid X receptors are essential for early mouse development and placentogenesis. *Proc Natl Acad Sci USA*. 1999;96(2):547–551.
213. Richardson BD, Cheng YH, Langland RA, Handwerger S. Differential expression of AP-2gamma and AP-2alpha during human trophoblast differentiation. *Life Sci*. 2001;69(18):2157–2165.
214. Hubert MA, Sherritt SL, Bachurski CJ, Handwerger S. Involvement of transcription factor NR2F2 in human trophoblast differentiation. *PLoS One*. 2010;5(2), e9417.
215. Kessler CA, Stanek JW, Stringer KF, Handwerger S. ETS1 induces human trophoblast differentiation. *Endocrinology*. 2015;156(5):1851–1859.
216. Cheng YH, Aronow BJ, Hossain S, Trapnell B, Kong S, Handwerger S. Critical role for transcription factor AP-2alpha in human trophoblast differentiation. *Physiol Genomics*. 2004;18(1):99–107.
217. Zhang Z, Zhang L, Jia L, et al. AP-2alpha suppresses invasion in BeWo cells by repression of matrix metalloproteinase-2 and -9 and up-regulation of E-cadherin. *Mol Cell Biochem*. 2013;381(1–2):31–39.
218. Kotani T, Iwase A, Ino K, et al. Activator protein-2 impairs the invasion of a human extravillous trophoblast cell line. *Endocrinology*. 2009;150(9):4376–4385.
219. Morasso MI, Grinberg A, Robinson G, Sargent TD, Mahon KA. Placental failure in mice lacking the homeobox gene Dlx3. *Proc Natl Acad Sci USA*. 1999;96(1):162–167.
220. Chui A, Pathirage NA, Johnson B, et al. Homeobox gene distal-less 3 is expressed in proliferating and differentiating cells of the human placenta. *Placenta*. 2010;31(8): 691–697.
221. Peng L, Huang Y, Jin F, Jiang SW, Payne AH. Transcription enhancer factor-5 and a GATA-like protein determine placental-specific expression of the Type I human 3beta-hydroxysteroid dehydrogenase gene, HSD3B1. *Mol Endocrinol*. 2004;18(8): 2049–2060.
222. Berghorn KA, Clark PA, Encarnacion B, et al. Developmental expression of the homeobox protein Distal-less 3 and its relationship to progesterone production in mouse placenta. *J Endocrinol*. 2005;186(2):315–323.
223. Roberson MS, Meermann S, Morasso MI, Mulvaney-Musa JM, Zhang T. A role for the homeobox protein Distal-less 3 in the activation of the glycoprotein hormone alpha subunit gene in choriocarcinoma cells. *J Biol Chem*. 2001;276(13):10016–10024.
224. Jacquemin P, Sapin V, Alsat E, Evain-Brion D, Dolle P, Davidson I. Differential expression of the TEF family of transcription factors in the murine placenta and during differentiation of primary human trophoblasts in vitro. *Dev Dyn*. 1998;212(3): 423–436.
225. Jacquemin P, Martial JA, Davidson I. Human TEF-5 is preferentially expressed in placenta and binds to multiple functional elements of the human chorionic somatomammotropin-B gene enhancer. *J Biol Chem*. 1997;272(20):12928–12937.
226. Barton K, Muthusamy N, Fischer C, et al. The Ets-1 transcription factor is required for the development of natural killer cells in mice. *Immunity*. 1998;9(4):555–563.
227. Takai N, Ueda T, Narahara H, Miyakawa I. Expression of c-Ets1 protein in normal human placenta. *Gynecol Obstet Invest*. 2006;61(1):15–20.
228. Yamamoto E, Ito T, Abe A, et al. Ikaros is expressed in human extravillous trophoblasts and involved in their migration and invasion. *Mol Hum Reprod*. 2005;11(11):825–831.

229. Ito T, Nomura S, Okada M, et al. Ap-2 and Ikaros regulate transcription of human placental leucine aminopeptidase/oxytocinase gene. *Biochem Biophys Res Commun.* 2002;290(3):1048–1053.
230. Blanchon L, Bocco JL, Gallot D, et al. Co-localization of KLF6 and KLF4 with pregnancy-specific glycoproteins during human placenta development. *Mech Dev.* 2001;105(1–2):185–189.
231. Racca AC, Camolotto SA, Ridano ME, Bocco JL, Genti-Raimondi S, Panzetta-Dutari GM. Kruppel-like factor 6 expression changes during trophoblast syncytialization and transactivates sshCG and PSG placental genes. *PLoS One.* 2011;6(7), e22438.
232. Racca AC, Ridano ME, Camolotto S, Genti-Raimondi S, Panzetta-Dutari GM. A novel regulator of human villous trophoblast fusion: the Kruppel-like factor 6. *Mol Hum Reprod.* 2015;21(4):347–358.
233. MacDonald BT, Tamai K, He X. Wnt/beta-catenin signaling: components, mechanisms, and diseases. *Dev Cell.* 2009;17(1):9–26.
234. Krivega M, Essahib W, Van de Velde H. WNT3 and membrane-associated beta-catenin regulate trophectoderm lineage differentiation in human blastocysts. *Mol Hum Reprod.* 2015;21(9):711–722.
235. Aoki M, Mieda M, Ikeda T, Hamada Y, Nakamura H, Okamoto H. R-spondin3 is required for mouse placental development. *Dev Biol.* 2007;301(1):218–226.
236. Ishikawa T, Tamai Y, Zorn AM, et al. Mouse Wnt receptor gene Fzd5 is essential for yolk sac and placental angiogenesis. *Development.* 2001;128(1):25–33.
237. Monkley SJ, Delaney SJ, Pennisi DJ, Christiansen JH, Wainwright BJ. Targeted disruption of the Wnt2 gene results in placentation defects. *Development.* 1996;122(11): 3343–3353.
238. Galceran J, Farinas I, Depew MJ, Clevers H, Grosschedl R. Wnt3a–/– like phenotype and limb deficiency in Lef1(–/–)Tcf1(–/–) mice. *Genes Dev.* 1999;13(6):709–717.
239. Matsuura K, Jigami T, Taniue K, et al. Identification of a link between Wnt/beta-catenin signalling and the cell fusion pathway. *Nat Commun.* 2011;2:548.
240. Sonderegger S, Husslein H, Leisser C, Knofler M. Complex expression pattern of Wnt ligands and frizzled receptors in human placenta and its trophoblast subtypes. *Placenta.* 2007;28(suppl A):S97–S102.
241. Getsios S, Chen GT, MacCalman CD. Regulation of beta-catenin mRNA and protein levels in human villous cytotrophoblasts undergoing aggregation and fusion in vitro: correlation with E-cadherin expression. *J Reprod Fertil.* 2000;119(1):59–68.
242. Pollheimer J, Loregger T, Sonderegger S, et al. Activation of the canonical wingless/T-cell factor signaling pathway promotes invasive differentiation of human trophoblast. *Am J Pathol.* 2006;168(4):1134–1147.
243. Knofler M, Pollheimer J. Human placental trophoblast invasion and differentiation: a particular focus on Wnt signaling. *Front Genet.* 2013;4:190.
244. Schreiber M, Wang ZQ, Jochum W, Fetka I, Elliott C, Wagner EF. Placental vascularisation requires the AP-1 component fra1. *Development.* 2000;127(22): 4937–4948.
245. Kubota K, Kent LN, Rumi MA, Roby KF, Soares MJ. Dynamic regulation of AP-1 transcriptional complexes directs trophoblast differentiation. *Mol Cell Biol.* 2015;35(18):3163–3177.
246. Kent LN, Rumi MA, Kubota K, Lee DS, Soares MJ. FOSL1 is integral to establishing the maternal-fetal interface. *Mol Cell Biol.* 2011;31(23):4801–4813.
247. Takeda K, Noguchi K, Shi W, et al. Targeted disruption of the mouse Stat3 gene leads to early embryonic lethality. *Proc Natl Acad Sci USA.* 1997;94(8):3801–3804.
248. San Martin S, Fitzgerald JS, Weber M, et al. Stat3 and Socs3 expression patterns during murine placenta development. *Eur J Histochem.* 2013;57(2), e19.

249. Roberts AW, Robb L, Rakar S, et al. Placental defects and embryonic lethality in mice lacking suppressor of cytokine signaling 3. *Proc Natl Acad Sci USA*. 2001;98(16): 9324–9329.
250. Takahashi Y, Carpino N, Cross JC, Torres M, Parganas E, Ihle JN. SOCS3: an essential regulator of LIF receptor signaling in trophoblast giant cell differentiation. *EMBO J*. 2003;22(3):372–384.
251. Takahashi Y, Dominici M, Swift J, Nagy C, Ihle JN. Trophoblast stem cells rescue placental defect in SOCS3-deficient mice. *J Biol Chem*. 2006;281(17):11444–11445.
252. Corvinus FM, Fitzgerald JS, Friedrich K, Markert UR. Evidence for a correlation between trophoblast invasiveness and STAT3 activity. *Am J Reprod Immunol*. 2003;50(4):316–321.
253. Busch S, Renaud SJ, Schleussner E, Graham CH, Markert UR. mTOR mediates human trophoblast invasion through regulation of matrix-remodeling enzymes and is associated with serine phosphorylation of STAT3. *Exp Cell Res*. 2009;315(10):1724–1733.
254. Poehlmann TG, Fitzgerald JS, Meissner A, et al. Trophoblast invasion: tuning through LIF, signalling via Stat3. *Placenta*. 2005;26(suppl A):S37–S41.

CHAPTER FOUR

The Phylogeny of Placental Evolution Through Dynamic Integrations of Retrotransposons

K. Imakawa*[,1], S. Nakagawa[†]
*Animal Resource Science Center, Graduate School of Agricultural and Life Science, The University of Tokyo, Kasama, Japan
[†]Biomedical Informatics Laboratory, Tokai University School of Medicine, Isehara, Japan
[1]Corresponding author: e-mail address: akaz@mail.ecc.u-tokyo.ac.jp

Contents

Abstract

Trophoblasts, a major constituent of the placenta, are known to express genes derived from various endogenous retroviruses (ERVs) as well as LTR retrotransposons. However, the evolutionary significance of *ERV*-derived genes involved in placental development has not been well characterized. In this review, we catalog the diverse morphology of placental structure among mammalian species with note of counterintuitive developments. We then detail the history of ancient placenta development with paternally expressed gene 10 (*Peg10/Sirh1*), *Peg11/Sirh2*, and *Sirh7/Ldoc1* as LTR retrotransposons, followed by independent captures of ERV-*env*-related genes such as *Syncytin-1*, *-2*, *-A*, *-B*, *-Rum1*, and *Fematrin-1* responsible for trophoblast cell fusion, resulting in multinucleate syncytiotrophoblast formation, and possibly morphological diversification of placentas. Because the endogenization of retroviral infections has occurred multiple times independently in different mammalian lineages, and some use the same molecules in their transcriptional activation, we speculate that *ERV* gene variants integrated into mammalian genomes in a locus-specific manner have replaced the genes previously responsible for cell fusion. Moreover, ERVs also work as transcriptional regulators of various genes such as interferon (*IFN*)-stimulated genes. The "baton pass" hypothesis suggests that

Progress in Molecular Biology and Translational Science, Volume 145
ISSN 1877-1173
http://dx.doi.org/10.1016/bs.pmbts.2016.12.004

evolutionary events caused by multiple successive retrotransposon integrations, possibly resulting in effective fusogenic activity, downstream gene transcription in a temporal and spatial manner, and/or increased diversity of placental structures.

1. PLACENTA: STRUCTURAL DIVERSITY

Following fertilization, the early embryonic cells begin to differentiate for the first time into an inner cell mass (ICM) and an outer trophectoderm, forming the blastocyst. The ICM differentiates into the embryo as well as the amnion, yolk sac, and allantois, whereas the trophectoderm develops into chorionic membrane, later becoming a major part of conceptus side of placenta. In most mammals, the allantois displaces the yolk sac from the trophoblast, resulting in chorioallantoic placentation.[1] However, in some other mammalian species such as rodents and lagomorphs, the yolk sac forms a maternal-facing absorptive epithelium, known as an inverted yolk sac placenta, and it persists until term.[2] This may reflect predominance of a yolk sac or inverted yolk sac placenta form of placentation in early mammals, after which chorioallantoic placentation developed and provided reproductive advantages.

1.1 Classification of Placenta

The placenta is a transient organ that provides an interface for metabolic exchange between the fetus and the mother. The placenta is composed of both the fetal chorion and the maternal uterine endometrium. Placentas are classified according to the distribution of chorionic villi: diffuse (pigs and horses), cotyledonary (ruminants), zonary (dogs and cats), and discoids (murines and primates) (Fig. 1); constituent cell types as well as anatomical structures vary considerably among mammalian species. Placentas are also classified by the number of placental cell layers that separate the fetal blood from the maternal blood: epitheliochorial, endotheliochorial, and hemochorial (Fig. 2). In an epitheliochorial placenta, the various cell types that form a layer between the maternal and fetal blood are: (1) the endothelium of the maternal capillary, (2) uterine endometrium (stroma and/or decidua), (3) the epithelial layer of the uterine endometrium, (4) the layer or layers of trophoblasts that make up the chorionic epithelium, (5) fetal connective tissues, and (6) the endothelium of the fetal capillary. Regardless of the number of cell types in between the mother and fetus, maternal nutrients and gases must traverse all intervening cell/tissue layers to reach fetal

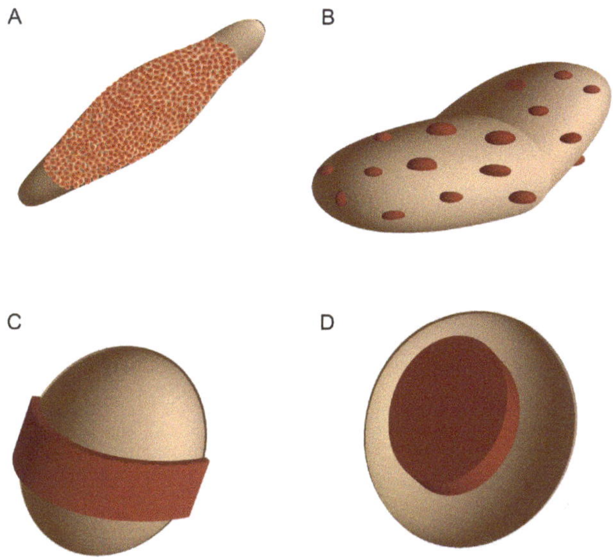

Fig. 1 Placental classification by distribution of chorionic villi. Gross anatomy showing diffuse (pig), cotyledonary (ruminants), zonary (dogs and cats), and discoid (rodents and primates). (A) Diffuse placentas have uniform distribution of chorionic villi that cover the surface of the chorion. (B) Cotyledonary placentas have numerous, discrete button-like structures called cotyledons. (C) Zonary placentas have a band-like zone of chorionic villi. (D) Discoid placentas form a regionalized disc-like structure.

circulation from which waste materials must then be expelled back to the maternal circulatory system.

In the epitheliochorial placenta, the uterine luminal epithelium is in direct contact with the chorionic trophoblast. This type of placentation is found in several orders including even-toed ungulates, whales, dolphins, and lower primates (Strepsirhini). In an endotheliochorial placenta, a loss of uterine epithelium and stromal thinning results in the endothelium of the maternal capillaries being located close to the trophoblast. This type of placenta is seen in carnivores, but it is also found in elephants (Proboscidea).[3] In a hemochorial placentation, maternal blood is directly in contact with the trophoblast, functioning without the capillary endothelium. This type of placentation is seen in many rodents and in higher primates including humans.

It was once thought that placental evolution had proceeded from the least invasive, epitheliochorial, to most invasive, hemochorial placentation. Using phylogenetic and statistical analyses of genomic and morphological data, Wildman *et al.* proposed that the ancestral eutherian placenta had a

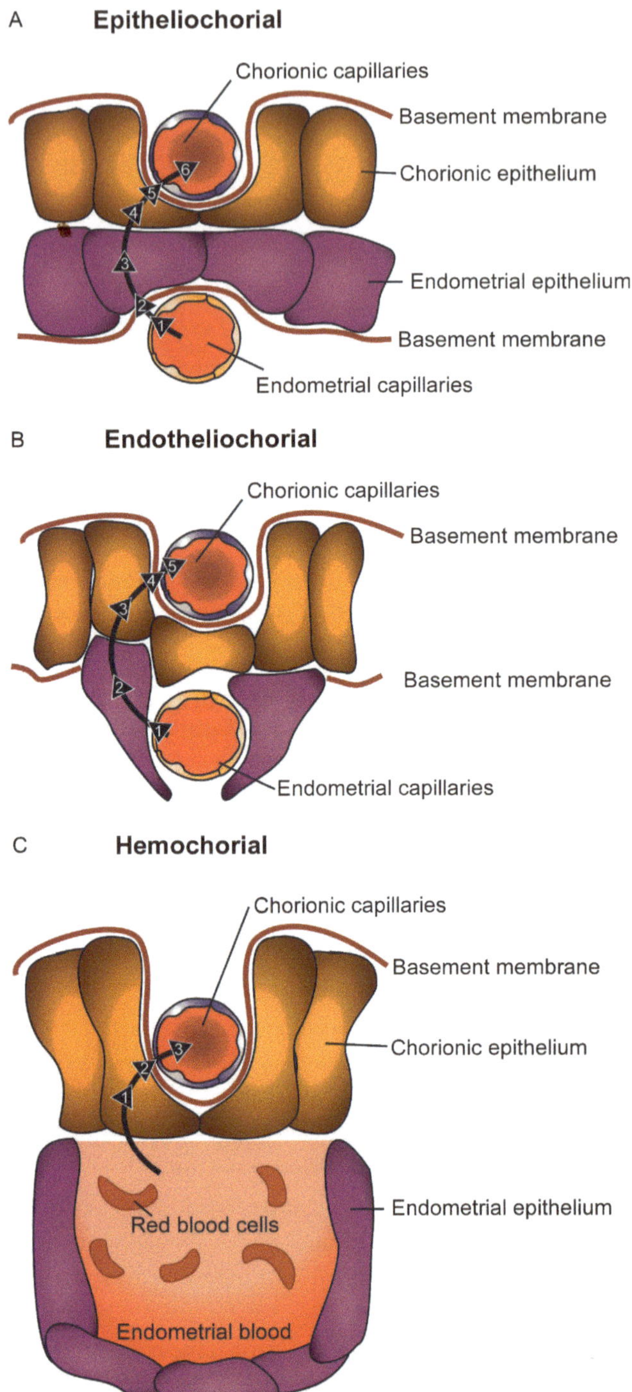

Fig. 2 See legend on opposite page.

hemochorial placental interface with discoid shape and labyrinthine maternofetal interdigitation.[4] However, recent phylogenetic analyses of molecular data sets do not fully support the notion that the highly invasive trophoblast is the most recently evolved or least refined form of placentation. For example, hyenas belong to the order carnivore, but unlike dogs and cats, the hyena undergoes hemochorial placentation.[5] These placental variations and similarities across separate phylogenic trees provide evidence for convergent evolution in mammalian placentation[6] (Table 1 and Fig. 3).

1.2 Evaluation of Evolution From Syncytiotrophoblast

Syncytiotrophoblast cells, components of the chorionic membrane, are generated from trophectodermal fusion and are located next to the maternal cell components. These cells manage efficient nutrient/gas exchange, enable the production of placental lactogen, and chorionic gonadotropin, and are also involved in immunotolerance of the conceptus by the maternal immune system.[11,12] In hemochorial placentation, syncytiotrophoblasts can be divided into three types depending on the number of trophoblast layers. The placenta of some murine species is characterized as monochorial placenta with a single layer of syncytiotrophoblast. Although the human placenta does not have a labyrinth zone like the murine species, still it can be classified as a monochorial placenta. In contrast, murines including Muridae and Cricetidae have three trophoblast layers and are therefore classified as trichorial placentation.[13] The first layer closest to the maternal stroma/decidua consists of mononucleate cytotrophoblast cells, and the second and third trophoblast layers are composed of syncytiotrophoblast cells. Beavers, rabbits, and bats have dichorial placentas, with the first layer of syncytiotrophoblast cells and the second of cytotrophoblast cells. Carnivora including Felidae and Canidae have two trophoblast layers, with the first

Fig. 2 Placental diversity based on cell/tissue layers between fetus and the mother. (A) In the epitheliochorial placenta (pig and horse), both the endometrial epithelium and the epithelium of chorionic villi are intact, and therefore, there are six-cell layers for maternal nutrients and gasses to reach the fetal blood: (1) endometrial cell layer of the maternal capillaries, (2) endometrial interstitium, (3) endometrial epithelial layer, (4) chorionic epithelium, (5) chorionic interstitium, and (6) endothelial cell layer of chorionic capillaries. (B) In the endotheliochorial placenta (cats and dogs), both the endometrial epithelium and underlying interstitium are eroded and maternal capillaries are directly exposed to epithelial cells of the chorion. (C) In the hemochorial placenta (rodents and primates), the chorionic villi are in direct apposition to maternal pools of blood. This results in direct exchanges of nutrients and gases, which move through only three tissue layers to reach the fetal blood.

Table 1 Placental Diversity and Animals That Could Be Assigned to That Category

		Epitheliochorial			Syndesmochorial	Endotheliochorial		Hemomonochorial			
Diffuse		Mono nucleate	Mono/bi nucleate	Multinucleate							
		Pig, Whale, Dolphin	Horse	Bush baby							
Cotyledonary					Mono/bi/ multinucleate						
					Cow, sheep						
Zonary						Mononucleate	Multinucleate	Mononucleate		Multinucleate	
						Elephant, kangaroo rat	Dog, cat, earless seal	Hyrax		Hyena	
Discoid	Bi					Multinucleate		Multinucleate			
						Treeshrew		Rhesus macaque, Japanese macaque			
	Mono					Star nosed mole		Mono (mononucleate)	Mono (multinucleate)	Bi (mono–multinucleate)	Bi (mono-multi–multinucleate)
								Tenrec, jerboa	Guinea pig, human (Primates)	Beaver, rabbit, Vespertilio sinensis	Mouse, rat, hamster

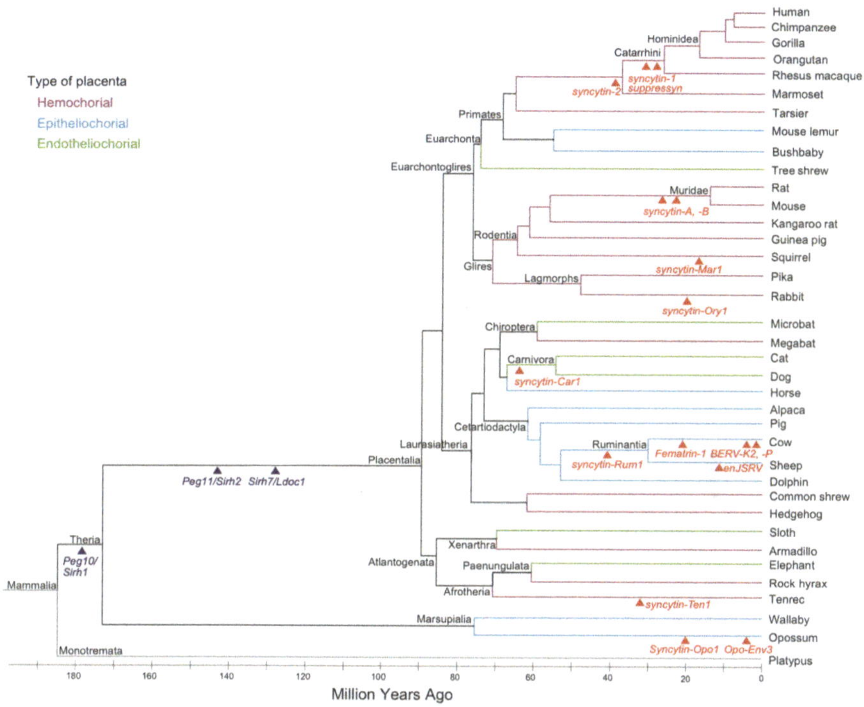

Fig. 3 Evolution and diversity of placenta in mammalian species and integration of genes derived from ERVs and LTR retrotransposons. This phylogenetic tree of mammalian species was based on the data shown in dos Reis *et al.*[7] and Imakawa *et al.*[8] An integrated gene known to be related to placental development that is derived from LTR retrotransposons or ERVs is shown in *blue* or *red*, respectively. The branch color designates the type of placenta[3,9,10] as shown in the *upper left panel.*

layer being cytotrophoblast and the second syncytiotrophoblast.[14] In the Bovidae family, syncytiotrophoblast cells are not formed; rather, through acytokinetic mitosis, the trophectoderm forms binucleate cells (BNCs), which migrate and fuse with epithelial cells, resulting in the formation of trinucleate cells (TNCs) localized in the uterine stroma.[15] These TNCs may serve functions similar to syncytiotrophoblasts in these species.[16] In animals with epitheliochorial placentation such as horses, camels, pigs, hippopotami, and cetaceans, syncytiotrophoblast layer is not formed; however, BNCs can be found in the horse placenta and *Galago* (bushbabies) possess syncytiotrophoblasts.[16] Although Procaviidae (Hyrax), *Tenrec ecaudatus*, and Dipodidae possess hemochorial placentation, syncytiotrophoblast cells are not formed in their placentas.[17–19] These observations indicate that although the placentas serve analogous functions of nutrition and gas exchange, their morphologies are far more diversified, and that their morphology can vary beyond the history of their specific section of the mammalian evolutionary tree (Table 1 and Fig. 3).

1.3 Significance of Syncytiotrophoblast Formation

Trophoblast cells, components of the outermost membrane, are located next to the maternal endometrium and must exhibit invasive properties to form an effective maternal–fetal vascular relationship. However, excessive invasiveness of trophoblast cells may cause too much stress to the uterus, resulting in uterine bleeding as well as inciting an immune response against the embryo. Although the uterine endometrium has means to control trophoblast invasiveness, trophoblasts themselves may have evolved to limit their own invasiveness into the uterine endometrium. One of these mechanisms appears to be slowing transit through their own cell cycles. Inhibition of cell cycles causes endoreduplication, forming trophoblast giant cells. In humans and murine species, syncytiotrophoblasts are formed through cytotrophoblast cell fusion, resulting in the cessation of cell cycles[20] and the regulation of trophoblast invasion of the maternal endometrium. Syncytialization may not be the only mechanisms through which trophoblasts control their invasiveness. Hyrax is known to have hemochorial placentation, but its trophoblasts do not form a syncytium. This independent appearance of similar functions in different clades is a prime example of convergent evolution.

2. PLACENTA ERVs

Transposable elements make up at least 45% and 40% of human and mouse genomes, respectively,[21,22] of which *ERVs* and long terminal repeat (LTR) retrotransposons account for a respective 8% and 10% of these genomes. Nucleotide structures of *ERVs* and LTR retrotransposons are quite similar to each other; both contain 5′- and 3′-LTRs in each terminus as regulatory elements and *gag*, *pro*, and *pol* protein-coding genes, but an *env*-coding gene that corresponds to a spike protein of viral envelope is included only in ERVs. In both retrotransposons, nucleotide structures largely consist of mutations including insertions and deletions. However, a number of protein-coding genes of ERVs and LTR retrotransposons are still actively transcribed in certain situations.

It has been estimated that a primitive placenta emerged in a mammalian ancestor 150–166 million years ago (MYA).[23] In considering LTR retrotransposon integrations into the mammalian genomes, there appeared to be occasional *ERV* endogenization events throughout mammalian evolution; it is hypothesized that earlier ones were necessary for structural formation of the placenta, while the latter facilitated morphological diversity of the placentas during mammalian evolution.

2.1 Sirh Family

A group of genes derived from the sushi-ichi LTR retrotransposon is called sushi-ichi-related retrotransposon homolog (*SIRH*) family, comprising 12 genes (*Sirh*1–12) in mammals. In this review, three *Sirh* family genes that are directly involved in placental structure and functioning will be discussed.

2.1.1 Peg10/Sirh1

Paternally expressed 10 (*Peg10/Sirh1*), a maternally imprinted gene, is expressed in trophoblasts and placentas.[24] *Peg10* contains *gag* and *pol* regions of retroviral genomes. A gene ablation study demonstrated that *Peg10* knockout mice suffered early embryonic death, resulting from severe placental defects with loss of spongiotrophoblasts and the labyrinth layer of the placenta. The labyrinth layer is essential for the exchange of nutrients and gases between maternal and fetal compartments. Integration of the gene is believed to have occurred 148 MYA, after the split with monotremes,[25] suggesting that this has contributed to the emergence of primitive placentation.

2.1.2 Peg11/Sirh2

Paternally expressed 11/retrotransposon-like 1 (*Peg11/Rtl1*) is a paternally imprinted gene with *gag* and *pol* regions of retroviral genomes.[26] The gene is believed to function in nutrient and gas exchanges at the allantochorion membranes; ablation of this gene causes early postnatal death resulting from malnutrition-like symptoms during gestation.[27] As this gene is conserved among placental mammals, this gene likely became integrated into the genome soon after marsupials diverged from eutherian mammals.

2.1.3 Sirh7/Ldoc1

Another LTR retrotransposon-derived *Sirh7/Ldoc* [sushi-ichi retrotransposon homolog 7/leucine zipper, downregulated in cancer 1, also called mammalian retrotransposon-derived 7 (*Mart7*)] has recently been characterized.[28] Ablation of this gene is associated with abnormal placental cell differentiation/maturation, leading to an overproduction of placental progesterone and placental lactogen (PL1) from trophoblast giant cells. Based on these observations, genes of the *Sirh* family likely contributed to the emergence of a primitive placenta. It should be emphasized, however, that the functions of *Sirh* family genes inferred to date are derived from gene ablation studies in present-day murines. Many years must have been required for these *Sirh* family genes to gain the functions we now recognize. Moreover, the observations that *Sirh* family genes are integrated into similar loci across mammalian species strongly indicate that these genes are highly conserved throughout mammalian evolution.

2.2 ERV-*env*

The effects of *Sirh* family gene integration into mammalian genomes are not sufficient to explain placental function and morphological diversity among mammalian placentas, because they are similarly integrated and fixed in the mammalian genomes. Integration of the *ERV* family genes occurred multiple times independently among mammalian species, which make these genes prime candidates for the emergence of efficient nutrient and gas exchange as well as for structural diversification of mammalian placentas. *Env* proteins of mammalian ERVs have been extensively studied, because *Env* proteins are required for viral infection to the host cells through specific receptors, and those of mammalian ERVs can also induce cell–cell fusion in host cells.[6,29] In humans, 18 ERV-*env* nucleotide structures have been identified, among which 16 genes have full coding *env* genes and are transcribed in several healthy tissues; however, only three of these 18 ERV-*env*s possess fusogenic activity.[30,31]

2.2.1 Syncytin-1 and -2

Syncytin-1 belongs to the HERV-W family, and *syncytin-2* belongs to the HERV-FRD family, both of which express env proteins and possess fusogenic activity.[32] In 2000, syncytin-1 was found in human syncytiotrophoblasts, and its fusogenic activity was demonstrated in cytotrophoblasts.[33,34] In the in vitro assay, *syncytin-1* mRNA is upregulated through an increase in cAMP levels when human choriocarcinoma BeWo cells are treated with forskolin, resulting in the fusogenic activity of this cell type. Cell fusion was also demonstrated in African green monkey COS-7, human rhabdomyosarcoma TE671, and human embryonic kidney-derived 293 T cells when these cells were transfected with *syncytin-1* expressing plasmids. In addition, the fusogenic activity of syncytin-1 was found in insect Sf9 cells. In syncytin-2-treated cells, fusogenic activity was found in feline

G355-5 and human 293 T cells.[32] Both syncytin-1 and -2, comprising 538 amino acids, contain surface (SU) and transmembrane (TM) subunits, which are cleaved by a furin protease, and their interaction is required for their fusogenic activity.[35,36]

A transcription factor, glial cell missing factor homolog 1 (*GCM1*), is required for the transcriptional regulation of *syncytin-1* and *-2* genes. The *GCM* family genes are present even in drosophila and are regarded as master regulators.[37] Previously, a multispanning transmembrane protein, CD9, was found as a regulator for the *GCM1* gene.[38] This protein is involved in the invasive behavior of cancer cells and in cell fusions between sperm and egg as well as myoblasts in muscle development.[39,40] *CD9* mRNA and protein are increased in BeWo cells when the cells are treated with forskolin. In addition, CD9 was downregulated by a protein kinase A (PKA) inhibitor, implicating regulation of GCM1 expression through the cAMP/PKA intracellular signaling system. It is possible that as *syncytin-2* is regulated by GCM1, *syncytin-2* is also regulated through the CD9 and cAMP/PKA signaling system. Furthermore, the transcription of *syncytin-2* gene is silenced epigenetically by CpG methylation in human 293 cells.[41] These findings suggest that *syncytin-1* and *-2* gene regulation has not been fully characterized and it awaits further investigation.

The expression of *syncytin-1* and *-2* transcripts differs in their cellular locations. In in situ hybridization studies, *syncytin-1* mRNA is found in syncytiotrophoblasts, whereas *syncytin-2* transcripts are found only in cytotrophoblasts. In addition, the abundance of *syncytin-1* transcripts is maintained throughout the entire gestational period; however, the level of *syncytin-2* decreases during the latter part of pregnancy.[42] Syncytin-1 receptor, RD114/mammalian type D retrovirus receptor (*ASCT2*), is expressed in both syncytiotrophoblast and cytotrophoblasts.[43] Syncytin-2 receptor, major facilitator superfamily domain containing 2 (*MFSD2*), is found only in cytotrophoblasts where syncytin-2 is expressed.[44] Upon forskolin treatment in BeWo cells, *syncytin-1* mRNA increases; however, its receptor *ASCT2* mRNA level decreases.[45] Based on these findings, it is possible that syncytin-1 may work not only on its primary syncytiotrophoblast target but also the secondary target of cytotrophoblasts. In contrast, MFSD2 expression is found only in cytotrophoblasts, and syncytin-2 may work only on cytotrophoblasts.[46] Recombinant protein produced from the *syncytin-2* gene possesses immunotolerizing activity, but the syncytin-1 protein does not, strongly suggesting that syncytin-1 and -2 may function differently in humans.

In the phylogenetic analysis, *syncytin-1* gene is found only in Catarrhini, but not in Platyrrhini, indicating that among primates, *syncytin-1* entered the primate lineages after Catarrhini separation from Platyrrhini lineages approximately 40 MYA.[47] In Catarrhini, the premature termination codon and/or frameshift mutations on *syncytin-1* genes provide evidence that functional syncytin has existed in primates after the separation from Old World monkeys approximately 25 MYA. It cannot be determined whether anthropoids were able to use inactivated *syncytin-1* gene or whether Old World monkeys were unable to use the activated *syncytin-1* gene. In any case, the function of syncytin-1 in the placental formation differs even among the Catarrhini lineages. On the contrary, syncytin-2 is conserved from Platyrrhini to humans (>87.9%) and regardless of lineage, it possesses fusogenic activity in a similar manner, suggesting that this gene has endured for more than 40 million years.[32]

Based on these observations, it is postulated that (a) *syncytin-2* entered the anthropoid lineages and acquired immunotolerance as well as fusogenic activity in cytotrophoblasts, generating primitive syncytiotrophoblasts and (b) *syncytin-1* then entered the Catarrhini lineages. In primates, syncytin-1 enables fusion between syncytiotrophoblasts and cytotrophoblasts. In humans, the fusion between cytotrophoblasts is initiated on days 7–11 days of gestation, and the fusion between syncytiotrophoblasts and cytotrophoblasts continues until the end of pregnancy.[48] Because syncytiotrophoblasts do not possess the ability to proliferate, it is possible that the syncytiotrophoblasts may maintain cellular activity through their fusion with cytotrophoblasts. If this is the case, syncytin-1 could be required to extend placental activity, resulting in a longer gestational period. A relatively short gestation period is seen in Platyrrhini and Old World monkeys; *syncytin-1* is not found in Platyrrhini, whereas the open reading frame of this gene is truncated and not functional in Old World monkeys.

2.2.2 Syncytin-A and -B

From the results of murine genome analysis, two *env* genes with their fusogenic activity in vitro were found and named *syncytin-A* and *-B*.[49] Mouse placenta consists of three trophoblast layers, and syncytin-A is found in the second layer, known as syncytiotrophoblast layer-I (ST-I), whereas syncytin-B is localized in the third or syncytiotrophoblast layer-II (ST-II).[50] Ablation of *syncytin-A* results in a lack of ST-I formation and these embryos die between days 11.5 and 13.5 of pregnancy.[51] Because syncytin-A exhibits fusogenic activity in Green monkey Vero and human 293 T cells, syncytin-A is involved in trophoblast cell fusion. Similar to human syncytins, GCM1 works as a transcription factor targeting the upstream region of the *syncytin-A* gene, regulating in the initiation of its transcription.[52] This is in agreement with the observation that *GCM1* gene ablation blocks the development of labyrinth zone in mouse placentas.[53] On the other hand, *syncytin-B* gene ablation does not result in embryonic death, although ST-II layer formation is insufficient and the number of pups born is smaller than that for control mice with functioning *syncytin-B*. In in vitro assay, syncytin-B exhibits fusogenic activity only in canine MDBK cells; however, it possesses strong immunosuppressive activity.[46] Findings for syncytin-B in murine trophoblasts closely resemble those for syncytin-2 in human cytotrophoblasts.

In a phylogenetic analysis, both *syncytin-A* and *-B* entered murine lineages approximately 20 MYA (Fig. 3). In Rodentia, *syncytin-A* and *-B* are found in Muroidea, but not in the rest of Myomorpha.[49] It should be noted that Muroidea possess a trichorial placenta and Myomorpha have monochorial placenta. It is thought that (a) an unknown gene X entered Rodentia, which was involved in the formation of monochorial placenta; (b) after Muridae's divergence from Myomorpha, Muridae incorporated *syncytin-A* and *-B*, which had fusogenic and immunosuppressive activity, respectively, and produced two layers of syncytiotrophoblasts; and (c) Muroidea with *syncytin-A* and *-B* lost the unknown gene X, resulting in the formation of the three trophoblast layers as we know them today. With regard to the function of each syncytin, the *syncytin-A* and *-B* genes are likely homologous to the human *syncytin-1* and *-2*.

2.2.3 ERVs in Ruminants

Unlike mammalian species discussed earlier, syncytiotrophoblasts do not exist in ruminant placentas. Instead, ruminant trophoblasts form cytotrophoblast BNCs, which fuse with maternal uterine epithelia, resulting in TNCs in the bovine or syncytial plaques in sheep and goats. It should be noted that these represent the only heterologous cell fusions between fetal and maternal cells in mammalian species. Using BLAST search on the bovine genome, two ORF regions derived from ERV-*env* genes were found and named as bovine endogenous retrovirus K1 (*BERV-K1*) and *BERV-K2*.[54] In addition to these ERVs, several other ERVs were recently found: *Syncytin-Rum1*,[55] *BERV-P*,[56] and endogenous Jaagsiekte sheep retroviruses (*enJSRV*).[57] Regions corresponding to *gag* and *pol* are lost from the nucleotide structures of *BERV-K1*, but its *env* region is well conserved.

On the other hand, *gag*, *pol*, and *env* regions are conserved in *BERV-K2*.[54] The expression of BERV-K1 mRNA and protein is much greater than those of BERV-K2.[58] In the comparison between nucleotide structures of *BERV-K1* and *BERV-K2*, integration of *BERV-K1* occurred much earlier than for *BERV-K2*. Nucleotide structure differences could be considered as reflecting relative degrees of endogenization of these genes. Recently, BERV-K1 was demonstrated as having strong fusogenic activity and identified as a main factor involved in TNC formation, and was therefore named as *Fematrin-1*.[58] It was also reported that *syncytin-Rum1* was inserted into ruminant genomes, including cattle and sheep, and was possibly involved in fetomaternal cell-to-cell fusion in both species.[55] However, *Fematrin-1* is integrated into the bovine genome, but not in the sheep genome, and it is now believed that *syncytin-Rum1* was integrated into ruminant genomes 20 MYA, while *Fematrin-1* was integrated into the bovine genome 11 MYA.[59]

3. HYPOTHESIS OF GENE EVOLUTION THROUGH BATON PASS

Genes originated from ERVs emerge mainly through viral infections and/or retrotranspositions; they do not typically form clusters in the genome. For example, *Fematrin-1* was integrated into the intron 18 of FAT tumor suppressor homolog 2 (*FAT2*), and no *ERVs* were found around the region.[54,58] Moreover, the integration of an exogenous virus is always associated with various components of genes (i.e., *gag*, *pol*, and *env*) as well as transcription regulatory sites (LTRs), and therefore their evolutionary pathways of *ERV* acquisition must operate in a functionality- and locus-specific manner. Indeed, a phylogenetic analysis of *env* genes derived from ERVs (ERV-*envs*), many of which are called "syncytins," clearly shows that those origins are phylogenetically distinct (Fig. 4).

In human *syncytin-1* and *-2* genes, the transcription factor GCM1 is the control element, regulating their expression.[37,41] CD9 is another upstream factor controlling GCM1 and *syncytin* genes.[38] Just as in humans, GCM1 also controls mouse *syncytin-A* and *-B* gene expression. During evolution, integration of one *ERV* could also be followed by another integration of *ERV*; newly acquired *ERVs* could function in trophoblast cell fusion and possibly placental morphogenesis with greater efficacy than the preexisting gene. Fematrin-1, for example, possessed much better fusogenicity than the earlier

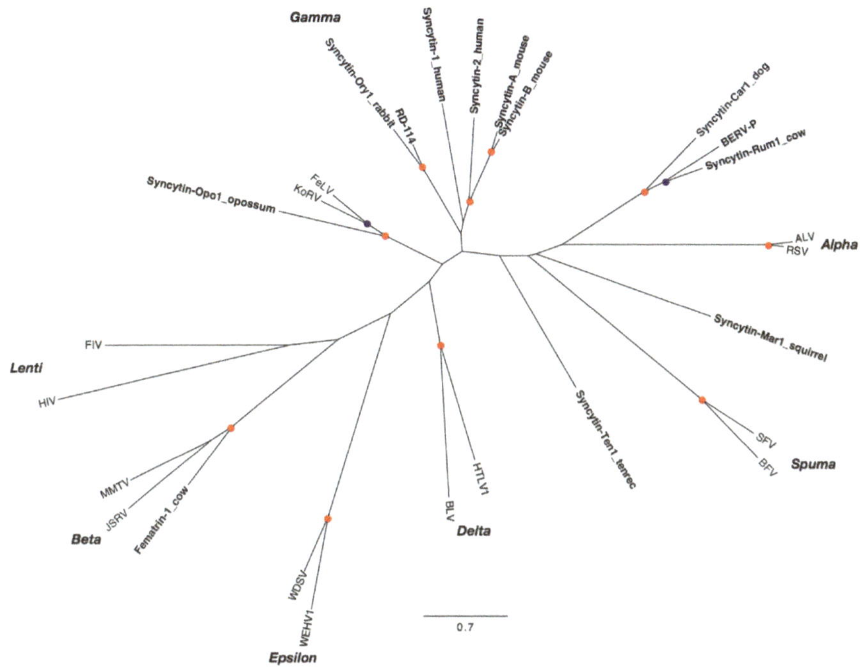

Fig. 4 Maximum likelihood (ML) tree of "syncytins" and Env sequences of various retroviruses. The ML tree was constructed as follows: (1) amino acid sequences of 12 Env sequences derived from endogenous retroviruses including *Syncytins* and 15 Env sequences of exogenous retroviruses that are representative in each genus of retroviruses—*Alpharetrovirus*, *Betaretrovirus*, *Gammaretrovirus*, *Deltaretrovirus*, *Epsilonretrovirus*, *Lentivirus*, and *Spumavirus*—were downloaded from NCBI database, (2) transmembrane regions were aligned using MAFFT L-INS-i[60] and gapped sites were removed using trimAl,[61] (3) LG amino acid replacement model[62] plus gamma distribution and invariant site (G+I) was selected using ProtTest3,[63] and (4) the phylogeny was inferred using RAxML program with 1000 rapid bootstrapping tests.[64] Twelve amino acid sequences of *Env* genes derived from ERVs used in this study were shown as follows: Syncytin-1 and -2 of humans, NP_055405.3 and NP_997465.1, respectively; Syncytin-A and -B of mice, NP_001013773.1 and NP_775596.1, respectively; Syncytin-Ory1 of rabbits, ACZ58381.1[65]; Syncytin-Car1 of dogs, AEX32761.1[66]; Fematrin-1,[58] Syncytin-Rum1,[55] and BERV-P[56] Env of cows, BAJ72717.1, NP_001292383.1, and BAN14723.1, respectively; Syncytin-Mar1 of squirrels, AHZ59674.1[67]; Syncytin-Ten1 of tenrecs, NP_001292515.1[68]; Syncytin-Opo1 of opossums, NP_001295306.1.[69] Fifteen Env amino acid sequences of exogenous retroviruses are shown as follows: ALV (Avian leukosis virus, *Alpharetrovirus*), AAU06813.1; RSV (Rous sarcoma virus, *Alpharetrovirus*), NP_056885.1; JSRV (Jaagsiekte sheep retrovirus, *Betaretrovirus*), AAD45228.2; MMTV (mouse mammary tumor virus, *Betaretrovirus*), BAA03768.1; FeLV (feline leukemia virus, *Gammaretrovirus*), AAA93093.1; KoRV (koala retrovirus, *Gammaretrovirus*), AAF15099.1; RD114 (RD114 retrovirus, *Gammaretrovirus*), YP_001497149.1; BLV (bovine leukemia virus,

(Continued)

Syncytin-Rum1 under physiological conditions.[55,58,70] These successive *ERV* acquisitions, therefore, are called a "baton pass": a new *ERV* replaced the preexisting *ERV* gene and acquired the role that that gene had played,[8,59,71] while the previous gene may either be lost or coopted for another function such as immunosuppressive ability.[6,72] In genes incorporated into reproductive processes through the baton pass, the integration of *ERVs* must be locus specific because they could be transcribed through their own LTRs or instead be transcribed along with placenta-specific genes.[58,73]

In our hypothesis of the baton pass, successive *ERVs* had replaced the preexisting genes and carried out their functions more effectively than their predecessors. The baton pass with *ERVs* in the placenta could also be seen in biological systems other than the process of placentation, such as innate immunity involving interferons (IFNs). IFNs are proinflammatory signaling molecules released upon infection, which promote transcription of IFN-stimulated genes (*ISGs*). When ISGs are activated, *cis*-regulatory elements of these genes are bound by the transcription factors IRF (interferon regulatory factor) and STAT signal transducer and activator of transcription.[74] Quite recently, Chuong *et al.* reported that *ERVs* could have shaped a transcriptional network of the IFN response, in which lineage-specific ERVs had dispersed numerous IFN-inducible enhancers in mammalian genomes. One *Gammaretrovirus*, *MER41*, was endogenized in the genome of an anthropoid primate ancestor 45–60 MYA, and six subfamilies (*MER41A*, *B*, *C*, *D*, *E*, and *G*) are now fixed in the human genome.[75] Deletion of these ERV elements in the human genome impaired expression of adjacent interferon gamma (IFNG)-induced genes, including activation of the gene *Absent in Melanoma* (*AIM2*) inflammasome. These authors also found ancestral sequences of *MER41*-like LTRs in lemuriformes, vesper bats, carnivores, and artiodactyls. Also, reconstructed *MER41*-like LTR in dogs and cows can drive robust IFNG-inducible reporter activity in HeLa cells,

Fig. 4—Cont'd *Deltaretrovirus*), AAO21338.2; HTLV1 (human T-cell lymphotropic virus type 1, *Deltaretrovirus*), NP_057865.1; WDSV (Walleye dermal sarcoma virus, *Epsilonretrovirus*), NP_045939.1; WEHV1 (Walleye epidermal hyperplasia virus types 1, *Epsilonretrovirus*), AAD30049.1; FIV (feline immunodeficiency virus, *Lentivirus*), NP_040976.1; HIV1 (human immunodeficiency virus 1, *Lentivirus*), AAC97548.1; BFV (bovine foamy virus, *Spumavirus*), AAN08117.1; SFV (Simian foamy virus, *Spumavirus*), AAA19979.1. *Env* genes derived from ERVs are shown in *bold letters*. *Red* and *blue circles* indicate that the clade was supported by $\geq$95% or $\geq$80% bootstrap values, respectively.

suggesting that *ERVs* may have independently expanded the IRF regulatory network in multiple mammalian lineages.

This hypothesis could also be applied to explain the variety of placental structures occurring among mammals. Although placental structures exhibit an abundance of diversity, trophoblast functions and their fusogenic activity exhibit more similarity than differences regardless of their invasive versus noninvasive trophoblast nature. Because of its distribution among mammalian genomes, *Peg10* may be a common gene initially required for the evolution of placental mammals.[25,26] While maintaining the same functions, placental structures have been diversified due to successive acquisition of new genes such as ERV-*envs*. When new *ERVs* that possess and provide reproductive advantages are integrated into the host genome, those new *ERVs* may then be retained in the population, eventually becoming common to the species.

4. CONCLUDING REMARKS

In this review, the expression of newly acquired and placenta-specific *ERVs* has been discussed pertaining to placental diversity and mammalian evolution. The main idea presented is that although *Peg10/Sirh1*, *Peg11/Sirh2*, and *Sirh7/Ldoc1* were integrated into primitive placentas, ERV-*envs* with fusogenic activity were independently and successively integrated into mammalian genomes. It should be noted that ERV-*envs* are not orthologous viral genes, but exhibit similar fusogenic functions. These independent integrations of ERV-*envs* must have caused structural diversification in a relatively short time. In the baton pass hypothesis, newly acquired ERV-*envs* replaced the preexisting genes and performed their functions much better than the predecessors.

What happened to old or preexisting genes once their function was taken over by the acquisition of new *ERVs*? Those genes tend to lose their function through mutations and/or deletions within a species. However, it is also possible that while those genes may have lost their original functions, they could have acquired alternative functions in other cell types or tissues. If this is the case, placental variations may still be proceeding through intermediate stages and their evolutionary history may still be far from over. Further experimentation should be carried out, through which other examples of the baton pass, function, and/or regulatory mechanisms of functional gene transcription, can be identified.

REFERENCES

1. Mess A, Carter AM. Evolutionary transformations of fetal membrane characters in Eutheria with special reference to Afrotheria. *J Exp Zool B Mol Dev Evol.* 2006;306:140–163.
2. Mess A. Evolutionary transformations of chorioallantoic placental characters in rodentia with special reference to hystricognath species. *J Exp Zool A Comp Exp Biol.* 2003;299:78–98.
3. Carter AM, Enders AC. Comparative aspects of trophoblast development and placentation. *Reprod Biol Endocrinol.* 2004;2:46.
4. Wildman DE, Chen C, Erez O, Grossman LI, Goodman RR. Evolution of the mammalian placenta revealed by phylogenetic analysis. *Proc Natl Acad Sci USA.* 2006;103:3203–3208.
5. Enders AC, Blankenship TN, Conley AJ, Jones CJ. Structure of the midterm placenta of the spotted hyena, Crocuta crocuta, with emphasis on the diverse hemophagous regions. *Cells Tissues Organs.* 2006;183:141–155.
6. Lavialle C, Cornelis G, Dupressoir A, et al. Paleovirology of 'syncytins', retroviral env genes exapted for a role in placentation. *Philos Trans R Soc Lond B Biol Sci.* 2013;368:20120507.
7. dos Reis M, Inoue J, Hasegawa M, Asher RJ, Donoghue PC, Yang Z. Phylogenomic data sets provide both precision and accuracy in estimating the timescale of placental mammal evolution. *Proc R Soc Lond B Biol Sci.* 2012;279:3491–3500.
8. Imakawa K, Nakagawa S, Miyazawa T. Baton pass hypothesis: successive incorporation of unconserved endogenous retroviral genes for placentation during mammalian evolution. *Genes Cells.* 2015;20:771–788.
9. Rezende LC, Barbeito CG, Favaron PO, Mess A, Miglino MA. The fetomaternal interface in the placenta of three species of armadillos (Eutheria, Xenarthra, Dasypodidae). *Reprod Biol Endocrinol.* 2012;10:38.
10. Carter AM, Mess A. Evolution of the placenta and associated reproductive characters in bats. *J Exp Zool Part B Mol Develop Evol.* 2008;310B:428–449.
11. Bischof P, Irminger-Finger I. The human cytotrophoblastic cell, a mononuclear chameleon. *Int J Biochem Cell Biol.* 2005;37:1–16.
12. Moffett A, Loke C. Immunology of placentation in eutherian mammals. *Nat Rev Immunol.* 2006;6:584–594.
13. Bhiwgade DA, Singh AB, Manekar AP, Menon SN. Ultrastructural development of chorioallantoic placenta in the Indian Miniopterus bat, Miniopterus schreibersii fuliginosus (Hodgson). *Acta Anat (Basel).* 1992;145:248–264.
14. Pfarrer C, Winther H, Leiser R, Dantzer V. The development of the endotheliochorial mink placenta: light microscopy and scanning electron microscopical morphometry of maternal vascular casts. *Anat Embryol (Berl).* 1999;199:63–74.
15. Wooding FB. Role of binucleate cells in fetomaternal cell fusion at implantation in the sheep. *Am J Anat.* 1984;170:233–250.
16. Hoffman LH, Wooding FB. Giant and binucleate trophoblast cells of mammals. *J Exp Zool.* 1993;266:559–577.
17. King BF, Mossman HW. The fetal membranes and unusual giant cell placenta of the jerboa (Jaculus) and jumping mouse (Zapus). *Am J Anat.* 1974;140:405–431.
18. Oduor-Okelo D, Musewe VO, Gombe S. Electron microscopic study of the chorioallantoic placenta of the rock hyrax (*Heterohyrax brucei*). *J Reprod Fertil.* 1983;68:311–318.
19. Carter AM, Blankenship TN, Künzle H, Enders AC. Structure of the definitive placenta of the tenrec, Echinops telfairi. *Placenta.* 2004;25:218–232.
20. Huppertz B, Kaufmann P, Kingdom J. Trophoblast turnover in health and disease. *Fetal Matern Med Rev.* 2002;13:103–118.

21. Lander ES, Linton LM, Birren B, et al. Initial sequencing and analysis of the human genome. *Nature*. 2001;409:860–921.
22. Chinwalla AT, Cook LL, Delehaunty KD, et al. Initial sequencing and comparative analysis of the mouse genome. *Nature*. 2002;429:520–562.
23. Luo ZX, Yuan CX, Meng QJ, Ji Q. A Jurassic eutherian mammal and divergence of marsupials and placentals. *Nature*. 2011;476:442–445.
24. Ono R, Kobayashi S, Wagatsuma H, et al. A retrotransposon-derived gene, PEG10, is a novel imprinted gene located on human chromosome 7q21. *Genomics*. 2001;73:232–237.
25. Suzuki S, Ono R, Narita T, et al. Retrotransposon silencing by DNA methylation can drive mammalian genomic imprinting. *PLoS Genet*. 2007;3: e55.
26. Ono R, Nakamura K, Inoue K, et al. Deletion of Peg10, an imprinted gene acquired from a retrotransposon, causes early embryonic lethality. *Nat Genet*. 2006;38:101–106.
27. Sekita Y, Wagatsuma H, Nakamura K, et al. Role of retrotransposon-derived imprinted gene, Rtl1, in the feto-maternal interface of mouse placenta. *Nat Genet*. 2008;40:243–248.
28. Naruse M, Ono R, Irie M, et al. Sirh7/Ldoc1 knockout mice exhibit placental P4 overproduction and delayed parturition. *Development*. 2014;141:4763–4771.
29. Dupressoir A, Lavialle C, Heidmann T. From ancestral infectious retroviruses to bona fide cellular genes: role of the captured syncytins in placentation. *Placenta*. 2012;33:663–671.
30. de Parseval N, Lazar V, Casella JF, Benit L, Heidmann T. Survey of human genes of retroviral origin: identification and transcriptome of the genes with coding capacity for complete envelop protein. *J Virol*. 2003;77:10414–10422.
31. Blaise S, de Parseval N, Heidmann T. Functional characterization of two newly identified human endogenous retroviruses coding envelop genes. *Retrovirology*. 2005;2:19.
32. Blaise S, de Parseval N, Bénit L, Heidmann T. Genomewide screening for fusogenic human endogenous retrovirus envelopes identifies syncytin-2, a gene conserved on primate evolution. *Proc Natl Acad Sci USA*. 2003;100:13013–13018.
33. Mi S, Lee X, Li X, et al. Syncytin is a captive envelope protein involved in human placental morphogenesis. *Nature*. 2000;403:785–789.
34. Kudaka W, Oda T, Jinno Y, Yoshimi N, Aoki Y. Cellular localization of placenta-specific human endogenous retrovirus (HERV) transcripts and their possible implication in pregnancy-induced hypertension. *Placenta*. 2008;29:282–289.
35. Chang C, Chen PT, Chang GD, Huang CJ, Chen H. Functional characterization of the placental fusogenic membrane protein syncytin. *Biol Reprod*. 2004;71:1956–1962.
36. Chen CP, Chen LF, Yang SR, et al. Functional characterization of the human placental fusogenic membrane protein syncytin 2. *Biol Reprod*. 2008;79:815–823.
37. Yu C, Shen K, Lin M, et al. GCMa regulates the syncytin-mediated trophoblastic fusion. *J Biol Chem*. 2002;277:50062–50068.
38. Muroi Y, Sakurai T, Hanashi A, Kubota K, Nagaoka K, Imakawa K. CD9 regulates transcription factor GCM1 and ERVWE1 expression through the cAMP/protein kinase A signaling pathway. *Reproduction*. 2009;138:945–951.
39. Tachibana I, Helmer ME. Role of transmembrane 4 superfamily (TM4SF) proteins CD9 and CD81 in muscle cell fusion and myotube maintenance. *J Cell Biol*. 1999;146:893–904.
40. Le Naour F, Rubinstein E, Jasmin C, Prenant M, Boucheix C. Severely reduced female fertility in CD9-deficient mice. *Science*. 2000;287:319–321.
41. Liang CY, Wang LJ, Chen CP, Chen LF, Chen YH, Chen H. GCM1 regulation of the expression of syncytin-2 and its cognate receptor MFSD2A in human placenta. *Biol Reprod*. 2010;83:387–395.

42. Okahara G, Matsubara S, Oda T, Sugimoto J, Jinno Y, Kanaya F. Expression analyses of human endogenous retroviruses (HERVs): tissue-specific and dedvelopmental stage-dependent expression of HERVs. *Genomics*. 2004;84:982–990.
43. Blond JL, Lavillette D, Cheynet V, et al. An envelope glycoprotein of the human endogenous retrovirus HERV-W is expressed in the human placenta and fuses cells expressing the type D mammalian retrovirus receptor. *J Virol*. 2000;74:3321–3329.
44. Hayward MD, Pötgens AJ, Drewlo S, Kaufmann P, Rasko JE. Distribution of human endogenous retrovirus type W receptor in normal human villous placenta. *Pathology*. 2007;39:406–412.
45. Kudo Y, Boyd CA. Changes in expression and function of syncytin and its receptor, amino acid transport system B (ASCT2), in human placental choriocarcinoma BeWo cells during syncytialization. *Placenta*. 2002;23:536–541.
46. Mangeney M, Renard M, Schlecht-Louf G, et al. Placental syncytins: genetic disjunction between the fusogenic and immunosuppressive activity of retroviral envelope proteins. *Proc Natl Acad Sci USA*. 2007;104:20534–20539.
47. Cáceres M, Comparative Sequencing Program, N.I.S.C., Thomas JW. The gene of retroviral origin Syncytin 1 is specific to hominoids and is inactive in Old World monkeys. *J Hered*. 2006;97:100–106.
48. Pötgens AJ, Schmitz U, Bose P, Versmold A, Kaufmann P, Frank HG. Mechanisms of syncytial fusion: a review. *Placenta*. 2002;23(suppl A):S107–S113.
49. Dupressoir A, Marceau G, Vernochet C, et al. Syncytin-A and syncytin-B, two fusogenic placenta-specific murine envelope genes of retroviral origin conserved in Muridae. *Proc Natl Acad Sci USA*. 2005;102:725–730.
50. Rawn SM, Cross JC. The evolution, regulation, and function of placenta-specific genes. *Annu Rev Cell Dev Biol*. 2008;24:159–181.
51. Dupressoir A, Vernochet C, Bawa O, et al. Syncytin-A knockout mice demonstrate the crucial role in placentation of a fusogenic, endogenous retrovirus-derived, envelop gene. *Proc Natl Acad Sci USA*. 2009;106:12127–12132.
52. Schubert SW, Lamoureux N, Kilian K, Klein-Hitpass L, Hashemolhosseini S. Identification of integrin-alpha4, Rb1, and syncytin-A as murine placental target genes of the transcription factor GCMa/Gcm1. *J Biol Chem*. 2008;283:5460–5465.
53. Schreiber J, Riethmacher-Sonnenberg E, Riethmacher D, et al. Placental failure in mice lacking the mammalian homolog of glial cells missing, GCMa. *Mol Cell Biol*. 2000;20:2466–2474.
54. Baba K, Nakaya Y, Shojima T, et al. Identification of novel endogenous Betaretroviruses which are transcribed in the bovine placenta. *J Virol*. 2011;85:1237–1245.
55. Cornelis G, Heidmann O, Degrelle SA, et al. Captured retroviral envelope syncytin gene associated with the unique placental structure of higher ruminants. *Proc Natl Acad Sci USA*. 2013;110:E828–837.
56. Nakagawa S, Bai H, Sakurai T, et al. Dynamic evolution of endogenous retrovirus-derived genes expressed in bovine conceptuses during the period of placentation. *Genome Biol Evol*. 2013;5:296–306.
57. Palmarini M, Hallwirth C, York D, et al. Molecular cloning and functional analysis of three type D endogenous retroviruses of sheep reveal a different cell tropism from that of the highly related exogenous Jaagsiekte sheep retrovirus. *J Virol*. 2000;74:8065–8076.
58. Nakaya Y, Koshi K, Nakagawa S, Hashizume K, Miyazawa T. Fematrin-1 is involved in fetomaternal cell-to-cell fusion in Bovinae placenta and has contributed to diversity of ruminant placentation. *J Virol*. 2013;87:10563–10572.
59. Nakaya Y, Miyazawa T. Dysfunction of bovine endogenous retrovirus K2 envelop glycoprotein is related to unsuccessful intracellular trafficking. *J Virol*. 2013;88:6896–6905.
60. Katoh K, Standley DM. MAFFT multiple sequence alignment software version 7: improvements in performance and usability. *Mol Biol Evol*. 2013;30:772–780.

61. Capella-Gutiérrez S, Silla-Martínez JM, Gabaldón T. trimAl: a tool for automated alignment trimming in large-scale phylogenetic analyses. *Bioinformatics*. 2009;25:1972–1973.
62. Le SQ, Gascuel O. An improved general amino acid replacement matrix. *Mol Biol Evol*. 2008;7:1307–1320.
63. Darriba D, Taboada GL, Doallo R, Posada D. ProtTest 3: fast selection of best-fit models of protein evolution. *Bioinformatics*. 2011;27:1164–1165.
64. Stamatakis A. RAxML version 8: a tool for phylogenetic analysis and post-analysis of large phylogenies. *Bioinformatics*. 2014;30:1312–1313.
65. Heidmann O, Vernochet C, Dupressoir A, Heidmann T. Identification of an endogenous retroviral envelope gene with fusogenic activity and placenta-specific expression in the rabbit: a new "syncytin" in a third order of mammals. *Retrovirology*. 2009;6:107.
66. Cornelis G, Heidmann O, Bernard-Stoecklin S, et al. Ancestral capture of syncytin-Car1, a fusogenic endogenous retroviral envelope gene involved in placentation and conserved in Carnivora. *Proc Natl Acad Sci USA*. 2012;109:E432–441.
67. Redelsperger F, Cornelis G, Vernochet C, et al. Capture of syncytin-Mar1, a fusogenic endogenous retroviral envelope gene involved in placentation in the Rodentia squirrel-related clade. *J Virol*. 2014;88:7915–7928.
68. Cornelis G, Vernochet C, Malicorne S, et al. Retroviral envelope syncytin capture in an ancestrally diverged mammalian clade for placentation in the primitive Afrotherian tenrecs. *Proc Natl Acad Sci USA*. 2014;111:E4332–E4341.
69. Cornelis G, Vernochet C, Carradec Q, et al. Retroviral envelope gene captures and syncytin exaptation for placentation in marsupials. *Proc Natl Acad Sci USA*. 2015;112: E487–E496.
70. Carter AM. Evolution of placental structure and function in ruminants. In: Juengel JL, Miyamoto A, Price C, Reynolds LP, Smith MF, Webb R, eds. *Reproduction in Domestic Ruminants VIII*. Leicestershire, UK: Context Products Ltd.; 2014:387–398.
71. Nakamura Y, Imakawa K. Retroviral endogenization and its role in the genital tract during mammalian evolution. *J Mamm Ova Res*. 2011;28:203–218.
72. Esnault C, Priet S, Ribet D, et al. A placenta-specific receptor for the fusogenic, endogenous retrovirus-derived, human syncytin-2. *Proc Natl Acad Sci USA*. 2008;105:17532–17537.
73. Dewannieux M, Heidmann T. Endogenous retroviruses: acquisition, amplification and taming of genome invaders. *Curr Opin Virol*. 2013;3:646–656.
74. Platanias LC. Mechanisms of type-I- and type-II-interferon-mediated signaling. *Nat Rev Immunol*. 2005;5:375–386.
75. Chuong EB, Elde NC, Feschotte C. Regulatory evolution of innate immunity through co-option of endogenous retroviruses. *Science*. 2016;351:1083–1087.

CHAPTER FIVE

Contribution of Syncytins and Other Endogenous Retroviral Envelopes to Human Placenta Pathologies

P.-A. Bolze*[,†,‡], M. Mommert[‡,§], F. Mallet[‡,§,1]
*University of Lyon 1, University Hospital Lyon Sud, Pierre Bénite, France
[†]French Reference Center for Gestational Trophoblastic Diseases, University Hospital Lyon Sud, Pierre Bénite, France
[‡]Joint Unit Hospices Civils de Lyon-bioMérieux, University Hospital Lyon Sud, Pierre Bénite, France
[§]EA 7526 Pathophysiology of Injury-Induced Immunosuppression, University of Lyon1-Hospices Civils de Lyon-bioMérieux, Hôpital Edouard Herriot, Lyon, France
[1]Corresponding author: e-mail address: francois.mallet@biomerieux.com

Contents

Abstract

Fusion, proliferation, angiogenesis, immune tolerance, and tissue survival are some of the critical functions involved in the physiological and pathological processes of placenta development. Strikingly, some of these properties are shared by envelope glycoproteins of retroviruses. Part of the overall retroviral world, the human retroviral heritage

Progress in Molecular Biology and Translational Science, Volume 145
ISSN 1877-1173
http://dx.doi.org/10.1016/bs.pmbts.2016.12.005

consists of hundred thousands of elements representing a huge amount of genetic material as compared to our 25,000 genes, whereas only a few tenths of retroviral loci still contain envelope genes exhibiting large open reading frames. Some of these envelopes, namely Syncytin-1, Syncytin-2, and ERV-3 Env, were shown to support essential functions in placenta development. First, in order to understand where these envelope genes originate and what are the critical mechanisms involved in transcription regulation and protein basic functions such as recognition of cellular receptor by viral envelopes, we will describe the retroviral life cycle and how repeated infections during species evolution led to the formation of retroviral families. We will emphasize how many envelope genes remain in our genome and in which organs they were found to be expressed. Second, Syncytin-1 will be used as a model to decipher essentially in placental context (i) the detailed modalities of transcriptional control including repressive histone marks and CpG methylation epigenetic mechanisms, involvement of tissue-specific transcription factors, and control of mRNA splicing, as well as (ii) the multiple steps required for protein maturation finally leading to a functional trimeric glycosylated protein. The extraordinary versatility of Syncytin-1 will permit to demonstrate that such proteins are likely involved in physiological processes not only in placenta but also in other organs, based on evidence of fusion/differentiation, immunomodulation, apoptosis, and proliferation properties. Third, we will describe extensively the altered behavior of the various levels of transcriptional control or of protein functions/localization/maturation displayed by Syncytins and other endogenous retroviral envelopes. We will exemplify how such altered states may contribute to human placenta pathologies, including Down syndrome, preeclampsia/hemolysis, elevated liver enzymes, and low platelets syndrome/intrauterine growth restriction, and gestational trophoblastic diseases including mole and choriocarcinoma. Similar deregulations will be respectively mentioned on this target of fetal invasion that is the endometrium, the reproductive organs that are the testis and the ovary, and in the breast nourisher of the newborn child. All these observations draw outlines of the symbiotic and conflicting mechanisms at work where the retrovirus world and the human world have converged.

ABBREVIATIONS

BaEV baboon endogenous virus
CHM complete hydatidiform mole
DC-SIGN dendritic cell-specific intercellular adhesion molecule-3-grabbing nonintegrin
DNMT3B DNA methyl transferase 3B
DRM detergent-resistant membrane
Env envelope
ERV endogenous retrovirus
FcEV Felis catus endogenous retrovirus
FP fusion peptide
FSFFV Friend spleen focus-forming virus
GCMa/1 glial cell missing a/1
GTN gestation trophoblastic neoplasia

HELLP hemolysis, elevated liver enzymes, and low platelets
HERV human endogenous retrovirus
HFV human foamy virus
HIV human immunodeficiency virus
HTLV human T-cell leukemia virus
ISD immunosuppressive domain
IUGR intrauterine growth restriction
JSRV Jaagsiekte sheep retrovirus
LPS lipopolysaccharide
LTR long terminal repeat
MALR mammalian apparent LTR retrotransposon
MLV murine leukemia virus
MMTV mouse mammary tumor virus
MSRV multiple sclerosis-associated retrovirus
OASIS old astrocytes specifically induced substance
ORF open reading frame
PBMC peripheral blood mononuclear cell
PBS primer-binding site
PCR polymerase chain reaction
PcRV Papio cynocephalus retrovirus
PE preeclampsia
PHM partial hydatidiform mole
RBD receptor-binding domain
RD114 a feline endogenous retrovirus
RT reverse transcriptase
SERV simian endogenous retrovirus
SETDB1 histone-lysine *N*-methyltransferase
SNV spleen necrosis virus
SP signal peptide
SRV simian retrovirus
SU surface unit
TLR4 toll-like receptor 4
TM transmembrane unit
tm transmembrane domain
URE upstream regulatory element

1. INTRODUCTION

Primitive cells were possibly very simple membrane-enclosed structures, containing RNA sequences, that underwent leakage and uptake of new molecules.[1] Cell division and particularly cell fusion created a dynamic setting in which to test new nucleic acid stretches, and consequently genetic exchanges became an early core evolutionary force.[2] In line with this, retroviruses can be seen as RNA shuttles ensuring genetic exchanges from one genome species to another. Temin formulated the hypothesis that retroviruses evolved from cellular moveable genetic elements, a key event of the process being the acquisition of an ancestral coding sequence leading to reverse transcriptase (RT)[3] whose characteristic motifs remain in the catalytic subunit of telomerase.[4] Subsequent to this association between mobile elements and precursor of RT, the structures of new entities have gained increasing complexity by adding regulatory sequences and sequences encoding enzymatic activities or structural proteins.[5] Last, infectious retroviruses appear to have emerged, escaping from the genomes of our distant ancestors by transcomplementation of cellular retrotransposons with viral envelope genes.[6] Such capture of a gene encoding a fusogenic envelope glycoprotein allowed the release and interspecies transmission of the genetic material of these newly born viruses, now called retroviruses.

Additional cross-species propagation may also result from another type of capture that exists between retroviruses of distant species, consisting in the swapping of envelopes. For example, the feline RD114 infectious endogenous virus comes from two genetic recombinations, i.e., the simian SERV *env* was first captured by the yellow baboon PcRV leading to the baboon BaEV and then the acquisition of BaEV *env* by feline FcEV led to the emergence of RD114 virus.[7] Due to the fusogenic capacities of their envelopes, retroviruses spread across species, occasionally infecting germ line cells where their genetic material became fixed as a provirus in the host genome, thereafter part of the genetic heritage of the host. During evolution, the founding-captured endogenous retrovirus (ERV) and later its descendant elements are mainly replicated by reinfection and retrotransposition mechanisms, and most of the elements contain disruptive substitutions, insertions, and deletions. Significantly, the preferential loss of the *env* gene of many ERV elements is a common phenomenon that may reflect the absence of selective pressure for retention of this gene once the species barrier of species is crossed.[8] Indeed, the retroviral genetic material is actually propagated via the reproductive activity of the host species. Today, in humans, mammalian apparent LTR retrotransposons (MaLR) and human endogenous retroviruses (HERVs) are the remnants of retrotransposon ancestors and of hundreds of retroviral infections, respectively. Altogether, retrovirus-like elements represent 8.3% of the human euchromatin,[9] comprising roughly 200,000 MaLR elements and 200,000 HERV elements, including 18 coding envelope genes that we will describe.

Why dwell on retroviruses while our main interest is the placenta and particularly placental pathologies? In the 1970s, electron microscopy described the presence of virus-related particles in placental chorionic villous tissues of humans and primates.[10] Further studies then revealed certain retroviral characteristics of these particles such as ultrastructural features and RT activity.[11] In addition to the presence of particles, mRNA expression from different HERV families was reported in placenta[5] and was followed by the detection of retroviral envelopes using immunohistochemical techniques in human[12] and in baboon.[13] Moreover, the human placenta is a transient organ whose contribution to fetal development involves various and complex functions such as local immunotolerance allowing initial blastocyst implantation and later maintenance of the fetal semiallograft, angiogenesis in response to hypoxia, massive proliferation, and finally differentiation including trophoblast fusion and prolonged cell survival postfusion. Interestingly, similar events are supported by classical/infectious retroviral envelopes and also by authentic genes encoding endogenous retroviral envelopes. We will describe all these properties through the most well-studied example, Syncytin-1, the envelope gene of the human endogenous retroviral locus ERVW-1 belonging to the HERV-W family. This gene, one of the most important genes involved in cell–cell fusion, was characterized at the protein and transcriptional levels initially in normal placenta and later in pathological pregnancies. Understanding these retroviral gene functions and their regulatory mechanisms is a prerequisite for elucidating their alterations in placental pathological contexts. In addition, these functions supported by retroviral envelope proteins are long known to be altered in placental pathology. This was exemplified by Ruebner concerning the fusion property: "Terminal differentiation of villous cytotrophoblasts (CTs) ends in formation of the multinucleated syncytiotrophoblast representing the fetal-maternal interface. Aberrations during this cell-fusion process are associated with IUGR, Preeclampsia (PE) and HELLP syndrome".[14] Moreover, alterations of the immune tolerance at the uteroplacental interface, apoptosis, and proliferation are important characteristics of placental syndromes such as PE and malignant gestational trophoblastic diseases. Last, abnormal proliferation and immune tolerance are associated with the development of reproduction-linked pathologies such as endometriosis, endometrial carcinoma, and tumoral testis. We will illustrate how modifications in the transcription regulation, protein synthesis, maturation, and cellular localization of endogenous retroviral envelopes are involved in these pathological contexts.

2. SO MANY HUMAN ENDOGENOUS RETROVIRAL LOCI, SO FEW GENES ENCODING ENVELOPES

2.1 From Infectious to Inheritable Retroviruses

The rare event that represents the infection of a germ line cell by an exogenous retrovirus leads to the integration into the host genome of a retroviral DNA, or provirus, thus becoming part of the genetic heritage of the host. Thereafter, this endogenous provirus is transmitted to the next generation in a Mendelian fashion. The parental infectious retrovirus is a diploid RNA virus whose 8–10-kb compact genome consists of three major genes, *gag*, *pol*, and *env*, encoding the proteins required for its replication life cycle, and flanked by 5′R-U5 and 3′U3-R untranslated regions. Fig.1A shows schematically the replication cycle of a simple infectious retrovirus in order to point out which HERV components, either proteins or long terminal repeat (LTR) regulatory elements involved in transcription initiation and termination, can fulfill a function by their contribution to a physiological or a pathological role. Briefly, following the entry into the cell, the viral RNA is reverse transcribed into DNA by the viral RT using a cell-specific tRNA as a primer hybridized with the PBS (primer-binding site) region located at the R-U5 and *gag* junction of the retroviral genome. The resulting double-stranded DNA, which contains at each end a noncoding LTR sequence derived from R-U5 and U3-R viral sequences, is integrated into the genome of the host cell through the action of the viral integrase. The expression of the proviral DNA then becomes dependent on the host cell machinery that provides the transcription factors required to activate the 5′LTR. The 5′LTR plays the role of promoter and enhancer sequence conferring tissue-specific expression. The distal 3′LTR contains the polyadenylation signal terminating the transcript. Classically, two major transcription products are produced: an unspliced genomic RNA containing all viral genes and supporting Gag Pol translation, and a spliced subgenomic RNA supporting Env translation. The *gag* gene encodes matrix, capsid, and nucleocapsid proteins necessary for viral RNA encapsidation and particle formation. The *pol* gene encodes the major viral enzyme machinery, including a protease, the RNA-dependent DNA polymerase that is RT, and a DNA integrase. These two nucleic acid-modifying enzymes are required sequentially for the successful conversion of viral RNA into double-stranded DNA and its subsequent integration as a provirus into the host cell genome. The protein modifying protease is required for cleavage

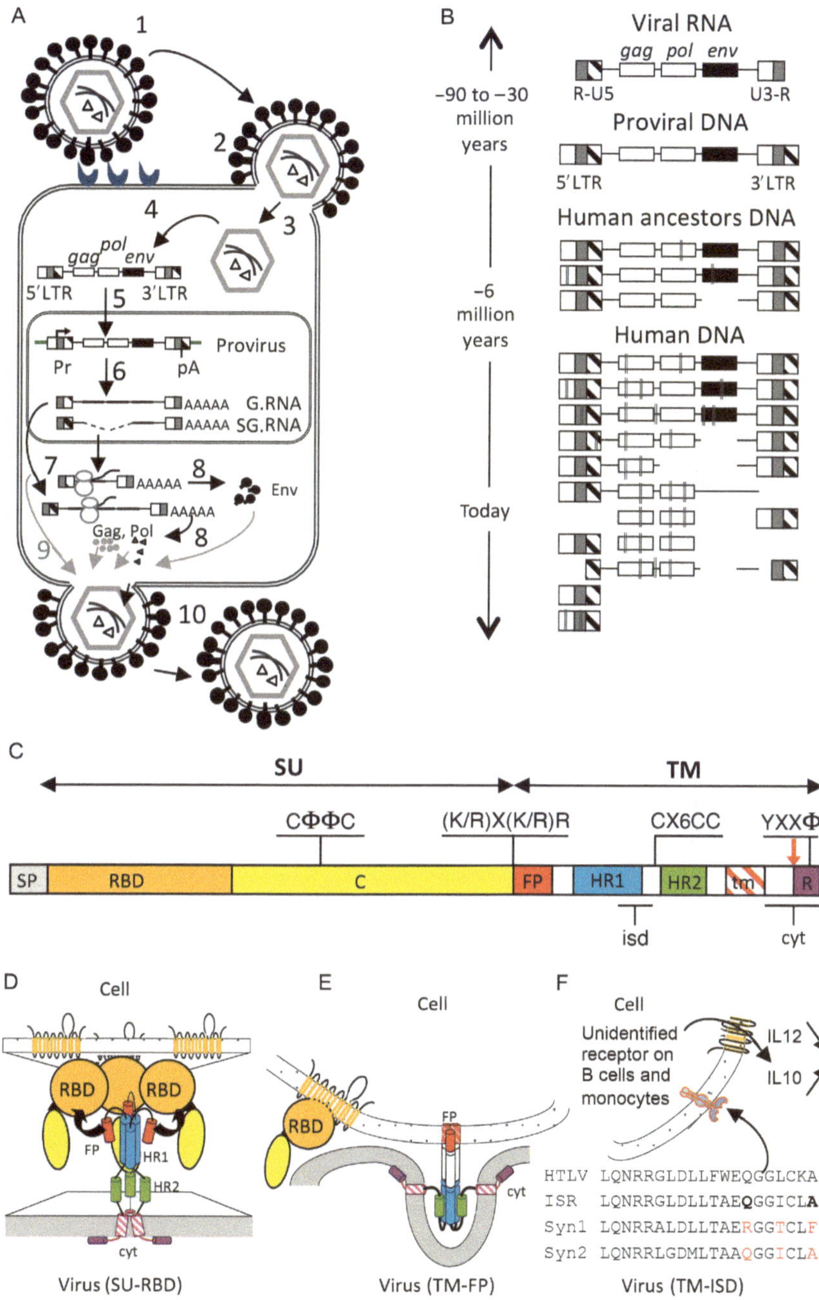

Fig. 1 See legend on next page.

of Gag-Pro-Pol and Gag polypeptide precursors, and also, in the case of the Gammaretroviridae genera (see Section 3.3), for the final Env maturation step leading to fusion competency. Last, the *env* gene encodes viral envelope glycoproteins that confer virus infectivity, i.e., receptor recognition via the SU (surface unit) subunit and virus–cell membrane fusion via the TM (transmembrane) subunit. In addition, the TM subunit contains motifs that are likely to confer immunosuppressive properties to retroviruses.

The process of endogenization is dependent upon the success of the retrovirus to overcome several cellular and evolutionary obstacles. These include notably the capacity of the retrovirus and its progeny to bypass

---

Fig. 1 From viral infection to domesticated envelope function in human host. (A) *Schematic representation of the replication life cycle of an infectious retrovirus,* depicting the functions sporadically preserved in endogenous retroviral remnants. (1) Binding to a cell receptor via the viral envelope surface subunit (SU, Env), (2) fusion of viral and cell membranes via the viral envelope transmembrane subunit (TM, Env), leading to (3) the entry of the capsid in the cytoplasm, (4) the conversion of the viral RNA to DNA by the reverse transcriptase (RT) and (5) the nuclear translocation and integration of the provirus flanked by two identical LTRs in the cellular DNA (provirus), (6) transcription (5′LTR promoter function, Pr; 3′LTR polyadenylation signal, pA) controlled by host cell transcription factors and production of genomic (G.RNA) and spliced subgenomic (SG.RNA) mRNAs (7) transport to the cytoplasm, (8) production of Gag (capsid, matrix, nucleocapsid) and Pol (RT, integrase, protease) polyproteins from G.RNA and Env from SG.RNA (9) assembly of the genomic RNA and processed viral proteins leading to (10) the budding and release of mature virions, which can infect new cells. (B) Constitution of an HERV family. The proviral DNA that derived from the RNA of the viral infectious ancestor integrated millions years ago into the DNA of a germ cell, spread mainly by reinfection and retrotransposition and the different offspring ERV elements went through a mutagenic process (*vertical lines*, deletion) during evolution. No contemporary HERV copy is infectious in human. Note the frequency of *env* gene deletion, the absence of U3 promoter region in 5′LTRs on certain elements, and the existence of solo LTR. (C) Structure of Syncytin-1, the only envelope conserved in the HERV-W family. SP, signal peptide. SU, surface unit contains RBD, receptor-binding domain, and C, C-terminal domain of SU with CΦΦC motif (Φ=L, I, V, F, M, or W), (K/R)X(K/R)R, furin cleavage site, TM, transmembrane unit contains FP, fusion peptide; leucine zipper motif with HR1 and HR2 heptad repeats followed by the CX6CC motif; tm, transmembrane anchorage domain; the ectodomain part of the TM contains a so-called immunosuppressive domain labeled isd; *cyt*, cytoplasmic tail with C-terminal R peptide containing YXXΦ motif. (D) Schematic representation of SU-RBD/receptor (hASCT2) interaction inducing conformational changes (E), which may lead to virus and cell membrane fusion mediated by the anchorage of the envelope fusion peptide (FP) located at N-term TM into the cell membrane. (F) Immunosuppressive domain (isd) of TM retroviral envelope (HTLV-1, immunosuppressive retroviruses state for MPMV, SRV SNV, BAEV, Syncytin-1, Syncytin-2) inhibits lymphocyte proliferation and alters cytokine release, possibly via the recognition of a specific receptor.[15]

the restrictive mechanisms common to all cells that act against retroviruses and retrotransposons, e.g., TRIM5α, OAS, RNaseL, and APOBEC3G factors,[16,17] and their need to infect reproductive tissues to reach the germ line and to infect one or more progenitor cells. Subsequently, the integrated virus should neither hamper the survival or the functional capacities of these germ cells to participate to the formation of a new zygote nor affect the fitness of the offspring so as to persist during the evolution of the host species.[18]

During the evolution, the founding-captured HERV provirus, and later elements of its progeny, is replicated by mechanisms such as reinfection and retrotransposition that essentially rely on transcription. Due to the overall absence of selection pressures, most if not all of the elements contain disruptive mutations, like substitutions, insertions, and deletions, in at least one of the structural genes of the provirus. Therefore, the preferential loss of the *env* gene of the overwhelming majority of HERV elements is a phenomenon that may reflect the expendability of this gene once the virus has crossed the species barrier.[8] Nevertheless, open reading frames (ORFs) can persist and lead to protein synthesis or even apparently matured but noninfectious particles.[19–21] In addition, each family contains numerous solitary LTRs, resulting from the recombination between the two distal LTRs flanking a proviral structure, resulting in the loss of internal coding sequences.[5] All these mechanisms lead to complex multicopy families, each consisting of heterogeneous elements (Fig. 1B). All loci of the contemporary HERV families are defective for complete retroviral life cycle, which means that they have lost their infectious properties and are exclusively engaged in a vertical mode of transmission. In addition to such significant level of complexity, the processes of spread within the genomes have generated a wide distribution of these sequences among chromosomes. A current nomenclature for HERV families is based on the PBS sequence located downstream of the 5′LTR, or its similarity to the infectious retroviruses PBS, which is recognized by a specific tRNA. The one letter code that refers to the amino acid specified by the tRNA is applied as a suffix, e.g., HERV-H exhibits a PBS which is recognized by a histidine (H) tRNA, the HERV-W PBS is homologous to the PBS of an avian retrovirus using tryptophan (W) tRNA. Multiple names sometimes exist for the same HERV, such as HERV-R initially known as ERV-3. This nomenclature can be misleading, e.g., the superfamily HERV-K contains 11 phylogenetically distinct subgroups referred to as HML-1 to HML-11,[18,22] or HML-5 members being primed by a methionine (M) or isoleucine (I) tRNA.[23,24] The International Committee on Taxonomy has established seven genera of Retroviridae based on sequence homologies of the *pol* region[25] and overlapping with the A-, B-, C-, or D-type classification based on virion morphology observed by electronic microscopy.[26] HERV elements are essentially present in Class I/gammaretroviruses, which correspond to mammalian type C retroviruses (MLV, murine leukemia virus), such as ERV3, HERV-E, HERV-FRD, HERV-H, and HERV-W families; Class II/betaretroviruses, which correspond to type B (MMTV, mouse mammary tumor virus) and type D (SRV-1, simian retrovirus) retroviruses, such as HML-1 and HML-2 family (HERV-K superfamily); and spumaviruses, which include HFV (human foamy virus) and the 90-million-year-old (H)ERV-L family, whose founding virus infected the germ line of a common ancestor prior to the divergence of placental mammals.[27,28]

A first complete view of the HERV landscape was provided by the publication of the human genome (International Human Genome Sequencing Consortium, 2001), revealing that LTR elements including endogenous retroviruses composed of 8.3% euchromatin. More precisely, the human genome contains 203,000 copies resulting from about 100 independent infectious events, although only about 40 groups have been studied,[29–33] and also contains some 240,000 MaLR elements. It is crucial to appreciate how these hundreds of thousands of retroviral loci constitute a mass significantly greater than the "classical" human genes, a number currently estimated to range between 20,000 and 25,000 (International Human Genome Sequencing Consortium 2004).

Given the huge amount, distribution, and retroviral and repetitive nature of these elements, several functions have gradually been conceived: (i) HERV may be involved in genomic plasticity during evolution as recombination sites within or between chromosomes[34]; (ii) such a recombinatorial property may induce germinal or somatic mutations giving rise to the loss of function of a cellular gene[35–37]; (iii) LTRs can modulate the expression of adjacent cellular genes[38–41]; and (iv) the expression of HERV envelopes with conventional retroviral functions like fusion or immunomodulation can impact physiological or pathological conditions of the host.[42–45]

2.2 The Retroviral Envelopes Heritage

Several studies proposed screening procedures to identify the coding envelope genes in the human genome. To first collect *env* genes, one main strategy was based on similarity search in the human genome using *env* gene consensus. These *env* consensuses were identified and picked from a recognized proviral element, built from an alignment of all the proviruses belonging to the same family, using characteristic Env stretches of amino acids such as the immunosuppressive domain (ISD). Such a study identified 476 potentially complete envelope genes, only 16 of which were found to potentially encode envelope proteins.[46] Another study was based on the systematic detection of HERV internal regions (*gag*, *pol*, and *env*) using both endogenous and replication-competent (with full ORFs) exogenous retroviral sequences from various host organisms. Scanning these elements for content of viral ORFs identified 134 *env* genes containing ORFs larger than 300 AA, and 73 and 29 loci containing ORFs larger than 400 and 500 AA, respectively.[47] Finally, three additional envelopes were identified, and one envelope initially selected was absent from the actual version of the human genome, leading to 18 ORFs similar to Env proteins of infectious viruses. It should be kept in mind that it is quite impossible to predict which HERV *env* gene may lead to the production of a protein envelope able to deliver a cellular signal. Thus, de Parseval selected *env* genes with an ORF beginning at the first Met codon of the consensus envelope gene, classically at the amino terminal part of the hydrophobic signal that allows the anchorage of the nascent chain to the endoplasmic reticulum (ER). Nevertheless, it is known that retroviruses may use nonconventional signals of translational initiation, frameshifting, and suppression of termination. All these mechanisms may support the production of alternative or larger ORFs than those predicted, ultimately smaller than a classical retroviral envelope, which may nevertheless contain functional signal-transducing motifs. This can be illustrated by the ERVW-2 locus of the HERV-W family, located on chromosome Xq22.3 that harbors an almost complete ORF for a full-length envelope protein but is interrupted by a stop codon downstream from the signal peptide. Hence, in vivo expression would require bypassing the stop codon or utilization of alternative initiation sites. Although no protein production was evidenced in vivo, the reversion of the stop codon in vitro led to the expression of a reconstituted, full-length HERV-W envelope protein sharing very similar posttranslational features with Syncytin-1.[48] In addition, although this locus lacks a 5′LTR promoter,[49] the transcription of the Xq22.3 env gene has been reported.[49–51]

Basically, assignment of putative functions of endogenous retroviral envelopes was based on known functions of envelopes of infectious retroviruses and the evolutionary preservation of associated canonical motifs. Nevertheless, in the absence of described functions and in view of our lack of understanding of the significance of truncated HERV envelope products,

Table 1 The 18 Coding Envelope Genes of the Human Genome

Gene[a] [Family (Class)]	Bibliographic Name[b]	Genomic Localization (Strand Sens)[c] (Cytogenetic Band)	AA[d]	CXXC (SU) and CX6CC, or CX7C (TM)[e]	Furin[f]	ISD[g]	Tissue[h]	Method[i]	References[j]
ERV3-1 [HERV-R (I)]	ERV3 envR	Chr7:64990324–64999936 (−) [7q11.21]	604	CWDC and CGKFNLTNCC	KSKR	YQNRLALDYLLAQE	Al, Bm, Br, Brt, Cl, Ht, Kd, Lv, Lg, Ov, PBL, **Pl** (SynT, CTv), Ps, Sk, Sp, Ts, Thy, Thd, Tr	RT-PCR, NB **WB**, **IHC**[k]	1, 2, **3**, 4, 5, 6, **7**, **8**
envF(c)1 [HERVF-(c)1 (I)]		chrX:95868842–95875915 (+) [Xq21.33]	584	CFLC and CMFLGEECC	RQKR	MQNRRALDLLTADK	Sk, Ts, Tr	RT-PCR	1, 9
ERVFC1-1 [HERVF(c)2 (I)]	envF(c)2	Chr7:153409434–153414153 (−) [7q36.2]	527	CFLC and CLFLQEECC	KSKW	AQNRQALDLLMAEK	Sk, Ts	RT-PCR	1, 9
ERVFRD-1 [HERV-FRD (I)]	Syncytin-2 envFRD	Chr6:11102749–11112216 (−) [6p24.1]	538	CWLC and CLALDEKCC	RVRR	LQNRRGLDMLFAAQ	Al, Bm, Br, Bst, Cl, Kd, Lg, Ov, PBL, **Pl** (SynT,CTv), Ps, Sk, Sp, Ts, Thy, Thd, Tr	RT-PCR, NB **WB**, **IHC**[k]	1, 9, **10**, 11, **12**
envH1 [HERV-H (I)]	envH/p62 H19	Chr2:166767244–166768998 (−) [2q24.3]	585	CWLC and CIFLNEECC	QQKR	LQNRRGLDLLTAEK	Sk, Ts	RT-PCR	1

envH2 [HERV-H (I)]	envH/p60	Chr3:167860265–167867997 (−) [3q26]	563	CWLC and mutated	RQKR	LQNRQGLDLLTAEK	Lg, Ts	RT-PCR	1
envH3 [HERV-H (I)]	envH/p59	Chr2:155931277–155932944 (+) [2q24.1]	555	CWLC and mutated	PQKR	LQNRRGLGLSILLN	Ts	RT-PCR	1
ERVK-21 [HML2 (II)]	HERV-K 74261 envK1	Chr12:58327459–58336915 (−) [12q14.1]	698	CDWNTSDFC	RSKR	LANQINDLRQTVIW	Al, Bm, Br, Bst, Cl, Kd, Ov, PBL, Pl, Ps, Sk, Sp, Ts, Thy, Thd, Tr	RT-PCR **WB**[l]	1, 9, **13**
ERVK-6 [HML2 (II)]	HML-2. HOM K (C7) HERV-K 108 envK2[m]	Chr7:4582426–4600400 (−) [7p22.1]	699	CDWNTSDFC	RSKR	LANQINDLRQTVIW	Al, Bm, Br, Bst, Cl, Kd, Ov, PBL, Pl, Ps, Sk, Sp, Ts, Thy, Thd, Tr	RT-PCR **WB**[l]	1, **13**, 14, 15, 16
ERVK-19 [HML2 (II)]	HERV-K (C19) HERV-K 17833 envK3	Chr19:32821287–32829201 (−) [19q12]	699	mutated	RSKR	LANQINDLRQTVIW	Al, Bm, Br, Bst, Cl, Kd, Ov, PBL, Pl, Ps, Sk, Sp, Ts, Thy, Thd, Tr	RT-PCR **WB**[l]	1, **13**, 15
ERVK-9 [HML2 (II)]	HERV-K109 envK4	Chr6:77716945–77726366 (−) [6q14.1]	698	CDWNTSDFC	RSKR	LANQINDLRQTVIW	Al, Bm, Br, Bst, Cl, Kd, Ov, PBL, Pl, Ps, Sk, Sp, Ts, Thy, Thd, Tr	RT-PCR **WB**[l]	1, **13**, 16

Continued

Table 1 The 18 Coding Envelope Genes of the Human Genome—cont'd

Gene [Family (Class)]	Bibliographic Name	Genomic Localization (Strand Sens) (Cytogenetic Band)	AA	CXXC (SU) and CX6CC, or CX7C (TM)	Furin	ISD	Tissue	Method	References
ERVK-8 [HML2 (II)]	HERV-K115 envK6	Chr8:7497875–7507337 (−) [8p23.1]	699	CDWNTSDFC	RSKR	LANQINDLRQTVIW	Al, Bm, Br, Bst, Cl, Kd, Ov, PBL, Pl, Ps, Sk, Sp, Ts, Thy, Thd, Tr	RT-PCR **WB**[l]	1, **13**, 17
ERVPABLB-1 [HERV-R(b) (I)]	envR(b)	Chr3:16763830–16772531 (+) [3p24.3]	514	CWVC and CALIKTECC	VVNQ RWVK[o]	LQNRMALDILTAAE	Pl, Ts	RT-PCR	1, 9
EnvP(b) [HERV-P(b) (II)]	ZFERV-like Env	Chr14:92621871–92630611 (−)	665	Absent/CYAC[n] CVMIGTQCC	RKTR	WENRMALDMILAEK	Al, Bm, Br, Bst, Cl, Ht, Kd, Lv, Lg, Ov, PBL, Pl, Ps, Sk, Sp, Ts, Thy, Thd, Tr	RT-PCR	18, 19
ERVS71-1 [HERV-T (I)]	envT	Chr19:20334642–20343232 (+) [19p13.11]	626	CWLC and CAALGESCC	RLHQ	LQNCRCLDLLFLS	Al, Bm, Br, Bst, Kd, Ov, Pl, Ps, Sk, Ts, Thd, Tr	RT-PCR	1, 9
ERVV-1 [HERV-V (II)]	HERV-W/FRD-like Env envV1	Chr19:58209156–58210586 (+) [19q13.41]	477	CWIC and CAVISKSCC	RQKR	MNNRLALDYLLAEQ	Pl	RT-PCR	18, 19, 20, 21

ERVV-2 [HERV-V (II)]	HERV-W/ FRD-like Env envV2	Chr19:58244317–58245921 (+) [19q13.41]	535	CWIC and CAVINKSCC	RQKR	MDNRLALDYLLAEQ	Pl	RT-PCR	18, 19, 20, 21
ERVW-1 [HERV-W (I)]	Syncytin-1 ERVWE1 envW	Chr7:92468380–92477986 (−) [7q21.2]	538	CWIC and CLFLGEECC	RNKR	LQNRRALDLLFAER	Al, Bm, Br, Bst, Cl, Kd, Ov, **Pl** (SynT, CTv, Ctev) Ps, Sk, Sp, Ts, Thy, Thd, Tr	RT-PCR, NB , cDNA **WB, IHC**[k]	1, **22**, **23**, 24

[a]Mayer *et al.*[52]; note that EnvK5 (HERV-113) at Chr19 is not present in GRCh38.p7 human genome version.
[b]Recapitulated from Villesen *et al.*[47], Blaise *et al.*,[53] and de Parseval *et al.*[46]
[c]Human genome version GRCh38.p7.
[d]Amino acid length.
[e]CXXC and CX6CC motifs involved in SU–TM interaction and protein folding in class I HERV and CX7C TM motif involved in protein folding in class II HERV.
[f]Furin cleavage site between SU and TM.
[g]ISD predicted immunosuppressive domain.
[h]Tissues where env transcript and protein were evidenced; *Al*, adrenal; *Bm*, bone marrow; *Br*, brain; *Bst*, breast; *Cl*, colon; *Kd*, kidney; *Ov*, ovary; *Pl*, placenta (*CTev*, extravillous cytotrophoblast; *CTv*, villous cytotrophoblast; *SynT*, syncytiotrophobast); *Ps*, prostate; *Sk*, skin; *Sp*, spleen; *Thd*, thyroid; *Thy*, thymus; *Tr*, trachea; *Ts*, testis. Detection by RT-PCR, except Northern blot and protein detections indicated by *underlined* and *bold letters*, respectively (reference numbers highlighted in column j).
[i]RT-PCR from de Parseval *et al.*[46] and Blaise *et al.*[53]
[j]Cited references are: (1) de Parseval *et al.*,[54] (2) Andersson *et al.*,[55] (3) Andersson *et al.*,[56] (4) Cohen *et al.*,[57] (5) Cohen *et al.*,[58] (6) Kato *et al.*,[59] (7) Larsson *et al.*,[60] (8) Venables *et al.*,[12](9) de Parseval *et al.*,[46] (10) Malassiné *et al.*,[61] (11) Blaise *et al.*,[62] (12) Chen *et al.*,[63] (13) Dewannieux *et al.*,[64] (14) Mayer *et al.*,[52] (15) Tonjes *et al.* 1999,[65] (16) Barbulescu *et al.*,[66] (17) Turner *et al.*,[67] (18) Villesen *et al.*,[47] (19) Blaise *et al.*,[53] (20) Vargas *et al.*,[68] (21) Kjeldbjerg *et al.*,[69] (22) Blond *et al.*,[70] (23) Frendo *et al.*,[71] (24) Pérot *et al.*[72]
[k]Immunohistochemistry.
[l]WB recognizes envK epitope and does not discriminate between envK1–4 and 6.
[m]The HML-2.HOM/K(C7)/HERV-K108 locus is organized as a tandem repeat in some individuals, both envelopes being 100% identical.[46]
[n]Absent from Blaise *et al.*,[53] CYAC this study.
[o]VVNQ in de Parseval *et al.*,[46] RWVK predicted.

we summarized the features of the 18 envelopes containing the larger ORFs in Table 1. All proteins contain the transmembrane hydrophobic domain close to the carboxyl terminal end of the TM subunit, allowing plasma membrane anchorage. Regarding gammaretroviruses (Class II), the CXXC motifs of SU subunits appear preserved in all envelopes, as do the complementary CX6CC motifs of TM subunits, except for envH2 and envH3 that lacked the two distal cysteine residues. Those two motifs are required for covalent SU–TM interaction as well as correct folding of the envelope. In betaretroviruses (Class I), which have noncovalently associated subunits, the CXXC motif is absent as expected.[73] In TM subunit, the CX7C motif required for SU folding was conserved in all HERV-K envelopes except EnvK3, which lacks the distal cysteine; 11 of the 18 Env proteins have a furin cleavage site identical to the canonical R-N-[K/R]-R motif, which suggests that they could be cleaved, although we cannot predict the functionality of other motifs notably ending with arginine, except EnvT, EnvR(b), and EnvF(c)2 that end with Q, Q or K, and W, respectively. Note that if inactivating mutations of the furin cleavage site abolish the fusogenic properties of Syncytin-1, the mutant protein can still reach the cell surface, albeit with a delayed kinetics.[74] Notably, Env3 and EnvP(b) lack a canonical fusion peptide typically located at the N-terminus of the TM subunit, suggesting that they are not competent for fusion.

Strikingly, EnvP(b) whose expression appears ubiquitous is highly fusogenic in vitro, but undetected in placenta.[75] These apparently conflicting observations deserve further investigations in order to identify envelope motifs and cellular-associated receptors that may drive fusion in other tissues. Conversely, EnvV(1) and EnvV(2) that present all required amino acid motifs are not fusogenic.[53] Finally, envelope fusogenic activity and discovery of cognate receptors were only formerly demonstrated for the Syncytin-1 and Syncytin-2, e.g., by coculture assays,[42,62] inhibiting key maturation steps,[74] using receptor interference for hASCT2 and hASCT1[42,76] or via the use of a human/hamster radiation hybrid panel for MFSD2A.[77] Expression of such envelopes in the placenta may protect against infection by closely related exogenous retroviruses via receptor interference. A receptor interference group is defined as a set of retroviruses that cannot infect a cell at the same time because they use the same receptor for entrance (fusion) and, following initial infection, receptors are trapped within the cell. To date, no other cellular receptors, which are potential fusion partners and possibly immunomodulatory actors (see

Section 3.4), have been identified for other human HERV Env. Finally, an apparent ISD is present in all envelopes. However, true demonstration of immunosuppressive properties has been a long and complicated path, based mainly on the absence of tumor rejection in a mouse model[78] and the inhibition of LPS-induced TLR4 stimulation.[79] The increasing amount of molecular data concerning Syncytins in humans and during evolution,[45,80] together with recent advances suggesting the existence of ISD receptors on monocytes and B lymphocytes,[15] will contribute to deciphering the multiple facets of induction of immunosuppression by envelopes. In addition, as observed for FSFFV, MMTV, and JSRV envelopes of infectious retroviruses,[81–83] some HERV envelopes were shown to be directly or indirectly involved in malignant transformation. HERV-K Env protein was overexpressed in breast cancers from US or Chinese women and proposed as a marker of disease progression and poor disease outcome[84] in childhood acute leukemia cells[85] and appeared to contribute to melanoma.[86] HERV-encoded sequences and notably Env should be considered as a new class of tumor-specific antigens, as exemplified by Env HERV-H elements.[87]

In summary, except for Syncytins (owing to the overall conservation of critical sites required for correct maturation and folding and the agreement between observed [in vitro] or predicted phenotypic properties and the presence of those canonical motifs), identification of the sites/organs of expression in the human body (Table 1) is a prerequisite prior to any evidence of a physiological or a pathological function. Globally, two situations occur: an HERV locus exhibiting either a wide range of tissue expression, e.g., envR, or a limited one such as envV, envH3, and envV transcripts being found exclusively in thymus and placenta, respectively. It should be noted that the observed transcriptional activity in some (if not most) tissues could be extremely low. Conversely, an extremely high level of transcription was observed for ERVW-1 in placenta and testis, ERVFRD-1 and envV in placenta, envR/ERV3 in placenta and adrenal gland, and envT in thyroid. Globally, few data are available concerning protein detection in vivo. The example of Syncytin-1 has allowed the reconciliation of molecular motifs, demonstrated functions and tissue tropism (Table 2), from which we can draw the most exhaustive information concerning regulation and deregulation, notably in placenta-associated pathologies, but also in other tissues where fusion property is required, e.g., myoblasts for muscles and osteoclasts for bones.

Table 2 Functions of Syncytin-1 Protein in Physiological and Pathological Contexts

Functions	Tissue[a]	Conditions[b]	Observations	References[c]
Fusion differentiation	Placenta syncytiotrophoblast	Normal	Specific splicing, detection of protein, functional tests (primary cytotrophoblasts)	1, 2, 3
	Placenta	Preeclampsia, HELLP HELLP, PE, IUGR human fetal growth restriction	Downregulation Protein redistribution Downregulation	4, 5, 6, 7, 8, 9
	Placental exosomes		Virus-like uptake by cells	10
	Muscle (myoblast)	Normal Motor neuron disease	Syncytin-1 and ASCT2 in primary myoblasts Functional assay	11, 12, 13
	Bone (osteoclast)	Normal	Synctyin-1 and ASCT2 in vitro Functional assay in vivo (human iliac crest)	14, 15
	Gametes	Normal	Synctyin-1 in sperm head ASCT2 in oocytes	16
	Breast	Cancer	Synctytin-1 in breast cancer ASCT2 in endothelium	17
Proliferation	Endometrium	Endometrial carcinoma	TGFβ1 and TGFβ3 increases the cholesterol efflux and modifies the membrane location and Syncytin-1 function	18
	B16F10 murine melanoma cells	Cancer cell line	Decreased of proliferation, migration, and invasion	19
Apoptosis	CHO Chinese hamster ovary cells	Cancer cell line: staurosporine-induced apoptosis model	Antiapoptotic (cell survival)	20

	Placenta	PE, HELLP	Increased apoptosis, cell death Activation of the calpain1-AIF-mediated apoptosis	6, 21
Senescence	Placenta	Normal	p53 and p16 pRb-dependent pathways	22
Immunosuppression	Placental exosomes/ microvesicles	LPS/PHA stimulation of PBMC	Inhibition of LPS/PHA-stimulated cytokine responses in human blood dampened PBMC responses to LPS challenge	23, 24
Cytotoxicity	Brain	Multiple sclerosis	Oligodendrocyte death or reduction of the oligodendroglial differentiation capacity Cytotoxic redox reactants increase iNOS Reduction of mitochondrial viability iNOS induction following interaction with TLR4	25, 26
Neurotoxicity	Sera	Schizophrenia	Syncytin 1 or env HERV-W-related protein in sera (ELISA method)	27

[a]Tissues or cell lines.
[b]Physiological or pathological contexts.
[c]Cited references are (1) Blond *et al.*,[70] (2) Mi *et al.*,[88] (3) Frendo *et al.*,[71] (4) Lee *et al.*,[89] (5) Chen *et al.*,[90] (6) Langbein *et al.*,[91] (7) Vargas *et al.*,[68] (8) Ruebner *et al.*,[14] (9) Pathirage *et al.*,[92] (10) Vargas *et al.*,[93] (11) Bjerregard *et al.*,[94] (12) Oluwole *et al.*,[95] (13) Redelsperger *et al.*,[96] (14) Søe *et al.*,[97] (15) Hobolt-Pedersen *et al.* 2014,[98] (16) Bjerregaard *et al.*,[99] (17) Bjerregaard *et al.*,[100] (18) Strick *et al.*,[101] (19) Mo *et al.*,[102] (20) Knerr *et al.*,[103] (21) Huang *et al.*,[104] (22) Chuprin *et al.*,[105] (23) Tolosa *et al.*,[79] (24) Holder *et al.*,[106] (25) Antony *et al.*,[107,108] (26) Kremer *et al.*,[109] (27) Perron *et al.*[110]

3. SYNCYTIN-1

3.1 Discovery of the ERVWE1 Bona Fide Gene

The molecular characterization of the HERV-W family relied on the isolation of placental cDNA clones, including one complete RU5-*env*-U3R-polyA sequence containing an *env* full-length viral ORF.[70] In 2000, protein truncation studies confirmed that this *env* ORF was unique in the genome and encodes a putative envelope gene,[111] in association with a functional U3 promoter.[112] Fusion events in the human rhabdomyosarcoma cell line (TE671) and choriocarcinoma cell lines (e.g., BeWo) were associated with a HERV-W envelope protein, thereby named Syncytin-1 by Mi in reference to the resulting syncytia.[42,88] Through a human genome-wide screening, Heidmann and colleagues identified a second envelope protein, Syncytin-2,[62] belonging to the HERV-FRD family and involved in cell–cell fusion events in human trophoblasts.[113,114] Early work also identified an ERV3 (HERV-R) envelope as a candidate for placental functions. The ERV3 Env expression affects proliferation and differentiation of BeWo cells in vitro.[115,116] Note that 1% of the Caucasian population has a mutation in ERV-3 *env* inducing a stop codon, resulting in a truncated envelope lacking both the noncanonical fusion peptide and the ISD.[117] In the murine genome, an in silico approach identified two coding envelope genes present as unique copies and with a placenta-specific pattern of expression: Syncytin-A and Syncytin-B.[118,119] To date, nine Syncytins with in vivo or ex vivo fusogenic properties have been identified in all eutherian mammal and marsupial clades including Haplorhini (Syncytin-1 and Syncytin–2), Rodentia (Syncytin-A and Syncytin–B), Tenrecidae (Syncytin-Ten1), Lagomorpha (Syncytin-Ory1), Ruminantia (Syncytin-Rum1), Carnivora (Syncytin-Car1), and Didelphimorphia (Syncytin-Opo1).[80,120] We will exemplify the involvement of endogenized retroviral envelopes through a focus on ERVW-1 locus/Syncytin-1 envelope, whose regulation and expression mechanisms are so far the most extensively described.

ERVW-1 is a 10.2-kb-long locus located between the 24th exon and the 5th exon of the PEX1 and ODAG genes, respectively, defining an LTR element-rich region of 30 kb in human chromosome 7q21.2 (Fig. 2A). ERVW-1 and downstream HERV-H-defective provirus are integrated within MaLR and HERV-P LTRs, respectively. The U3 region of the ERVW-1 5′LTR possesses the promoter activity. ERVW-1 can produce

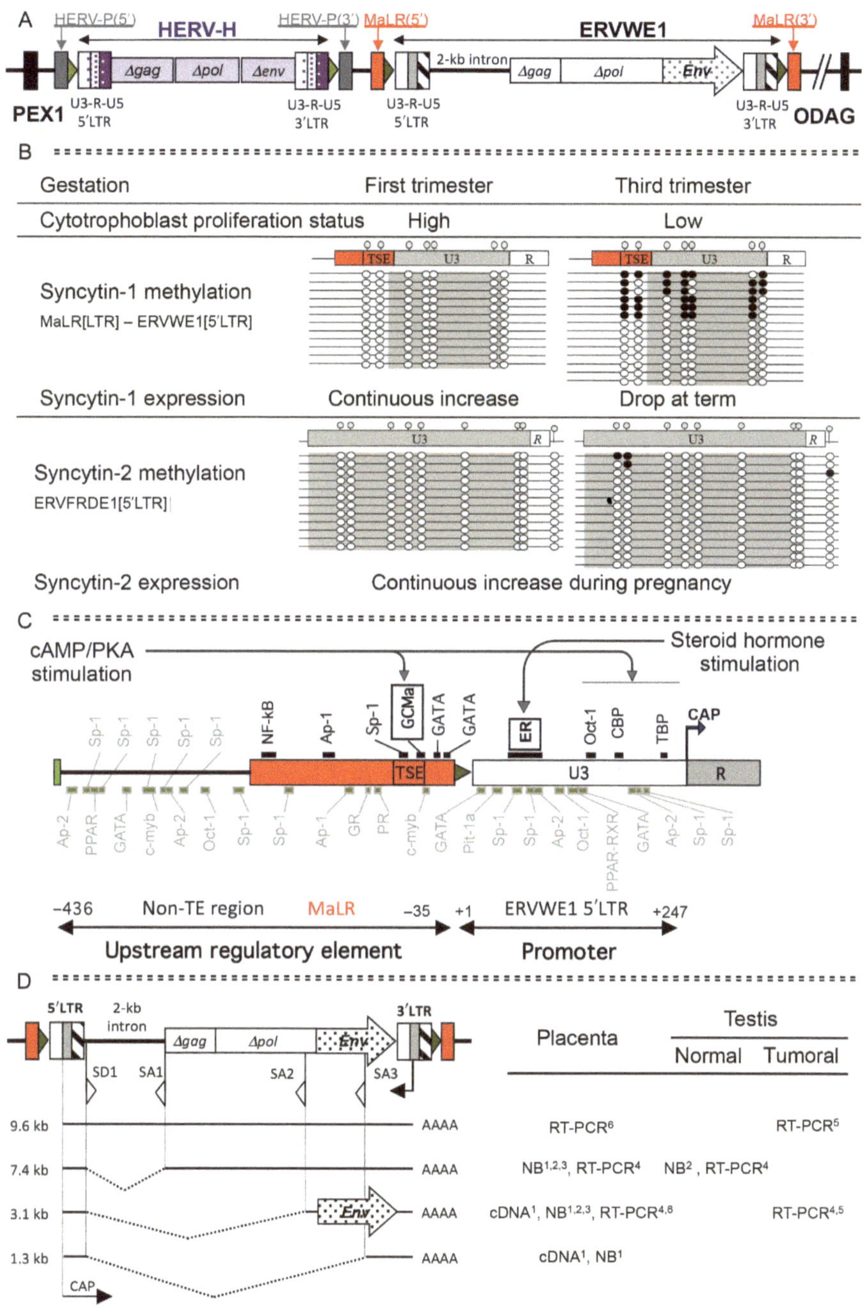

Fig. 2 See legend on next page.

different spliced transcripts, one of which encodes the envelope dubbed Syncytin-1. The genomic integration context, the control of the transcription, and the protein maturation are evidences of a domestication process of Syncytin-1, whose locus ERVW-1 is now considered as a bona fide human gene.

3.2 Transcriptional Regulation

3.2.1 Histone Marks and CpG Methylation

The expression of Syncytin-1 from the ERVW-1 locus is regulated at multiple levels involving epigenetic modification, transcription factors, and splicing strategy. Trejbalová *et al.* studied chromatin modifications in 5′LTR and associated H3K9 trimethylations at a transcriptionally inactive ERVW-1 locus in HeLa cells. The authors also associated the high density of H3K36 trimethylation along the intron–exon boundary of the Syncytin-1 envelope with high expression and efficient splicing of the envelope gene transcripts.[123] In nonplacental cell lines, infection with influenza

Fig. 2 Epigenetic, transcriptional, and posttranscriptional multilevels of control of the ERVWE1 locus expression. (A) *Schematic representation of the retroviral-enriched PEX1-ODAG intergenic region* (24th and 5th exons, respectively; *black boxes*). ERVW-1 provirus encoding only the envelope dubbed Syncytin-1 *(dotted arrow)*, and downstream HERV-H-defective provirus are integrated within MaLR and HERV-P LTRs, respectively. The U3, R, and U5 regions of HERV-H and ERVWE1 proviruses are depicted. (B) *Transcriptional and epigenetic control of Syncytin-1 and -2* during human gestation. 228-bp MalR enhancer, 247-bp U3 LTR promoter, and R transcription initiation site regions are indicated as *boxes*. MalR contains a trophoblast-specific enhancer (TSE). CpG methylation is determined by bisulfite sequencing PCR in cytotrophoblasts at two times of gestation. Each *line* represents an independent molecule. Methylated CpGs are schematized by *black circles* and unmethylated CpGs by *white circles*. (C) *ERVWE1/Syncytin-1 transcriptional regulatory element* is a bipartite element consisting of a cyclic AMP-inducible LTR retroviral promoter (ERVWE1 5′LTR U3 region) adjacent to an upstream regulatory element (URE) of composite origin including a 228-bp MaLR LTR (*box*) containing a trophoblast-specific enhancer (TSE). True (*top black boxes*) and putative (*bottom gray boxes*) transcription factor-binding sites along ERVWE1 5′LTR and URE are indicated. (D) *ERVWE1 splicing strategy in placenta and normal and tumoral testis. Left panel*: The CAP transcription initiation site (*right arrow*) and the polyadenylation signal (*left arrow*) are located at the 5′end and the 3′end of the R region of the 5′- and 3′LTRs, respectively. ERVWE1 appears to produce four single-spliced transcripts, a genomic 9.6 kb, the subgenomic 7.4- and 3.1-kb mRNAs, and the fully spliced 1.3-kb mRNA. Only the 3.1-kb variant is accountable for Syncytin-1 translation. Splice donor (SD) and acceptor (SA) sites are indicated by *right* and *left arrows*, respectively. *Right panel*: These four transcripts have been evidenced by Northern blot (NB), by RT-PCR, or as almost complete cDNA clones. References: 1, Blond *et al.*[70]; 2, Mi *et al.*[88]; 3, Smallwood *et al.*[121]; 4, Gimenez *et al.*[122]; 5, Trejbalova *et al.*[123]; 6, Li *et al.*[124]

A/WSN/33 virus induced the transcription of the ERVW-1 locus and one of its transcription factors, GCMa/1. This was associated with a decrease of H3K9 trimethylation and histone methyltransferase SETDB1 levels. Unexpectedly, these changes in the histone code were not associated with modifications of promoter region CpG methylation.[124] Methylation pattern studies of the ERVW-1 5′LTR revealed an inverse correlation between CpG methylation and locus expression. This indicates that general demethylation of the 5′promoter is a prerequisite for Syncytin-1 expression in trophoblasts cells.[125,126] Conversely, the promoter region is highly methylated in other tissues including placental fibroblasts, fetal cord blood, and maternal PBL.[125,126] Changes in methylation patterns within ERVW-1 during pregnancy were also studied by a comparison of first- and third-trimester samples[125] (Fig. 2B). Methylation of the ERVW-1 5′LTR reaches 40% at term, while it is completely absent at the beginning of the pregnancy. Thus, the selective and temporal demethylation of the ERVW-1 locus in placenta during the first trimester may allow Syncytin-1-mediated cell differentiation and fusion, while, in contrast, increased methylation at term may reduce Syncytin-1 production and consequently cell fusion or putative antiapoptotic protection,[103] in accordance with limited cytotrophoblast fusion and higher apoptosis rate at term.[127] Interestingly, ERVFRD-1/Syncytin-2 and ERV3 proviruses, which are involved in fusion/immunomodulation and proliferation, respectively,[45,56,59,62] exhibit different and independent methylation patterns compared to the ERVW-1 locus in the placenta. This may reflect complementary and ordered physiological functions for these three provirus sequences.[125]

3.2.2 The Ying and Yang of Transcription Factors: GCMa/1 vs TGFβ

As for conventional retroviruses, the ERVW-1 locus displays within its LTR all the signals required for transcription initiation and regulation, with the U3 region of the ERVW-1 5′LTR possessing the promoter activity. The core promoter domain within the U3 region contains the CAAT box and the TATA box located upstream of the CAP site, marking the beginning of the R region.[112] The 5′end of the U3 region harbors multiple binding sites contributing to overall promoter efficiency, including sites for the transcriptional regulators GATA, Sp-1, AP-2, Oct-1, and peroxisome proliferator-activated receptor-γ/retinoid X receptor-α (PPAR-γ/RXR-α). PPAR-γ/RXR-α were shown to bind the ERVWE1 5′LTR and seemed to be required to regulate Syncytin-1 expression (and fusion) in primary trophoblast cultures from human placenta.[128] Syncytin-1 regulatory elements include not only the 5′LTR but also a so-called upstream regulatory element (URE), a cellular 436-bp sequence located immediately upstream the Syncytin-1 proviral integration site, that, together with the 5′LTR, defines a bipartite control element[112] (Fig. 2C). This URE is composed of two main domains: (i) a distal regulatory region, including the previously mentioned putative binding sites found in the promoter core, as well as binding sites for NF-κB and AP-1[129]; (ii) an MaLR ancestor retrotransposon with putative binding sites for steroid hormone receptors, which features a trophoblast-specific enhancer with putative sequences for ubiquitous Ap-2 and Sp-1 but also placenta-specific GCMa/1-binding sites.[112] Glial cell missing a (GCMa/1) is a transcription factor primarily expressed in the human placenta and highly expressed in the labyrinth trophoblast cells in mice[130] but weakly expressed in other tissues.[131] Two binding sites by which GCMa/1 can specifically transactivate Syncytin-1 have been described,[132] and functional GCMa/1-binding sites were also identified in Syncytin-2 and MFSD2A related receptor promoters.[133] GCMa/1 regulation has been linked to cAMP/protein kinase A signaling pathways[134,135] and hypoxia levels.[136] Interestingly, Syncytin-A was shown to be downregulated in murine GCMa/1-deficient placenta,[137] and siRNA-mediated GCMa/1 inhibition in BeWo cells led to a decrease in syncytialization activity manifested as fusion events.[138] All these data support a crucial role for GCMa/1 in the regulation of Syncytin-1 expression and therefore placental growth and maintenance.[139,140]

In the U3 region of the ERVW-1 5′LTR, a binding site for estrogen receptors acting as transcriptional regulators factor was identified as involved in Syncytin-1-dependent proliferation under TGFβ influence in pathological contexts.[101] Although Syncytin-1 was suggested to be involved in myoblast and osteoclast differentiation, how Syncytin-1 is regulated at epigenetic and transcriptional levels outside the placenta remains unknown. Interestingly, in a nonplacental cell line, changes in the histone code, such as decrease in H3K9 trimethylation following influenza A/WSN/33 infection, were a prerequisite for the production of spliced ERVW-1 transcripts, while CpG methylation of ERVW-1 promoter region remained unchanged.[141]

3.2.3 Alternative Splicing as a Rule of Thumb

ERVW-1 is a 10.2-kb-long locus that can produce differentially spliced transcripts depending on the context. Four single-spliced transcripts have been detected in the placenta.[70,88,121,124,125] The genomic 9.6-kb nascent transcript contains the intron, the *gag* and *pol* pseudogenes, and the *env* gene.

The 7.4-kb subgenomic transcript contains the *gag* and *pol* pseudogenes and the *env* gene. A second subgenomic 3.1-kb transcript includes only the ORF for the envelope protein Syncytin-1. Additionally, early Northern blot experiments detected a fully spliced 1.3-kb transcript in the placenta.[70] Along with the placenta, different ERVW-1 mRNAs were identified in testis in physiological and pathological contexts.[88,122,123] Although the biological significance of noncoding splice forms of ERVW-1 in cancers is subject to debate, the fact that appearance of the 3.1-kb env-coding mRNA is normally restricted to the organ(s) in which a physiological function exists but can reappear in pathological contexts argues that splicing variations may represent an additional level of control of the function of endogenized retroviral sequences, balanced by other epigenetic mechanisms (Fig. 2D). As stated previously, splicing of ERVW-1 locus may be regulated by epigenetic modification, i.e., the histone-specific methylation pattern may influence alternative splicing. Indeed, Trejbalová *et al.* have reported that in a testis germ cell tumor (seminoma), efficient splicing, and expression of ERVW-1 were accompanied by a high level of H3K36Me3 at the intron and at the intron/exon boundary.[123]

3.3 Maturation of the Trimeric Glycoprotein

Syncytin-1 is synthesized in the lumen of ER as a gPr73-glycosylated precursor. The maturation process of Syncytin-1 glycoprotein is indistinguishable from that of an infectious retroviral envelope, excepting the protease-dependent step involved in fusion competency of gammaretroviruses. After folding and disulfide bond formation, the protein is assembled as homotrimers involving a leucine zipper-like motif ($LX_6LX_6NX_6LX_6L$). Syncytin-1 is then transported to the Golgi apparatus and is specifically cleaved at a consensus RNKR furin cleavage site, giving rise to two mature subunits, gp50 SU and gp24 TM.[74] The mature envelope protein possesses seven N-glycosylation sites, including high-mannose N-glycans on the six glycosylation sites of the carboxy-terminal domain of the SU and one in the TM subunit, which is essential for correct envelope protein folding and function.[142] Finally, a disulfide bond is established between CϕϕC (ϕ is an amino acid with a bulky hydrophobic side chain [Leu, Ile, Phe, Val, or Met]) and CX_6CC motifs located on the SU and TM subunits. The SU is composed of a signal peptide, a 124-amino acid domain in the N-terminus involved in interaction with two receptors hASCT1 and hASCT2, and a C-terminal domain containing the CϕϕC motif.

The cytoplasmic tail region of numerous retroviral envelopes plays a critical role in triggering fusion. In the retrovirus life cycle, the presence of an R peptide basically prevents fusion from occurring, notably because of the presence of the YXXφ motif absent from Syncytin-1. Accordingly, Syncytin-1 fusion competence does not involve an R-like peptide cleavage but relies instead upon the first 16 residues of the intracytoplasmic tail. Constitutive fusogenicity results from a 12-bp (corresponding to four LQMV amino acids) deletion in its cytoplasmic tail,[143] leading to an R peptide region with loss of fusion inhibition control.[74] Ultimately, the Syncytin-1 glycoprotein is synthetized as a glycosylated gp73 precursor, cleaved into a gp50 surface unit containing the receptor-binding domain and a gp24 TM subunit containing the fusion peptide and an ISD, with SU and TM subunits associated as homotrimers.[74]

Syncytin-1 is expressed under physiological conditions in every trophoblast subtype throughout the human placental development process, i.e., villous cytotrophoblasts and syncytiotrophoblast. Moreover, Syncytin-1 is also expressed in extravillous cytotrophoblasts, in particular in uterine spiral arteries.[71,144] Two convergent studies noted that Syncytin-1 is located on specific cholesterol-rich membrane areas and termed detergent-resistant membrane (DRMs) or rafts.[74,101] In addition to their localization at the cell surface of trophoblasts, Syncytin-1 is also located on the surface of various extracellular microvesicules of placental origin, including exosomes.[79,93,106]

3.4 A Compendium of Functions

Syncytin-1 is one of the major genes involved in cytotrophoblast differentiation to syncytiotrophoblast through cell–cell fusion events. In primary cultures of human villous cytotrophoblasts, expression of Syncytin-1 mRNA and protein as well as the syncytium formation by cell fusion events was significantly reduced by using specific antisense oligonucleotides and siRNA strategies.[71,113] The endocrine function of syncytiotrophoblast was also impaired by Syncytin-1 knockdown, as illustrated by a decrease in hCG secretion.[71] Syncytin-1-dependent cell–cell fusion in humans involves two sodium-dependent neutral amino acid transporters such as receptors, hASCT2 (SLC1A5), and the related hASCT1 (SLC1A4).[42,76] The level of Syncytin-1 protein increases during early pregnancy but remarkably decreases in the late third trimester.[121,145] More recently, studies have suggested the involvement of Syncytin-1 in very early cell–cell fusion events of human reproduction, since its expression was detected in sperm head and around the equatorial segment at mRNA and protein levels, while hASCT2 is expressed on oocytes, consistent with a possible role of Syncytin-1 in fusion between oocyte and sperm.[99] A recent study showed the expression of Syncytin-1 from the preimplantation stage blastocyst, particularly on the cell surface of trophoblastic cells and consistently on the trophectoderm underlying the inner cell mass, which suggests that Syncytin-1 expression is a prerequisite for embryo implantation and placentation.[146]

Syncytin-1 protein expression has also been shown outside its privileged tissue, in myoblasts and osteoclasts. Recent studies report that, in vitro, Syncytin-1 is directly involved in human myotube differentiation through cell–cell fusion events[94,96] and that murine Syncytins contribute to myoblast fusion and muscle repair after cardiotoxin-induced injury.[96] Interestingly, in vitro studies also suggest that Syncytin-1 is involved in osteoclast fusion.[97]

Using a similar targeting strategy against Syncytin-2, Vargas *et al.* showed that Syncytin-2 inhibition in cytotrophoblast primary cell cultures leads to a decrease in the fusion index that is more pronounced than for Syncytin-1.[113] Syncytin-2, which is expressed in villous cytotrophoblast and syncytiotrophoblast,[63] interacts with the carbohydrate transporter MFSD2A expressed in syncytiotrophoblast.[77] In mice, Syncytin-A and Syncytin-B are able to trigger cell–cell fusion in ex vivo transfection assays,[118] and double knockout mice exhibit premature death of embryos,[119] thereby suggesting a critical role of Syncytin-A and Syncytin-B in murine syncytiotrophoblast differentiation. Similarly in sheep, the envelope of endogenous Jaagsiekte sheep retrovirus (enJSRV) regulates trophectoderm growth and differentiation in the periimplantation conceptus. Experiments of envelope knockdown with morpholino antisense oligonucleotides showed impaired trophectoderm differentiation and induced pregnancy loss.[147]

Syncytin-1 might also modulate the immune system. A first mechanism of Syncytin-mediated immunosuppressive activity may stem from the presence in the TM subunit of a putative immunosuppressive region (LQNRRGLDLLTAE**Q**GGICL**A** [see Fig. 1F; Table 1]) conserved among murine, feline, and human retroviruses[148] referred to hereafter as the MPMV, SRV, SNV, and BaEV immunosuppressive retroviruses. In contrast, the analysis of this domain for the human and mouse Syncytins in a mouse model of transplant rejection revealed an immunosuppressive activity for Syncytin-2 and Syncytin-B but not for the Syncytin-1 and Syncytin-A.[45] Two amino acids (indicated by bold letters in sequence above) have been described as commutator points that can be alternatively turned "on" or "off" in substitution experiments and trigger a switch from immunosuppressive to nonimmunosuppressive activity. However, the expression of Syncytin-1 at the direct maternal–fetal interface on cytotrophoblasts of maternal spiral arteries suggests that it is involved in control of placental tolerance by maternal immune cells. In HIV infection, the envelope TM subunit and the ISD were shown to inhibit lymphocyte proliferation and alter cytokine release; the ISD was shown to bind peripheral blood mononuclear cells (PBMCs), preferentially monocytes and B cells, suggesting that a particular receptor might be involved in immunosuppression.[15]

A second mechanism of Syncytin-1-dependent immunomodulation was proposed after endotoxin-response experiments. When challenged by lipopolysaccharide (LPS), the activation of PBMCs was reduced in the presence of Syncytin-1-containing exosomes.[79,106] Moreover, LPS stimulation in the presence of recombinant SU of Syncytin-1 dampened PBMC response to LPS stimulation, suggesting that the immunomodulatory property of Syncytin-1 is also partly supported by the SU subunit.[106] Interestingly, infection of cells with Syncytin-1 phylogenetically related RD114 and simian immunosuppressive type D retroviruses resulted in impaired amino acid transport. This mechanism was proposed to mediate viral immunosuppression.[149] In line with this, glutamine, an amino acid accepted by both ASCT1 and ASCT2 transporters, was shown to influence the balance within the T lymphocyte subpopulations, potentially modulating the host response.[150] The constant interaction of Syncytin-1 with hASCT1 and hASCT2 might therefore influence the level of extracellular glutamine and therefore disturb the development of allogeneic maternal T cells. Moreover, Syncytin-1 might also target T-cell activation indirectly by modulating the stimulatory activity of dendritic cells.[151] These observations should be linked with in vitro studies, suggesting that an envelope from HERV-W family (MSRV Env) interacts with toll-like receptor 4 (TLR4)[152] and that Syncytin-1 interacts with dendritic cell-specific intercellular adhesion molecule-3-grabbing nonintegrin (DC-SIGN).[74] Interestingly, in contrast to immunosuppressive properties, proinflammatory features of Syncytin-1 have been proposed. Syncytin-1 induces in astrocytes the expression of OASIS, an ER stress sensor, which increases inducible nitric oxide synthase expression but concurrently downregulates hASCT1.[107] This results in diminished myelin protein production comparable to lesions observed in multiple sclerosis brains.

Syncytin-1 is also involved in cell proliferation and antiapoptotic processes. In vitro studies implicated Syncytin-1 in both the fusion and the proliferation of endometrial carcinoma cells.[101] Syncytin-1 upregulation via the cAMP pathway leads to cell–cell fusion, while induction by steroid hormones (estradiol, progesterone) leads to proliferation. This molecular switch is apparently controlled by TGFβ1 and TGFβ3 induced by steroid hormones and may override Syncytin-1-mediated cell–cell fusions.[101] ERV-3, whose expression was specifically detected in the multinucleated syncytiotrophoblast in vivo,[12] is also involved in choriocarcinoma cell line proliferation.[115,116] The treatment of Syncytin-1-transfected CHO cells with staurosporine showed a reduced apoptosis rate compared to controls at 18 and 24 h. The delayed apoptosis was associated with a reduction of active caspase-3. The antiapoptotic marker Bcl-2 was increased in Syncytin-1-transfected CHO cells at baseline and following staurosporine treatment.[103]

4. DEREGULATION OF SYNCYTIN-1 IN HUMAN PLACENTA PATHOLOGIES

In light of the critical biological functions of human endogenous retroviruses in placental morphogenesis, it is not surprising that a number of studies have identified alterations of HERV expressions in placental pathological contexts (Figs. 3 and 4).

4.1 Down Syndrome

Down syndrome (trisomy of chromosome 21) is a major known cause for mental retardation. The incidence is strongly influenced by maternal age at conception and occurs in about 1:450 pregnancies in Europe.[156] Since the placenta also carries this aneuploidy, trophoblast development is affected and characterized by a defect in syncytiotrophoblast formation.[157] In vitro, primary cultured cytotrophoblasts from trisomy 21 placenta show impaired and delayed cell–cell fusion.[158,159] Levels of ERVW-1 and ERVFRD-1 mRNA expression were unchanged in these cells after 72 h of culture, whereas transcription of ERVW-1 increased and ERVFRD-1 decreased during normal cytotrophoblast culture.[160] The synthesis of pregnancy-associated hormones by syncytiotrophoblast is altered,[161] and an abnormally hyperglycosylated form of human chorionic gonadotropin is secreted in trisomy 21 placenta.[162] Interestingly, the fusion defect observed in syncytiotrophoblast from trisomy 21 is reversible in vitro after addition of recombinant hCG and cAMP.[160]

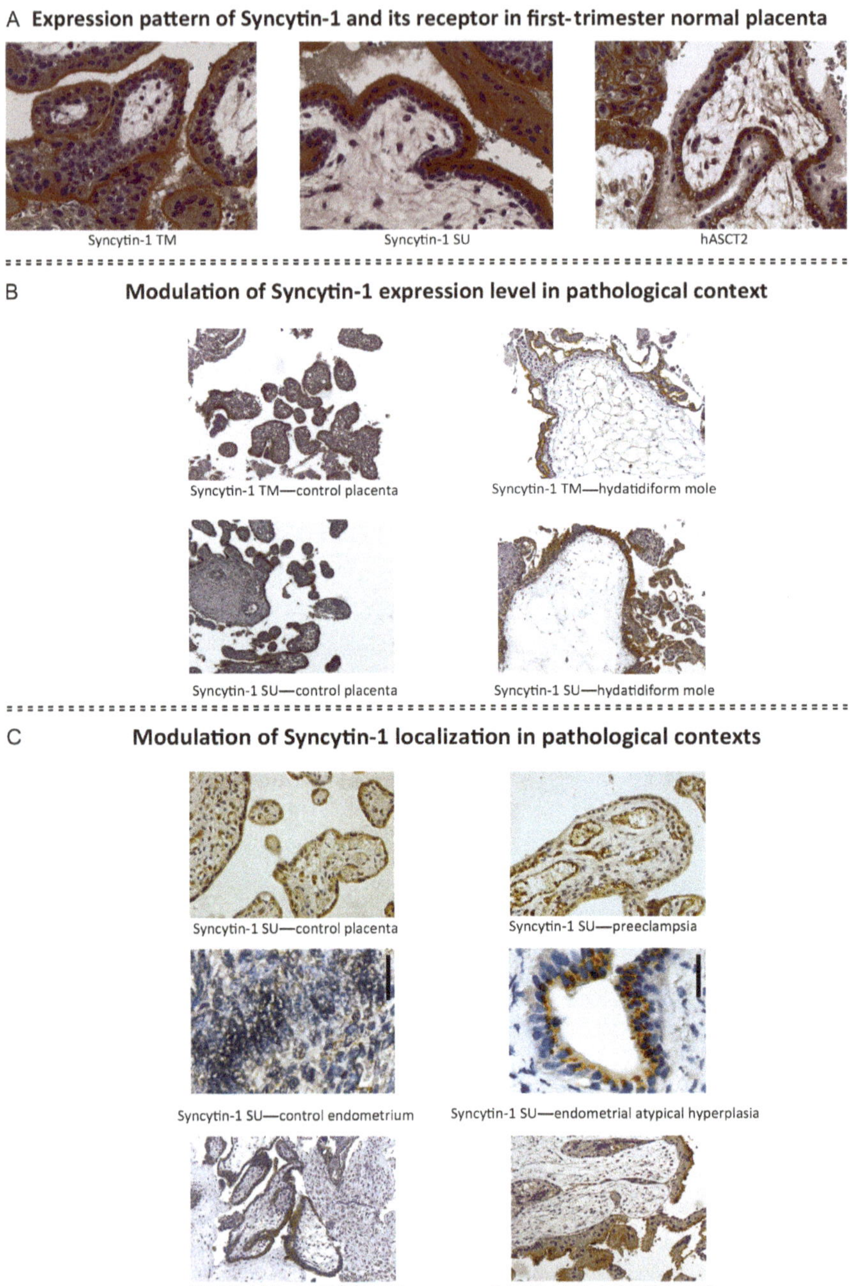

Fig. 3 See legend on opposite page.

4.2 PE, HELLP, and IUGR

PE, defined as new-onset hypertension and proteinuria occurring after 20 weeks of gestation, affects between 1.4% and 4.0% of pregnant women from Europe, Northern America, and Australia.[163] This syndrome is thought to develop in two stages: an earlier placental problem and a later maternal inflammatory response. The two major life-threatening consequences of PE are HELLP syndrome on the maternal side and intrauterine growth restriction (IUGR) on the fetal side. The deep invasion by cytotrophoblasts and replacement of endothelial cells in the spiral arteries of the inner third myometrium frequently are altered in PE. Moreover, abnormal syncytiotrophoblast differentiation is also observed in preeclamptic placentas, with increased proportion of syncytial knots and an early hypermaturation of chorionic villi.[164,165] Although the exact causes of PE remain unknown, the observed defects affect the unique immune tolerance phenomena involving convergence of three different genomes: the mother, the father, and the fetus. In light of the placental functions altered in PE, the study of endogenous retroviral envelope deregulations has provided contributive data to the understanding of this disease.

Fig. 3 Modulation of Syncytin-1 glycoprotein expression in physiological and pathological contexts. Immunohistochemistry on formalin-fixed and paraffin-embedded samples. (A) Physiological expression on human first-trimester normal pregnancy. Expression of Syncytin-1 in syncytiotrophoblast cytoplasm using 6A2B2 and 1F11B10 primary antibodies raised against transmembrane (TM, *left panel*) and surface unit (SU, *middle panel*) of human Syncytin-1 glycoprotein, respectively; expression of Syncytin-1 receptor hASCT2 at the cytoplasmic membrane of villous and extravillous cytotrophoblasts using rabbit polyclonal anti-D-type mammalian retrovirus receptor primary antibody (rp21C, *right panel*) (magnification ×100, unpublished data). (B) Modulation of Syncytin-1 level of expression in a 5-month twin pregnancy composed of a healthy fetus linked to a normal placenta associated to a complete hydatidiform mole. Syncytin-1 TM and SU subunits are immunostained with 6A2B2 (*top panels*) and 1F11B10 (*bottom panels*) primary antibodies (magnification ×20, Bolze *et al.*[153]). (C) Modulation of Syncytin-1 localization in pathological contexts. Immunostaining of Syncytin-1 surface unit on fetal endothelial cells during preeclampsia compared to gestational age-matched normal pregnancy control (*top panels*, magnification ×40, with permission from Holder *et al.*[154]). Syncytin-1 cytosolic protein expression as "punctate" foci ("Golgi-like") localizing paranuclear or apically at the membrane of epithelial cells of endometrial glands in premalignant endometrial hyperplasia compared to normal endometrium (*middle panels*, magnification ×100, with permission from Strissel *et al.*[155]). Enhancement of Syncytin-1 expression on paddle-shape syncytiotrophoblast outgrowths in complete hydatidiform mole using anti-Syncytin-1 TM subunit 6A2B2 primary antibody compared to first-trimester normal pregnancy control (*bottom panels*, magnification ×20, Bolze *et al.*[153]).

		Physiological Context		Pathological Context			
		Placenta	**Other Tissues**	**Preeclampsia**	**Other Pathologies**		
					Endometriosis Endometrial Carcinoma	Testis Germ Cell Tumor	Influenza Infected CCF-STTG1 Cell Line
Histone code	5'LTR	*BeWo* **High H3K9Ac** **Low H3K9Me3**	*HeLa* **Low H3K9Ac** **High H3K9Me3**				**Low H3K9Me3**
	Splicing	**High H3K36Me3**[a]					
CpG methylation	(MaLR enhancer & 5'LTR promoter regions)	*Primary CT* 1st T: **fully unmethylated** 3rt T: **partially remethylated**	*Pla fib, mat PBL, Sk, HeLa, PBMC, testis* **Methylated**[b]	*Primary CT* **Remethylated**	**Unmethylated**	Adjacent: **partial demethylation of MaLR**[c] Tumor: **unmethylated**	**Partial unmethylation of MaLR**[d]
SETDB1							**Downregulated**
DNMT3B				**Upregulated**	*Endometriosis* **Downregulated**		
Transcription factors	GCMa/1	**Highly expressed**	**Weakly expressed**[e]	**Downregulated**[f]			**Upregulated**[g]
	PPAR-γ	**Highly expressed**[h]	**Ubiquitous**[i]	**Downregulated**[j]			
	TGFß			induces SMAD2 phosphorylation **then GCM1 decrease**	*Endometrial carcinoma* + cAMP or + steroids[k] **induces proliferation**[l]		
Splicing	envelope coding mRNA	**Highly produced**	**–** (except: myoblasts, osteoclasts & sperm)	**Poorly produced**	**+**[m]	**+**	**+**
Syncytin-1	glycoprotein	**Highly expressed**	**–** (except: myoblasts, osteoclasts & sperm)	**+**	**+**	**Not known**	**Not known**

Fig. 4 See legend on opposite page.

Langbein *et al.* showed that cultured cytotrophoblasts from PE and HELLP-associated IUGR correlated with a pronounced lower cell–cell fusion index, fewer nuclei per syncytiotrophoblast, a significantly decreased β-hCG secretion, and a reduction of Syncytin-1 expression compared with controls.[91] Similarly, Vargas *et al.* described a pronounced deficiency in cellular fusion in trophoblast cells from patients with PE when compared to controls.[68] This was concomitant with a reduced expression of mRNA and protein levels of Syncytin-1 and Syncytin-2 in PE placentas compared to controls. The reduction was more pronounced for Syncytin-2 and was correlated with PE severity. These observations of reduced Syncytin-1 protein levels in placenta from PE and HELLP syndrome were corroborated by other teams.[89,90,166] Various cellular mechanisms were suggested to explain the association of reduced Syncytin-1 expression and decreased fusion events in syncytiotrophoblast formation. Low ambient oxygen conditions of cytotrophoblast cultures, intended to mimic hypoxic conditions of PE, are associated with downregulation of ERVW-1 and hASCT2 transcripts. This deregulation of Syncytin-1 may suppress the normal process of cell fusion necessary for syncytiotrophoblast formation and contributes to syncytiotrophoblast abnormalities characteristic of PE.[167] A reduced placental GCMa/1 expression has been reported as a causative factor in defective syncytiotrophoblast differentiation in human PE.[168] Under hypoxic conditions associated with PE, GCMa/1 was ubiquitinated and degraded.[169] GCMa/1 polymorphisms carried by the fetus were reported to induce hypertension

Fig. 4 Comparative regulation of Syncytin-1 expression in normal vs pathological contexts. *CT*, cytotrophoblast; *mat PBL*, maternal peripheral blood lymphocytes; *PBMCs*, peripheral blood mononuclear cells; *Pla fib*, placental fibroblasts; *Sk*, skin. [a]H3K36me3 intron/exon boundary increases env mRNA splicing and is also active in nonplacental cells transfected with ERVW-1. [b]Noninvestigated in myoblast, osteoclasts, and sperm. [c]The two CpGs present in MaLR trophoblast-specific enhancer are unmethylated in almost 50% of the sequenced DNA molecules. [d]partial unmethylation of MaLR element is similar to the demethylation profile observed in the tissue adjacent to the tumor in testis and is identical in infected and uninfected cells (Mallet F *et al.*, unpublished data). [e]GCMa/1 gene expression evidenced by Northern blot analysis.[131] [f]Downregulation may also result from gene polymorphism.[139] [g]Insufficient by itself to increase the level of Syncityn-1 in cell lines of nonplacental origin. [h]Specifically expressed in villous and extravillous cytotrophoblast and syncytiotrophoblast[128] and plays a role in human placentogenesis influencing both villous and extravillous cytotrophoblast. [i]Ubiquitous but expression levels vary widely between tissues. [j]Associated with reduced p38α (MAPK14) kinase phosphorylation.[128] [k]Estradiol and progesterone.[101] [l]TGFβ induced by steroids or added to forskolin (cAMP inducer) produces a shift from fusion to proliferation.[101] [m]Presence observed, comparative quantitation unreachable.

during pregnancy.[139] In parallel, epigenetic hypermethylation of the ERVW-1 promoter region was shown to be linked with decreased Syncytin-1 expression in placental syndromes such as PE, HELLP, and IUGR, as well as in monozygotic dichorionic discordant twins.[14,104,170,171] The reduction of fusion events observed in PE might also be linked with abnormal Syncytin-1 protein localization and conformation. While Syncytin-1 expression in trophoblasts was generally localized at the apical border of the syncytiotrophoblast, Lee *et al.* described a redistribution from the basal side in control placentas to the apical side in PE.[89] Moreover, Syncytin-1 protein assembly was described to be altered in PE, with a higher level of Syncytin-1 SU and its appearance on fetal endothelium. This suggests a shedding of SU from syncytiotrophoblast into fetal circulation where it may bind the endothelium.[154]

Another characteristic of PE and HELLP syndrome placentas is the increased cytotrophoblast apoptotic rate. Syncytin-1 gene expression in primary placental tissues of PE/IUGR and HELLP/IUGR is lower along with a significant increase in the apoptosis rate compared with controls.[91] As noted previously, Knerr *et al.* challenged CHO cells overexpressing Syncytin-1 in CHO cells with apoptosis-induced staurosporine.[103] It was found that Syncytin-1 ectopic overexpression resulted in a delayed and reduced cell death. The reduced activation of caspase-3 and increased expression of Bcl-2 appeared to be among the mechanisms underlying the antiapoptotic effects of Syncytin-1. These findings in support of the protective role of Syncytin-1 appear to be consistent with the decreased Syncytin-1 expression and increased trophoblast apoptosis/deportation observed under preeclamptic condition.

Cytotrophoblasts from IUGR placentas, not related to PE or HELLP syndrome, demonstrated a lower cell fusion index and nuclei per syncytiotrophoblast in vitro. Syncytin-1 and Syncytin-2 gene expression were downregulated and Syncytin-1 protein expression was reduced in PE.[172] hASCT1 and hASCT2 gene expression were not modified in IUGR or PE,[90,172] but the receptor of Syncytin-2, MFSD2a, was downregulated in IUGR placenta[172] and PE.[114] Interestingly, in a murine model of PE, induced overexpression of STOX1, a transcription factor involved in trophoblast proliferation and migration, reproduced clinical symptoms and placental features of PE.[173] When STOX1 was overexpressed in BeWo choriocarcinoma cell line, the transcription levels of Syncytin-1 were altered along with GCMa/1 and hCG, supporting a key role in the modulation of PE.[174] A disturbed PPAR-γ/RXR-α pathway could also contribute to pathological human pregnancies. Significant reduction of Syncytin-1 in PE and IUGR was found associated with reduced PPAR-γ/RXR-α expression levels in PE and reduced p38α protein phosphorylation in PE and IUGR.[128] This appears consistent with the demonstrated activation pathways of PPAR-γ, either direct such as p38 α (MAPK14) kinase or indirect via stimulation of adenylate cyclase and PKA phosphorylation by forskolin.[175] Interestingly, although a direct link between Syncytin-1 expression in endometriosis lesions and PPAR-γ has not yet been identified, PPAR-γ was suggested to be involved in growth and survival of endometriosis lesions.[176,177] TGFβ has also been implicated in the pathogenesis of PE.[178] Recent data indicate that TGFβ signaling, through SMAD 2/7 and SMAD-independent pathways, plays a role in trophoblast growth and proliferation. In particular, altered SMAD regulation is associated with downregulation of GCMa/1 and Syncytin-1, thereby suggesting altered cell–cell fusion.[179] This study reinforces the hypothesis of a pivotal role of TGFβ in the modulation of Syncytin-1 functions in placental pathologies. Finally, a recent study has identified DNA polymorphisms in ERVW-1 regulatory regions, but no significant association was found between the development of PE and such polymorphisms in the studied population.[180] These conclusions merit further investigation on larger population cohorts.

4.3 Gestational Diabetes

Fetal growth may also be affected by placental dysfunctions of diabetic origin. Gestational diabetes is defined by glucose intolerance appearing during pregnancy. Increased glucose levels may cause higher rates of early pregnancy loss and maternal hypertensive disorders. The fetus is then exposed to increased hypoxic stress.[181–185] Characteristic features of gestational diabetes in placenta are a dysmature or relatively immature villous structure with an increased number of villous trophoblastic cells, pointing to an increased trophoblast proliferation.[186] Recently, Soygur *et al.* have studied the expression of Syncytin-1 and -2 with their receptors in gestational diabetes placental samples.[187] They reported a decreased expression of Syncytin-2 protein and its receptor MFSD2A in diabetic term placentas compared with normal term placentas. Immunoreactive Syncytin-1 protein was not decreased, but ERVW-1 mRNA was downregulated. They hypothesized that altered Syncytin-2 production might be involved in placental pathogenesis of gestational diabetes.

4.4 Hydatidiform Moles and Gestational Trophoblastic Neoplasia

In contrast with placental dysfunctions of the PE spectrum, gestational trophoblastic diseases form a group of placental disorders characterized by proliferative and metastatic potential. They encompass a premalignant stage termed hydatidiform moles in which no viable fetus develops and a malignant stage, gestational trophoblastic neoplasia (GTN), consisting of solid tumors derived from gestational trophoblasts and thus derived from a fertilized egg.[188,189] Complete hydatidiform mole (CHM) and partial hydatidiform mole (PHM) occur in 0.6–2.4 per 1000 pregnancies in western countries. CHMs are androgenetic diploid products of conception with pronounced trophoblastic proliferation, while PHMs are diandric triploid products of conception with mild proliferation and an abnormal or absent embryo. GTN encompasses invasive mole, choriocarcinoma, placental site trophoblastic tumor, and epithelioid trophoblastic tumor. GTN develops after 15%–20% of CHM and 1% of PHM (postmolar GTN), abortion, or term pregnancy.[188,190] The 5-year overall survival rate of patients developing GTN is nearly 98% when a management in accordance to international guidelines is provided.[191,192]

Our team recently described that Syncytin-1 protein was highly expressed in syncytiotrophoblast of hydatidiform moles with an apical enhancement when compared to normal first-trimester placenta.[153] Moreover, SU and TM subunits of Syncytin-1 showed different immunoreactivities. In moles with further malignant transformation, Syncytin-1 SU C-terminus was strongly stained when compared to moles with spontaneous remission. This suggests posttranslational modifications such as reduction of disulfide bonds and altered glycosylation, which may induce a different level of glycoprotein assembly and epitope unmasking. The local strong staining of Syncytin-1 TM ISD on syncytiotrophoblast outgrowths suggests that it contributes to the immune tolerance of CHM. Last, given the potential interaction of MSRV Env with TLR4 resulting in the production of major proinflammatory cytokines by monocytes,[152] it is then interesting to note the downregulation of TLR4 in CHM when compared to normal first-trimester placenta, suggesting an altered sensor role of TLR4 in gestational trophoblastic diseases despite the strong expression of Syncytin-1.[153]

When assessing other placentally expressed endogenous retroviral envelopes, mRNA levels of ERVFRD-1 and ERV3 were dramatically reduced in CHM, PHM, and GTN, while ERVW-1 was reduced only in GTN. The mechanism involved is unknown, but, given the epigenetic regulation described in PE, the hypothesis of an acquired methylation of promoter regions of these loci leading to a pathological condition such as GTN must be further investigated. Alternatively, such downregulation could, at least for ERVW-1 and ERVFRD-1, be due to an alteration of the upstream placental-shared GCMa/1 transcription factor. At the protein level, since no modification in GCMa/1 transcript levels was observed, ubiquitination and protein degradation of GCMa/1 (as observed in hypoxic condition such as PE[169]) might influence ERVW-1 and ERVFRD-1 downregulation.

The expression of Syncytin-1 in CHM, which generally contains only paternal chromosomes,[190] suggests that Syncytin-1 could be a paternally expressed gene. This hypothesis emerged with the observations that ERVW-1 maps closely to two maternally imprinted genes (*SGCE* and *PEG10*) and that their regulation was temporally coordinated.[121] However, we cannot exclude that the persistence of Syncytin-1 expression in CHM in the absence of maternal chromosomes is a consequence of a deregulation of the normal methylation patterns of imprinted genes in hydatidiform moles.[193]

4.5 Other Reproduction-Related Pathologies

The expression of human endogenous retrovirus envelope proteins was shown in various cancers, all linked by their involvement in reproduction. Most of the time, the expression only involves mRNA but sometimes also encompass envelope proteins.

4.5.1 Endometrial Carcinoma and Endometriosis

Endometrial cancers encompass two main histological entities, endometrial carcinoma driven from premalignant stages to invasive malignant stages by steroids such as estrogens and progesterone. Strick *et al.* have shown that ERVW-1 mRNA and Syncytin-1 protein levels increased along with the transition from premalignant to malignant stages of endometrial proliferation compared to normal postmenopausal endometrium.[155] The treatment of endometrial carcinoma cell lines with steroid hormones induced Syncytin-1 expression due to an LTR HERV-W estrogen response element (Fig. 2A) and resulted in increased proliferation. In contrast, the activation of cAMP pathway by forskolin treatment also induced Syncytin-1, but resulted in cell–cell fusion as observed in syncytiotrophoblast. They showed that the switch from cell–cell fusion to proliferation was driven by TGFβ1 and TGFβ3. This suggests that Syncytin-1 is involved early in the tumorigenic process of endometrial carcinoma.

Beyond premalignant lesions, other reproductive tract conditions such as endometriosis exhibit tumorigenic features and may be associated with HERV deregulation. Endometriosis is defined by ectopic endometrium, i.e., the presence of normal endometrium outside the uterine cavity. Several mechanisms are proposed to explain the development of endometriosis, and the most plausible is the retrograde migration through Fallopian tubes during menstruation with following endometrial implantation in pelvic organs. Opplet *et al.*, studying the expression of ERVW-1 in eutopic and ectopic endometrium, showed that ERVW-1 mRNA levels were increased only in eutopic endometrium from patients with endometriosis but not in their endometriosis foci.[194] Interestingly, Zhou *et al.* have suggested that epigenetic modifications drive the ectopic expression of Syncytin-1 in endometriosis through hypomethylation of the 5′LTR region of ERVW-1 along with reduced levels of DNA methyl transferase 3B.[195] This suggests that Syncytin-1 may also influence the development of endometriosis during very early steps of pathogenesis. Other HERV loci are activated in endometriosis lesions, mainly from HERV-E family.[196]

4.5.2 Testis Germ Cell Tumor, Ovarian Tumor, and Breast Cancer

Synctyin-1, Synctin-2, and ERV3 protein expression was evident neither in normal nor in pathological ovary and testis tissues. However, in addition to the normal testis where the 7.4-kb spliced ERVW-1 transcript was identified by Northern blot and RT-PCR,[88,122] the development of testis germ cell tumor such as seminoma was associated with the presence of 3.1-kb spliced ERVW-1 transcript involved in Syncytin-1 envelope translation.[123,125] Syncytin-1 glycoprotein has never been described in the tumor context. The presence of the 3.4-kb ERVW-1 transcript in seminoma was associated with epigenetic modifications, i.e., CpG demethylation in the MaLR enhancer and 5′LTR promoter regions.[122,123] Interestingly, the tissue adjacent to seminoma was partially demethylated, suggesting a permissiveness state.[122]

HERV-W and HERV-E mRNA were found expressed in normal ovarian tissue, indeed at levels higher than in ovarian cancer.[196] Conversely, env-specific HERV-E, HERV-K, and ERV3 mRNAs were detected in ovarian cancer, and, strikingly, anti-HERV antibodies including anti-ERV3, anti-HERV-E, and anti-HERV-K were detected in patients but not in normal female controls. Interestingly, Env HERV-K protein was observed on the surface and in the cytoplasm of ovarian cancer cells. An increase was observed in tumors with low malignant potential and low grade, when compared to expression in normal ovarian tissues.[197] HERV-K env and ERV-3 env were detected in the baboon ovary.[198]

Along with physiological cell–cell fusion processes involved in oocyte fertilization, and the differentiation of syncytiotrophoblasts, myoblasts, and osteoclasts, cell–cell fusion events have also been reported in cancer, resulting either in subsequent apoptosis or conversely in unlimited proliferative ability[199] and aggravated malignancy.[200–202] Bjerregaard *et al.* showed that Syncytin-1-dependent cell–cell fusions occurred between breast cancer cell lines and endothelial cells upon interaction with hASCT2 expressed by endothelial cells.[100] Moreover, in vivo expression of Syncytin-1 was reported for 38% of examined breast cancer specimen. Syncytin-1 expression was cytoplasmic with sometimes an intracytoplasmic punctuate enhancement.[203] The authors associated the expression of Syncytin-1 in vivo with a positive prognosis on recurrence-free survival.

5. CONCLUSION

Our appreciation of the friendly cohabitation of the retroviruses and the placenta is in its adolescence. The HERV-W family containing the ERVWE1 locus encoding the protein known as Syncytin-1 was experimentally depicted 17 years ago, and the receptor-dependent fusogenic property of the protein together with its placental immunolocalization was described in vitro and in vivo, respectively, between 16 and 13 years ago.[42,71,88] This has led to the publication of about 400 articles among which 54% concern the placenta and 15% focused on placenta-related pathologies. Today, much of the interest in placental pathologies concerns Syncytins, although ERV3, EnvP(b), and EnvV clearly deserve further investigation due to their observed properties in vitro and modes of expression in vivo. In addition to these hypothesis-driven approaches, systematic "omics" approaches may reveal the involvement of other retroviral elements, e.g., short HERV-K peptides in *env* regions as observed in melanoma cells,[204] HML-2 Gag and Rec retroviral regulatory protein evidenced during normal human embryogenesis beginning with embryonic genome activation at the eight-cell stage,[205] and also noncoding HERV-H sequences, as observed in embryonic stem cell and signing pluripotency.[206] In addition to Env immunological properties, noncoding sequences show evidence of involvement in immune tuning.[207–209] HERV deregulation may thus convert this site of suspended conflicts that is the mother–child interface into a hostile zone, as occurs in many of the pathologies described herein. Hence, understanding the role of envelopes and other retroviral proteins as well as of the tightly regulated retroviral DNA/RNA may lead to the development of new diagnostic tools allowing earlier and specific detection of placental pathologies and also drive the use of conventional retroviral therapies and DNA methylation inhibitors in placenta-related pathologies as suggested in the cancer field.[208,210]

ACKNOWLEDGMENT

We thank Beth Holder, John Aplin, Reiner Strick, and Pamela Strissel for kindly providing photographs.

REFERENCES

1. Brosius J, Gould SJ. On "genomenclature": a comprehensive (and respectful) taxonomy for pseudogenes and other "junk DNA". *Proc Natl Acad Sci USA.* 1992;89(22):10706–10710.
2. Brosius J. Echoes from the past—are we still in an RNP world? *Cytogenet Genome Res.* 2005;110(1–4):8–24. http://dx.doi.org/10.1159/000084934.
3. Temin HM. Origin of retroviruses from cellular moveable genetic elements. *Cell.* 1980;21(3):599–600.
4. Lingner J, Hughes TR, Shevchenko A, Mann M, Lundblad V, Cech TR. Reverse transcriptase motifs in the catalytic subunit of telomerase. *Science.* 1997;276(5312):561–567.
5. Lower R, Lower J, Kurth R. The viruses in all of us: characteristics and biological significance of human endogenous retrovirus sequences. *Proc Natl Acad Sci USA.* 1996;93(11):5177–5184.
6. Malik HS, Henikoff S, Eickbush TH. Poised for contagion: evolutionary origins of the infectious abilities of invertebrate retroviruses. *Genome Res.* 2000;10(9):1307–1318.
7. Kim FJ, Battini J-L, Manel N, Sitbon M. Emergence of vertebrate retroviruses and envelope capture. *Virology.* 2004;318(1):183–191. http://dx.doi.org/10.1016/j.virol.2003.09.026.
8. Coffin JM, Hughes SH, Varmus HE, Boeke JD, Stoye JP. *Retrotransposons, Endogenous Retroviruses, and the Evolution of Retroelements.* New York, NY: Cold Spring Harbor Laboratory Press; 1997.
9. Lander ES, Linton LM, Birren B, et al. Initial sequencing and analysis of the human genome. *Nature.* 2001;409(6822):860–921.
10. Kalter SS, Heberling RL, Helmke RJ, et al. A comparative study on the presence of C-type viral particles in placentas from primates and other animals. *Bibl Haematol.* 1975;40:391–401.
11. Lyden TW, Johnson PM, Mwenda JM, Rote NS. Ultrastructural characterization of endogenous retroviral particles isolated from normal human placentas. *Biol Reprod.* 1994;51(1):152–157.
12. Venables PJ, Brookes SM, Griffiths D, Weiss RA, Boyd MT. Abundance of an endogenous retroviral envelope protein in placental trophoblasts suggests a biological function. *Virology.* 1995;211(2):589–592.
13. Langat DK, Johnson PM, Rote NS, et al. Characterization of antigens expressed in normal baboon trophoblast and cross-reactive with HIV/SIV antibodies. *J Reprod Immunol.* 1999;42(1):41–58.

14. Ruebner M, Strissel PL, Ekici AB, et al. Reduced syncytin-1 expression levels in placental syndromes correlates with epigenetic hypermethylation of the ERVW-1 promoter region. *PLoS One*. 2013;8(2). e56145http://dx.doi.org/10.1371/journal.pone.0056145.
15. Mühle M, Kroniger T, Hoffmann K, Denner J. The immunosuppressive domain of the transmembrane envelope protein gp41 of HIV-1 binds to human monocytes and B cells. *Immunol Res*. 2016;64(3):721–729. http://dx.doi.org/10.1007/s12026-015-8776-4.
16. Douville RN, Hiscott J. The interface between the innate interferon response and expression of host retroviral restriction factors. *Cytokine*. 2010;52(1–2):108–115. http://dx.doi.org/10.1016/j.cyto.2010.04.010.
17. Goodier JL. Restricting retrotransposons: a review. *Mob DNA*. 2016;7(1):16. http://dx.doi.org/10.1186/s13100-016-0070-z.
18. Blikstad V, Benachenhou F, Sperber GO, Blomberg J. Evolution of human endogenous retroviral sequences: a conceptual account. *Cell Mol Life Sci*. 2008;65(21): 3348–3365. http://dx.doi.org/10.1007/s00018-008-8495-2.
19. Beimforde N, Hanke K, Ammar I, Kurth R, Bannert N. Molecular cloning and functional characterization of the human endogenous retrovirus K113. *Virology*. 2008;371(1):216–225.
20. Boller K, Schonfeld K, Lischer S, et al. Human endogenous retrovirus HERV-K113 is capable of producing intact viral particles. *J Gen Virol*. 2008;89(pt 2):567–572.
21. Contreras-Galindo R, Kaplan MH, Leissner P, et al. Human endogenous retrovirus K (HML-2) elements in the plasma of people with lymphoma and breast cancer. *J Virol*. 2008;82(19):9329–9336. http://dx.doi.org/10.1128/JVI.00646-08.
22. Subramanian RP, Wildschutte JH, Russo C, Coffin JM. Identification, characterization, and comparative genomic distribution of the HERV-K (HML-2) group of human endogenous retroviruses. *Retrovirology*. 2011;8(1):90. http://dx.doi.org/10.1186/1742-4690-8-90.
23. Gifford R, Tristem M. The evolution, distribution and diversity of endogenous retroviruses. *Virus Genes*. 2003;26(3):291–315.
24. Lavie L, Medstrand P, Schempp W, Meese E, Mayer J. Human endogenous retrovirus family HERV-K(HML-5): status, evolution, and reconstruction of an ancient betaretrovirus in the human genome. *J Virol*. 2004;78(16):8788–8798. http://dx.doi.org/10.1128/JVI.78.16.8788-8798.2004.
25. van Regenmortel MH, Mayo MA, Fauquet CM, Maniloff J. Virus nomenclature: consensus versus chaos. *Arch Virol*. 2000;145:2227–2232.
26. Coffin JM. Genetic diversity and evolution of retroviruses. *Curr Top Microbiol Immunol*. 1992;176:143–164.
27. Benit L, Lallemand JB, Casella JF, Philippe H, Heidmann T. ERV-L elements: a family of endogenous retrovirus-like elements active throughout the evolution of mammals. *J Virol*. 1999;73(4):3301–3308.
28. Belshaw R, Katzourakis A, Paces J, Burt A, Tristem M. High copy number in human endogenous retrovirus families is associated with copying mechanisms in addition to reinfection. *Mol Biol Evol*. 2005;22(4):814–817.
29. Benit L, Dessen P, Heidmann T. Identification, phylogeny, and evolution of retroviral elements based on their envelope genes. *J Virol*. 2001;75(23):11709–11719.
30. Jurka J, Kapitonov VV, Pavlicek A, Klonowski P, Kohany O, Walichiewicz J. Repbase Update, a database of eukaryotic repetitive elements. *Cytogenet Genome Res*. 2005;110(1–4):462–467. http://dx.doi.org/10.1159/000084979.
31. Mager DL, Medstrand P. Retroviral repeat sequences. In: Cooper D, ed. *Nature Encyclopedia of the Human Genome*. vol 5: Hampshire, England: Nature Publishing Group; 2003:57–63.
32. Tempel S, Jurka M, Jurka J. VisualRepbase: an interface for the study of occurrences of transposable element families. *BMC Bioinf*. 2008;9(1):345. http://dx.doi.org/10.1186/1471-2105-9-345.

33. Tristem M. Identification and characterization of novel human endogenous retrovirus families by phylogenetic screening of the human genome mapping project database. *J Virol.* 2000;74(8):3715–3730.
34. Hughes JF, Coffin JM. Evidence for genomic rearrangements mediated by human endogenous retroviruses during primate evolution. *Nat Genet.* 2001;29(4):487–489. http://dx.doi.org/10.1038/ng775.
35. Blanco P, Shlumukova M, Sargent CA, Jobling MA, Affara N, Hurles ME. Divergent outcomes of intrachromosomal recombination on the human Y chromosome: male infertility and recurrent polymorphism. *J Med Genet.* 2000;37(10):752–758.
36. Kamp C, Hirschmann P, Voss H, Huellen K, Vogt PH. Two long homologous retroviral sequence blocks in proximal Yq11 cause AZFa microdeletions as a result of intrachromosomal recombination events. *Hum Mol Genet.* 2000;9(17):2563–2572.
37. Sun C, Skaletsky H, Rozen S, et al. Deletion of azoospermia factor a (AZFa) region of human Y chromosome caused by recombination between HERV15 proviruses. *Hum Mol Genet.* 2000;9(15):2291–2296.
38. Cohen CJ, Lock WM, Mager DL. Endogenous retroviral LTRs as promoters for human genes: a critical assessment. *Gene.* 2009;448:105–114.
39. Long Q, Bengra C, Li C, Kutlar F, Tuan D. A long terminal repeat of the human endogenous retrovirus ERV-9 is located in the 5′ boundary area of the human beta-globin locus control region. *Genomics.* 1998;54(3):542–555. http://dx.doi.org/10.1006/geno.1998.5608.
40. Schulte AM, Lai S, Kurtz A, Czubayko F, Riegel AT, Wellstein A. Human trophoblast and choriocarcinoma expression of the growth factor pleiotrophin attributable to germ-line insertion of an endogenous retrovirus. *Proc Natl Acad Sci USA.* 1996;93(25):14759–14764.
41. Ting CN, Rosenberg MP, Snow CM, Samuelson LC, Meisler MH. Endogenous retroviral sequences are required for tissue-specific expression of a human salivary amylase gene. *Genes Dev.* 1992;6(8):1457–1465.
42. Blond J-L, Lavillette D, Cheynet V, et al. An envelope glycoprotein of the human endogenous retrovirus HERV-W is expressed in the human placenta and fuses cells expressing the type D mammalian retrovirus receptor. *J Virol.* 2000;74(7):3321–3329.
43. Boese A, Sauter M, Galli U, et al. Human endogenous retrovirus protein cORF supports cell transformation and associates with the promyelocytic leukemia zinc finger protein. *Oncogene.* 2000;19(38):4328–4336.
44. Magin C, Lower R, Lower J. cORF and RcRE, the Rev/Rex and RRE/RxRE homologues of the human endogenous retrovirus family HTDV/HERV-K. *J Virol.* 1999;73(11):9496–9507.
45. Mangeney M, Renard M, Schlecht-Louf G, et al. Placental syncytins: genetic disjunction between the fusogenic and immunosuppressive activity of retroviral envelope proteins. *Proc Natl Acad Sci USA.* 2007;104(51):20534–20539. http://www.ncbi.nlm.nih.gov/entrez/query.fcgi?cmd=Retrieve&db=PubMed&dopt=Citation&list_uids=18077339.
46. de Parseval N, Lazar V, Casella JF, Benit L, Heidmann T. Survey of human genes of retroviral origin: identification and transcriptome of the genes with coding capacity for complete envelope proteins. *J Virol.* 2003;77(19):10414–10422. http://dx.doi.org/10.1128/JVI.77.19.10414-10422.2003.
47. Villesen P, Aagaard L, Wiuf C, Pedersen FS. Identification of endogenous retroviral reading frames in the human genome. *Retrovirology.* 2004;1:32.
48. Roebke C, Wahl S, Laufer G, et al. An N-terminally truncated envelope protein encoded by a human endogenous retrovirus W locus on chromosome Xq22.3. *Retrovirology.* 2010;7(1):69. http://dx.doi.org/10.1186/1742-4690-7-69.

49. Nellaker C, Yao Y, Jones-Brando L, Mallet F, Yolken RH, Karlsson H. Transactivation of elements in the human endogenous retrovirus W family by viral infection. *Retrovirology*. 2006;3:44. http://dx.doi.org/10.1186/1742-4690-3-44.
50. Laufer G, Mayer J, Mueller BF, Mueller-Lantzsch N, Ruprecht K. Analysis of transcribed human endogenous retrovirus W env loci clarifies the origin of multiple sclerosis-associated retrovirus env sequences. *Retrovirology*. 2009;6:37.
51. Schmitt K, Richter C, Backes C, Meese E, Ruprecht K, Mayer J. Comprehensive analysis of human endogenous retrovirus group HERV-W locus transcription in multiple sclerosis brain lesions by high-throughput amplicon sequencing. *J Virol*. 2013;87(24): 13837–13852. http://dx.doi.org/10.1128/JVI.02388-13.
52. Mayer J, Blomberg J, Seal RL. A revised nomenclature for transcribed human endogenous retroviral loci. *Mob DNA*. 2011;2(1):7. http://dx.doi.org/10.1186/1759-8753-2-7.
53. Blaise S, de Parseval N, Heidmann T. Functional characterization of two newly identified Human Endogenous Retrovirus coding envelope genes. *Retrovirology*. 2005;2:19.
54. de Parseval N, Casella J, Gressin L, Heidmann T. Characterization of the three HERV-H proviruses with an open envelope reading frame encompassing the immunosuppressive domain and evolutionary history in primates. *Virology*. 2001;279(2): 558–569. http://dx.doi.org/10.1006/viro.2000.0737.
55. Andersson AC, Merza M, Venables P, et al. Elevated levels of the endogenous retrovirus ERV3 in human sebaceous glands. *J Invest Dermatol*. 1996;106(1):125–128.
56. Andersson AC, Yun Z, Sperber GO, Larsson E, Blomberg J. ERV3 and related sequences in humans: structure and RNA expression. *J Virol*. 2005;79(14):9270–9284.
57. Cohen M, Powers M, O'Connell C, Kato N. The nucleotide sequence of the env gene from the human provirus ERV3 and isolation and characterization of an ERV3-specific cDNA. *Virology*. 1985;147(2):449–458.
58. Cohen M, Kato N, Larsson E. ERV3 human endogenous provirus mRNAs are expressed in normal and malignant tissues and cells, but not in choriocarcinoma tumor cells. *J Cell Biochem*. 1988;36(2):121–128. http://dx.doi.org/10.1002/jcb.240360203.
59. Kato N, Pfeifer-Ohlsson S, Kato M, et al. Tissue-specific expression of human provirus ERV3 mRNA in human placenta: two of the three ERV3 mRNAs contain human cellular sequences. *J Virol*. 1987;61(7):2182–2191.
60. Larsson E, Andersson AC, Nilsson BO. Expression of an endogenous retrovirus (ERV3 HERV-R) in human reproductive and embryonic tissues—evidence for a function for envelope gene products. *Ups J Med Sci*. 1994;99(2):113–120.
61. Malassiné A, Blaise S, Handschuh K, et al. Expression of the fusogenic HERV-FRD Env glycoprotein (syncytin 2) in human placenta is restricted to villous cytotrophoblastic cells. *Placenta*. 2007;28(2–3):185–191. http://dx.doi.org/10.1016/j.placenta.2006.03.001.
62. Blaise S, de Parseval N, Bénit L, Heidmann T. Genomewide screening for fusogenic human endogenous retrovirus envelopes identifies syncytin 2, a gene conserved on primate evolution. *Proc Natl Acad Sci USA*. 2003;100(22):13013–13018. http://dx.doi.org/10.1073/pnas.2132646100.
63. Chen CP, Chen LF, Yang SR, et al. Functional characterization of the human placental fusogenic membrane protein syncytin 2. *Biol Reprod*. 2008;79(5):815–823.
64. Dewannieux M, Blaise S, Heidmann T. Identification of a functional envelope protein from the HERV-K family of human endogenous retroviruses. *J Virol*. 2005;79(24):15573–15577.
65. Tonjes RR, Czauderna F, Kurth R. Genome-wide screening, cloning, chromosomal assignment, and expression of full-length human endogenous retrovirus type K. *J Virol*. 1999;73(11):9187–9195.

66. Barbulescu M, Turner G, Seaman MI, Deinard AS, Kidd KK, Lenz J. Many human endogenous retrovirus K (HERV-K) proviruses are unique to humans. *Curr Biol.* 1999;9(16):861–868.
67. Turner G, Barbulescu M, Su M, Jensen-Seaman MI, Kidd KK, Lenz J. Insertional polymorphisms of full-length endogenous retroviruses in humans. *Curr Biol.* 2001;11(19):1531–1535.
68. Vargas A, Toufaily C, LeBellego F, Rassart É, Lafond J, Barbeau B. Reduced expression of both syncytin 1 and syncytin 2 correlates with severity of preeclampsia. *Reprod Sci.* 2011;18(11):1085–1091. http://dx.doi.org/10.1177/1933719111404608.
69. Kjeldbjerg AL, Villesen P, Aagaard L, Pedersen FS. Gene conversion and purifying selection of a placenta-specific ERV-V envelope gene during simian evolution. *BMC Evol Biol.* 2008;8(1):266. http://dx.doi.org/10.1186/1471-2148-8-266.
70. Blond JL, Beseme F, Duret L, et al. Molecular characterization and placental expression of HERV-W, a new human endogenous retrovirus family. *J Virol.* 1999;73(2):1175–1185.
71. Frendo JL, Olivier D, Cheynet V, et al. Direct involvement of HERV-W Env glycoprotein in human trophoblast cell fusion and differentiation. *Mol Cell Biol.* 2003;23(10):3566–3574.
72. Pérot P, Bolze PA, Mallet F. From viruses to genes: syncytins. In: Witzany G, ed. *Viruses: Essential Agents of Life*. Dordrecht, Netherlands: Springer Science + Business Media; 2012:325–361 [chapter 17]. http://dx.doi.org/10.1007/978-94-007-4899-6_17.
73. Henzy JE, Coffin JM. Betaretroviral envelope subunits are noncovalently associated and restricted to the mammalian class. *J Virol.* 2013;87(4):1937–1946. http://dx.doi.org/10.1128/JVI.01442-12.
74. Cheynet V, Ruggieri A, Oriol G, et al. Synthesis, assembly, and processing of the Env ERVWE1/syncytin human endogenous retroviral envelope. *J Virol.* 2005;79(9): 5585–5593. http://dx.doi.org/10.1128/JVI.79.9.5585-5593.2005.
75. Vargas A, Thiery M, Lafond J, Barbeau B. Transcriptional and functional studies of Human Endogenous Retrovirus envelope EnvP(b) and EnvV genes in human trophoblasts. *Virology.* 2012;425(1):1–10. http://dx.doi.org/10.1016/j.virol.2011.12.015.
76. Lavillette D, Marin M, Ruggieri A, Mallet F, Cosset FL, Kabat D. The envelope glycoprotein of human endogenous retrovirus type W uses a divergent family of amino acid transporters/cell surface receptors. *J Virol.* 2002;76(13):6442–6452.
77. Esnault C, Priet S, Ribet D, et al. A placenta-specific receptor for the fusogenic, endogenous retrovirus-derived, human syncytin-2. *Proc Natl Acad Sci USA.* 2008;105(45):17532–17537.
78. Mangeney M, Heidmann T. Tumor cells expressing a retroviral envelope escape immune rejection in vivo. *Proc Natl Acad Sci USA.* 1998;95(25):14920–14925.
79. Tolosa JM, Schjenken JE, Clifton VL, et al. The endogenous retroviral envelope protein syncytin-1 inhibits LPS/PHA-stimulated cytokine responses in human blood and is sorted into placental exosomes. *Placenta.* 2012;33(11):933–941. http://dx.doi.org/10.1016/j.placenta.2012.08.004.
80. Lavialle C, Cornelis G, Dupressoir A, et al. Paleovirology of "syncytins", retroviral env genes exapted for a role in placentation. *Philos Trans R Soc B.* 2013;368(1626): 20120507. http://dx.doi.org/10.1098/rstb.2012.0507.
81. Umehara D, Kawamura M, Odahara Y, et al. Role of N-terminal sequences of the tyrosine kinase sf-Stk in transformation of rodent fibroblasts by variants of Friend spleen focus-forming virus. *Int J Cancer.* 2012;131(5):1083–1094. http://dx.doi.org/10.1002/ijc.27330.
82. Feldman D, Roniger M, Bar-Sinai A, et al. The signal peptide of mouse mammary tumor virus-env: a phosphoprotein tumor modulator. *Mol Cancer Res.* 2012;10(8):1077–1086. http://dx.doi.org/10.1158/1541-7786.MCR-11-0581.

83. Hull S, Lim J, Hamil A, Nitta T, Fan H. Analysis of jaagsiekte sheep retrovirus (JSRV) envelope protein domains in transformation. *Virus Genes*. 2012;45(3):508–517. http://dx.doi.org/10.1007/s11262-012-0793-y.
84. Zhao J, Rycaj K, Geng S, et al. Expression of human endogenous retrovirus type K envelope protein is a novel candidate prognostic marker for human breast cancer. *Genes Cancer*. 2011;2(9):914–922. http://dx.doi.org/10.1177/1947601911431841.
85. Januszkiewicz-Lewandowska D, Nowicka K, Rembowska J, et al. Env gene expression of human endogenous retrovirus-k and human endogenous retrovirus-w in childhood acute leukemia cells. *Acta Haematol*. 2013;129(4):232–237. http://dx.doi.org/10.1159/000345407.
86. Krone B, Grange JM. Melanoma, Darwinian medicine and the inner world. *J Cancer Res Clin Oncol*. 2010;136(12):1787–1794. http://dx.doi.org/10.1007/s00432-010-0949-x.
87. Mullins CS, Linnebacher M. Human endogenous retroviruses and cancer: causality and therapeutic possibilities. *World J Gastroenterol*. 2012;18(42):6027–6035. http://dx.doi.org/10.3748/wjg.v18.i42.6027.
88. Mi S, Lee X, Li X, et al. Syncytin is a captive retroviral envelope protein involved in human placental morphogenesis. *Nature*. 2000;403(6771):785–789. http://dx.doi.org/10.1038/35001608.
89. Lee X, Keith Jr JC, Stumm N, et al. Downregulation of placental syncytin expression and abnormal protein localization in pre-eclampsia. *Placenta*. 2001;22(10):808–812. http://dx.doi.org/10.1053/plac.2001.0722.
90. Chen CP, Wang KG, Chen CY, Yu C, Chuang HC, Chen H. Altered placental syncytin and its receptor ASCT2 expression in placental development and pre-eclampsia. *BJOG*. 2006;113(2):152–158. http://dx.doi.org/10.1111/j.1471-0528.2005.00843.x.
91. Langbein M, Strick R, Strissel PL, et al. Impaired cytotrophoblast cell–cell fusion is associated with reduced Syncytin and increased apoptosis in patients with placental dysfunction. *Mol Reprod Dev*. 2007;75(1):175–183. http://dx.doi.org/10.1002/mrd.20729.
92. Pathirage NA, Cocquebert M, Sadovsky Y, et al. Homeobox gene transforming growth factor β-induced factor-1 (TGIF-1) is a regulator of villous trophoblast differentiation and its expression is increased in human idiopathic fetal growth restriction. *Mol Hum Reprod*. 2013;19(10):665–675. http://dx.doi.org/10.1093/molehr/gat042.
93. Vargas A, Zhou S, Ethier-Chiasson M, et al. Syncytin proteins incorporated in placenta exosomes are important for cell uptake and show variation in abundance in serum exosomes from patients with preeclampsia. *FASEB J*. 2014;28(8):3703–3719. http://dx.doi.org/10.1096/fj.13-239053.
94. Bjerregard B, Ziomkiewicz I, Schulz A, Larsson L-I. Syncytin-1 in differentiating human myoblasts: relationship to caveolin-3 and myogenin. *Cell Tissue Res*. 2014;357(1):355–362. http://dx.doi.org/10.1007/s00441-014-1930-9.
95. Oluwole SO, Yao Y, Conradi S, Kristensson K, Karlsson H. Elevated levels of transcripts encoding a human retroviral envelope protein (syncytin) in muscles from patients with motor neuron disease. *Amyotroph Lateral Scler*. 2007;8(2):67–72.
96. Redelsperger F, Raddi N, Bacquin A, et al. Genetic evidence that captured retroviral envelope syncytins contribute to myoblast fusion and muscle sexual dimorphism in mice. *PLoS Genet*. 2016;12(9). e1006289http://dx.doi.org/10.1371/journal.pgen.1006289.
97. Søe K, Andersen TL, Hobolt-Pedersen A-S, Bjerregaard B, Larsson L-I, Delaissé J-M. Involvement of human endogenous retroviral syncytin-1 in human osteoclast fusion. *Bone*. 2011;48(4):837–846. http://dx.doi.org/10.1016/j.bone.2010.11.011.
98. Hobolt-Pedersen A-S, Delaissé J-M, Søe K. Osteoclast fusion is based on heterogeneity between fusion partners. *Calcif Tissue Int*. 2014;95(1):73–82. http://dx.doi.org/10.1007/s00223-014-9864-5.

99. Bjerregaard B, Lemmen JG, Petersen MR, et al. Syncytin-1 and its receptor is present in human gametes. *J Assist Reprod Genet*. 2014;31(5):533–539. http://dx.doi.org/10.1007/s10815-014-0224-1.
100. Bjerregaard B, Holck S, Christensen IJ, Larsson LI. Syncytin is involved in breast cancer-endothelial cell fusions. *Cell Mol Life Sci*. 2006;63(16):1906–1911. http://dx.doi.org/10.1007/s00018-006-6201-9.
101. Strick R, Ackermann S, Langbein M, et al. Proliferation and cell-cell fusion of endometrial carcinoma are induced by the human endogenous retroviral Syncytin-1 and regulated by TGF-beta. *J Mol Med (Berl)*. 2007;85(1):23–38. http://dx.doi.org/10.1007/s00109-006-0104-y.
102. Mo H, Ouyang D, Xu L, Gao Q, He X. Human endogenous retroviral syncytin exerts inhibitory effect on invasive phenotype of B16F10 melanoma cells. *Chin J Cancer Res*. 2013;25(5):556–564. http://dx.doi.org/10.3978/j.issn.1000-9604.2013.10.06.
103. Knerr I, Schnare M, Hermann K, et al. Fusiogenic endogenous-retroviral syncytin-1 exerts anti-apoptotic functions in staurosporine-challenged CHO cells. *Apoptosis*. 2007;12(1):37–43. http://dx.doi.org/10.1007/s10495-006-0329-9.
104. Huang Q, Chen H, Li J, et al. Epigenetic and non-epigenetic regulation of syncytin-1 expression in human placenta and cancer tissues. *Cell Signal*. 2014;26(3):648–656. http://dx.doi.org/10.1016/j.cellsig.2013.11.002.
105. Chuprin A, Gal H, Biron-Shental T, et al. Cell fusion induced by ERVWE1 or measles virus causes cellular senescence. *Genes Dev*. 2013;27(21):2356–2366. http://dx.doi.org/10.1101/gad.227512.113.
106. Holder BS, Tower CL, Forbes K, Mulla MJ, Aplin JD, Abrahams VM. Immune cell activation by trophoblast-derived microvesicles is mediated by syncytin 1. *Immunology*. 2012;136(2):184–191. http://dx.doi.org/10.1111/j.1365-2567.2012.03568.x.
107. Antony JM, Ellestad KK, Hammond R, et al. The human endogenous retrovirus envelope glycoprotein, syncytin-1, regulates neuroinflammation and its receptor expression in multiple sclerosis: a role for endoplasmic reticulum chaperones in astrocytes. *J Immunol*. 2007;179(2):1210–1224.
108. Antony JM, van Marle G, Opii W, et al. Human endogenous retrovirus glycoprotein-mediated induction of redox reactants causes oligodendrocyte death and demyelination. *Nat Neurosci*. 2004;7(10):1088–1095. http://dx.doi.org/10.1038/nn1319.
109. Kremer D, Schichel T, Förster M, et al. Human endogenous retrovirus type W envelope protein inhibits oligodendroglial precursor cell differentiation. *Ann Neurol*. 2013;74(5):721–732. http://dx.doi.org/10.1002/ana.23970.
110. Perron H, Mekaoui L, Bernard C, Veas F, Stefas I, Leboyer M. Endogenous retrovirus type W GAG and envelope protein antigenemia in serum of schizophrenic patients. *Biol Psychiatry*. 2008;64(12):1019–1023.
111. Voisset C, Bouton O, Bedin F, et al. Chromosomal distribution and coding capacity of the human endogenous retrovirus HERV-W family. *AIDS Res Hum Retrovir*. 2000;16(8):731–740.
112. Prudhomme S, Oriol G, Mallet F. A retroviral promoter and a cellular enhancer define a bipartite element which controls env ERVWE1 placental expression. *J Virol*. 2004;78(22):12157–12168. http://dx.doi.org/10.1128/JVI.78.22.12157-12168.2004.
113. Vargas A, Moreau J, Landry S, et al. Syncytin-2 plays an important role in the fusion of human trophoblast cells. *J Mol Biol*. 2009;392(2):301–318.
114. Toufaily C, Vargas A, Lemire M, Lafond J, Rassart E, Barbeau B. MFSD2a, the Syncytin-2 receptor, is important for trophoblast fusion. *Placenta*. 2013;34(1):85–88. http://dx.doi.org/10.1016/j.placenta.2012.10.012.
115. Lin L, Xu B, Rote NS. Expression of endogenous retrovirus ERV-3 induces differentiation in BeWo, a choriocarcinoma model of human placental trophoblast. *Placenta*. 1999;20(1):109–118.

116. Lin L, Xu B, Rote NS. The cellular mechanism by which the human endogenous retrovirus ERV-3 env gene affects proliferation and differentiation in a human placental trophoblast model, BeWo. *Placenta*. 2000;21(1):73–78.
117. de Parseval N, Heidmann T. Physiological knockout of the envelope gene of the single-copy ERV-3 human endogenous retrovirus in a fraction of the Caucasian population. *J Virol*. 1998;72(4):3442–3445.
118. Dupressoir A, Marceau G, Vernochet C, et al. Syncytin-A and syncytin-B, two fusogenic placenta-specific murine envelope genes of retroviral origin conserved in Muridae. *Proc Natl Acad Sci USA*. 2005;102(3):725–730.
119. Dupressoir A, Vernochet C, Harper F, et al. A pair of co-opted retroviral envelope syncytin genes is required for formation of the two-layered murine placental syncytiotrophoblast. *Proc Natl Acad Sci USA*. 2011;108(46):E1164–E1173. http://dx.doi.org/10.1073/pnas.1112304108.
120. Cornelis G, Vernochet C, Carradec Q, et al. Retroviral envelope gene captures and syncytin exaptation for placentation in marsupials. *Proc Natl Acad Sci USA*. 2015;112(5):E487–E496. http://dx.doi.org/10.1073/pnas.1417000112.
121. Smallwood A. Temporal regulation of the expression of syncytin (HERV-W), maternally imprinted PEG10, and SGCE in human placenta. *Biol Reprod*. 2003;69(1):286–293. http://dx.doi.org/10.1095/biolreprod.102.013078.
122. Gimenez J, Montgiraud C, Pichon J-P, et al. Custom human endogenous retroviruses dedicated microarray identifies self-induced HERV-W family elements reactivated in testicular cancer upon methylation control. *Nucleic Acids Res*. 2010;38(7):2229–2246. http://dx.doi.org/10.1093/nar/gkp1214.
123. Trejbalová K, Blazková J, Matousková M, et al. Epigenetic regulation of transcription and splicing of syncytins, fusogenic glycoproteins of retroviral origin. *Nucleic Acids Res*. 2011;39(20):8728–8739. http://dx.doi.org/10.1093/nar/gkr562.
124. Li F, Nellaker C, Sabunciyan S, et al. Transcriptional derepression of the ERVWE1 locus following influenza A virus infection. *J Virol*. 2014;88(8):4328–4337. http://dx.doi.org/10.1128/JVI.03628-13.
125. Gimenez J, Montgiraud C, Oriol G, et al. Comparative methylation of ERVWE1/syncytin-1 and other human endogenous retrovirus LTRs in placenta tissues. *DNA Res*. 2009;16(4):195–211. http://dx.doi.org/10.1093/dnares/dsp011.
126. Matouskova M, Blazkova J, Pajer P, Pavlicek A, Hejnar J. CpG methylation suppresses transcriptional activity of human syncytin-1 in non-placental tissues. *Exp Cell Res*. 2006;312(7):1011–1020.
127. Chen YX, Allars M, Maiti K, et al. Factors affecting cytotrophoblast cell viability and differentiation: evidence of a link between syncytialisation and apoptosis. *Int J Biochem Cell Biol*. 2011;43(5):821–828. http://dx.doi.org/10.1016/j.biocel.2011.02.007.
128. Ruebner M, Langbein M, Strissel PL, et al. Regulation of the human endogenous retroviral Syncytin-1 and cell-cell fusion by the nuclear hormone receptors PPARγ/RXRα in placentogenesis. *J Cell Biochem*. 2012;113(7):2383–2396. http://dx.doi.org/10.1002/jcb.24110.
129. Mameli G, Astone V, Khalili K, Serra C, Sawaya BE, Dolei A. Regulation of the syncytin-1 promoter in human astrocytes by multiple sclerosis-related cytokines. *Virology*. 2007;362(1):120–130. http://dx.doi.org/10.1016/j.virol.2006.12.019.
130. Basyuk E, Cross JC, Corbin J, et al. Murine Gcm1 gene is expressed in a subset of placental trophoblast cells. *Dev Dyn*. 1999;214(4):303–311. http://dx.doi.org/10.1002/(SICI)1097-0177(199904)214:4<303::AID-AJA3>3.0.CO;2-B.
131. Janatpour MJ, Utset MF, Cross JC, et al. A repertoire of differentially expressed transcription factors that offers insight into mechanisms of human cytotrophoblast differentiation. *Dev Genet*. 1999;25(2):146–157. http://dx.doi.org/10.1002/(SICI)1520-6408(1999)25:2<146::AID-DVG9>3.0.CO;2-K.

132. Yu C, Shen K, Lin M, et al. GCMa regulates the syncytin-mediated trophoblastic fusion. *J Biol Chem*. 2002;277(51):50062–50068.
133. Liang C-Y, Wang L-J, Chen C-P, Chen L-F, Chen Y-H, Chen H. GCM1 regulation of the expression of syncytin 2 and its cognate receptor MFSD2A in human placenta. *Biol Reprod*. 2010;83(3):387–395. http://dx.doi.org/10.1095/biolreprod.110.083915.
134. Chang C-W, Chuang H-C, Yu C, Yao T-P, Chen H. Stimulation of GCMa transcriptional activity by cyclic AMP/protein kinase A signaling is attributed to CBP-mediated acetylation of GCMa. *Mol Cell Biol*. 2005;25(19):8401–8414. http://dx.doi.org/10.1128/MCB.25.19.8401-8414.2005.
135. Knerr I, Schubert SW, Wich C, et al. Stimulation of GCMa and syncytin via cAMP mediated PKA signaling in human trophoblastic cells under normoxic and hypoxic conditions. *FEBS Lett*. 2005;579(18):3991–3998. http://dx.doi.org/10.1016/j.febslet.2005.06.029.
136. Klase Z, Winograd R, Davis J, et al. HIV-1 TAR miRNA protects against apoptosis by altering cellular gene expression. *Retrovirology*. 2009;6:18.
137. Schubert SW, Lamoureux N, Kilian K, Klein-Hitpass L, Hashemolhosseini S. Identification of integrin-alpha4, Rb1, and syncytin a as murine placental target genes of the transcription factor GCMa/Gcm1. *J Biol Chem*. 2008;283(9):5460–5465.
138. Baczyk D, Drewlo S, Proctor L, Dunk C, Lye S, Kingdom J. Glial cell missing-1 transcription factor is required for the differentiation of the human trophoblast. *Cell Death Differ*. 2009;16(5):719–727. http://dx.doi.org/10.1038/cdd.2009.1.
139. Wilson ML, Brueggmann D, Desmond DH, Mandeville JE, Goodwin TM, Ingles SA. A fetal variant in the GCM1 gene is associated with pregnancy induced hypertension in a predominantly hispanic population. *Int J Mol Epidemiol Genet*. 2011;2(3):196–206.
140. Muroi Y, Sakurai T, Hanashi A, Kubota K, Nagaoka K, Imakawa K. CD9 regulates transcription factor GCM1 and ERVWE1 expression through the cAMP/protein kinase A signaling pathway. *Reproduction*. 2009;138(6):945–951. http://dx.doi.org/10.1530/REP-09-0082.
141. Li F, Karlsson H. Expression and regulation of human endogenous retrovirus W elements. *APMIS*. 2016;124(1–2):52–66. http://dx.doi.org/10.1111/apm.12478.
142. Cheynet V, Oriol G, Mallet F. Identification of the hASCT2-binding domain of the Env ERVWE1/syncytin-1 fusogenic glycoprotein. *Retrovirology*. 2006;3:41. http://dx.doi.org/10.1186/1742-4690-3-41.
143. Bonnaud B, Bouton O, Oriol G, Cheynet V, Duret L, Mallet F. Evidence of selection on the domesticated ERVWE1 env retroviral element involved in placentation. *Mol Biol Evol*. 2004;21(10):1895–1901.
144. Malassine A, Handschuh K, Tsatsaris V, et al. Expression of HERV-W Env glycoprotein (syncytin) in the extravillous trophoblast of first trimester human placenta. *Placenta*. 2005;26(7):556–562. http://dx.doi.org/10.1016/j.placenta.2004.09.002.
145. Muir A. Human endogenous retrovirus-W envelope (syncytin) is expressed in both villous and extravillous trophoblast populations. *J Gen Virol*. 2006;87(7):2067–2071. http://dx.doi.org/10.1099/vir.0.81412-0.
146. Soygur B, Moore H. Expression of Syncytin 1 (HERV-W), in the preimplantation human blastocyst, embryonic stem cells and trophoblast cells derived in vitro. *Hum Reprod*. 2016;31:1455–1461. http://dx.doi.org/10.1093/humrep/dew097.
147. Dunlap KA, Palmarini M, Varela M, et al. Endogenous retroviruses regulate periimplantation placental growth and differentiation. *Proc Natl Acad Sci USA*. 2006;103(39):14390–14395. http://dx.doi.org/10.1073/pnas.0603836103.
148. Cianciolo GJ, Copeland TD, Oroszlan S, Snyderman R. Inhibition of lymphocyte proliferation by a synthetic peptide homologous to retroviral envelope proteins. *Science*. 1985;230(4724):453–455.

149. Rasko JE, Battini JL, Gottschalk RJ, Mazo I, Miller AD. The RD114/simian type D retrovirus receptor is a neutral amino acid transporter. *Proc Natl Acad Sci USA*. 1999;96(5):2129–2134.
150. Chang WK, Yang KD, Shaio MF. Effect of glutamine on Th1 and Th2 cytokine responses of human peripheral blood mononuclear cells. *Clin Immunol*. 1999;93(3):294–301. http://dx.doi.org/10.1006/clim.1999.4788.
151. Hummel J, Kämmerer U, Müller N, Avota E, Schneider-Schaulies S. Human endogenous retrovirus envelope proteins target dendritic cells to suppress T-cell activation. *Eur J Immunol*. 2015;45(6):1748–1759. http://dx.doi.org/10.1002/eji.201445366.
152. Rolland A, Jouvin-Marche E, Viret C, Faure M, Perron H, Marche PN. The envelope protein of a human endogenous retrovirus-W family activates innate immunity through CD14/TLR4 and promotes Th1-like responses. *J Immunol*. 2006;176(12):7636–7644. http://www.ncbi.nlm.nih.gov/entrez/query.fcgi?cmd=Retrieve&db=PubMed&dopt=Citation&list_uids=16751411.
153. Bolze P-A, Patrier S, Cheynet V, et al. Expression patterns of ERVWE1/Syncytin-1 and other placentally expressed human endogenous retroviruses along the malignant transformation process of hydatidiform moles. *Placenta*. 2016;39:116–124. http://dx.doi.org/10.1016/j.placenta.2016.01.011.
154. Holder BS, Tower CL, Abrahams VM, Aplin JD. Syncytin 1 in the human placenta. *Placenta*. 2012;33(6):460–466. http://dx.doi.org/10.1016/j.placenta.2012.02.012.
155. Strissel PL, Ruebner M, Thiel F, et al. Reactivation of codogenic endogenous retroviral (ERV) envelope genes in human endometrial carcinoma and prestages: emergence of new molecular targets. *Oncotarget*. 2012;3(10):1204–1219. http://dx.doi.org/10.18632/oncotarget.679.
156. Loane M, Morris JK, Addor M-C, et al. Twenty-year trends in the prevalence of Down syndrome and other trisomies in Europe: impact of maternal age and prenatal screening. *Eur J Hum Genet*. 2012;21(1):27–33. http://dx.doi.org/10.1038/ejhg.2012.94.
157. Roberts L, Sebire NJ, Fowler D, Nicolaides KH. Histomorphological features of chorionic villi at 10–14 weeks of gestation in trisomic and chromosomally normal pregnancies. *Placenta*. 2000;21(7):678–683. http://dx.doi.org/10.1053/plac.2000.0553.
158. Frendo JL, Vidaud M, Guibourdenche J, et al. Defect of villous cytotrophoblast differentiation into syncytiotrophoblast in Down's syndrome. *J Clin Endocrinol Metabol*. 2000;85(10):3700–3707. http://dx.doi.org/10.1210/jcem.85.10.6915.
159. Massin N, Frendo JL, Guibourdenche J, et al. Defect of syncytiotrophoblast formation and human chorionic gonadotropin expression in Down's syndrome. *Placenta*. 2001;22(suppl A):S93–S97. http://dx.doi.org/10.1053/plac.2001.0658.
160. Malassiné A, Pidoux G, Gerbaud P, Frendo J-L, Evain-Brion D. Human trophoblast in trisomy 21: a model for cell-cell fusion dynamic investigation. *Adv Exp Med Biol*. 2011;-714(chapter 4):103–112. http://dx.doi.org/10.1007/978-94-007-0782-5_4.
161. Pidoux G, Gerbaud P, Laurendeau I, et al. Large variability of trophoblast gene expression within and between human normal term Placentae. *Placenta*. 2004;25(5):469–473. http://dx.doi.org/10.1016/j.placenta.2003.10.016.
162. Frendo JL, Guibourdenche J, Pidoux G, et al. Trophoblast production of a weakly bioactive human chorionic gonadotropin in trisomy 21-affected pregnancy. *J Clin Endocrinol Metabol*. 2004;89(2):727–732. http://dx.doi.org/10.1210/jc.2003-030668.
163. Roberts CL, Ford JB, Algert CS, et al. Population-based trends in pregnancy hypertension and pre-eclampsia: an international comparative study. *BMJ Open*. 2011;1(1). e000101http://dx.doi.org/10.1136/bmjopen-2011-000101.
164. Benirschke K. Remarkable placenta. *Clin Anat*. 1998;11(3):194–205. http://dx.doi.org/10.1002/(SICI)1098-2353(1998)11:3<194::AID-CA8>3.0.CO;2-T.

165. Devisme L, Merlot B, Ego A, Houfflin-Debarge V, Deruelle P, Subtil D. A case–control study of placental lesions associated with pre-eclampsia. *Int J Gynaecol Obstet.* 2013;120(2):165–168. http://dx.doi.org/10.1016/j.ijgo.2012.08.023.
166. Knerr I, Beinder E, Rascher W. Syncytin, a novel human endogenous retroviral gene in human placenta: evidence for its dysregulation in preeclampsia and HELLP syndrome. *Am J Obstet Gynecol.* 2002;186(2):210–213. http://dx.doi.org/10.1067/mob.2002.119636.
167. Kudo Y, Boyd CA, Sargent IL, Redman CW. Hypoxia alters expression and function of syncytin and its receptor during trophoblast cell fusion of human placental BeWo cells: implications for impaired trophoblast syncytialisation in pre-eclampsia. *Biochim Biophys Acta.* 2003;1638(1):63–71.
168. Bainbridge SA, Minhas A, Whiteley KJ, Qu D. Effects of reduced Gcm1 expression on trophoblast morphology, fetoplacental vascularity, and pregnancy outcomes in mice. *Hypertension.* 2012;59(3):732–739. http://dx.doi.org/10.1161/HYPERTENSIONAHA.111.183939/-/DC1.
169. Chiang M-H, Liang F-Y, Chen C-P, et al. Mechanism of hypoxia-induced GCM1 degradation: implications for the pathogenesis of preeclampsia. *J Biol Chem.* 2009;284(26):17411–17419. http://dx.doi.org/10.1074/jbc.M109.016170.
170. Zhuang X-W, Li J, Brost BC, et al. Decreased expression and altered methylation of syncytin-1 gene in human placentas associated with preeclampsia. *Curr Pharm Des.* 2014;20(11):1796–1802.
171. Gao Y, He Z, Wang Z, et al. Increased expression and altered methylation of HERVWE1 in the human placentas of smaller fetuses from monozygotic, dichorionic, discordant twins. *PLoS One.* 2012;7(3). e33503http://dx.doi.org/10.1371/journal.pone.0033503.t004.
172. Ruebner M, Strissel PL, Langbein M, et al. Impaired cell fusion and differentiation in placentae from patients with intrauterine growth restriction correlate with reduced levels of HERV envelope genes. *J Mol Med (Berl).* 2010;88(11):1143–1156. http://dx.doi.org/10.1007/s00109-010-0656-8.
173. Doridot L, Passet B, Méhats C, Rigourd V. Preeclampsia-like symptoms induced in mice by fetoplacental expression of STOX1 are reversed by aspirin treatment. *Hypertension.* 2013;61:662–668. http://dx.doi.org/10.1161/HYPERTENSIONAHA.111.202994/-/DC1.
174. Rigourd V, Chauvet C, Chelbi ST, et al. STOX1 overexpression in choriocarcinoma cells mimics transcriptional alterations observed in preeclamptic placentas. *PLoS One.* 2008;3(12):e3905–e3909. http://dx.doi.org/10.1371/journal.pone.0003905.
175. Lazennec G, Canaple L, Saugy D, Wahli W. Activation of peroxisome proliferator-activated receptors (PPARs) by their ligands and protein kinase A activators. *Mol Endocrinol.* 2000;14(12):1962–1975. http://dx.doi.org/10.1210/mend.14.12.0575.
176. Lebovic DI, Kavoussi SK, Lee J, Banu SK, Arosh JA. PPARγ activation inhibits growth and survival of human endometriotic cells by suppressing estrogen biosynthesis and PGE2 signaling. *Endocrinology.* 2013;154(12):4803–4813. http://dx.doi.org/10.1210/en.2013-1168.
177. Sidell N, Han SW, Parthasarathy S. Regulation and modulation of abnormal immune responses in endometriosis. *Ann N Y Acad Sci.* 2002;955:159–173. Discussion 199–200, 396–406.
178. Shaarawy M, Meleigy El M, Rasheed K. Maternal serum transforming growth factor beta-2 in preeclampsia and eclampsia, a potential biomarker for the assessment of disease severity and fetal outcome. *J Soc Gynecol Investig.* 2001;8(1):27–31.
179. Xu J, Sivasubramaniyam T, Yinon Y, et al. Aberrant TGFβ signaling contributes to altered trophoblast differentiation in preeclampsia. *Endocrinology.* 2016;157(2):883–899. http://dx.doi.org/10.1210/en.2015-1696.

180. Priscakova P, Konkolova J, Petrovic R, et al. ERVW-1 gene polymorphisms related to preeclampsia. *Bratisl Lek Listy*. 2016;117(06):340–344. http://dx.doi.org/10.4149/BLL_2016_067.
181. Mondestin MAJ, Ananth CV, Smulian JC, Vintzileos AM. Birth weight and fetal death in the United States: the effect of maternal diabetes during pregnancy. *Am J Obstet Gynecol*. 2002;187(4):922–926.
182. Cheng X, Chapple SJ, Patel B, et al. Gestational diabetes mellitus impairs Nrf2-mediated adaptive antioxidant defenses and redox signaling in fetal endothelial cells in utero. *Diabetes*. 2013;62(12):4088–4097. http://dx.doi.org/10.2337/db13-0169.
183. Hajj El N, Pliushch G, Schneider E, et al. Metabolic programming of MEST DNA methylation by intrauterine exposure to gestational diabetes mellitus. *Diabetes*. 2013;62(4):1320–1328. http://dx.doi.org/10.2337/db12-0289.
184. Ruchat S-M, Houde A-A, Voisin G, et al. Gestational diabetes mellitus epigenetically affects genes predominantly involved in metabolic diseases. *Epigenetics*. 2013;8(9):935–943. http://dx.doi.org/10.4161/epi.25578.
185. West NA, Kechris K, Dabelea D. Exposure to maternal diabetes in utero and DNA methylation patterns in the offspring. *Immunometabolism*. 2013;1:1–9. http://dx.doi.org/10.2478/immun-2013-0001.
186. Benirschke K, Burton GJ, Baergen RN. *Pathology of the Human Placenta;* New York, NY: Springer; 2012:959.
187. Soygur B, Sati L, Demir R. Altered expression of human endogenous retroviruses syncytin-1, syncytin-2 and their receptors in human normal and gestational diabetic placenta. *Histol Histopathol*. 2016;31:1037–1047. http://dx.doi.org/10.14670/HH-11-735.
188. Lurain JR. Gestational trophoblastic disease I: epidemiology, pathology, clinical presentation and diagnosis of gestational trophoblastic disease, and management of hydatidiform mole. *Am J Obstet Gynecol*. 2010;203(6):531–539. http://dx.doi.org/10.1016/j.ajog.2010.06.073.
189. Lurain JR. Gestational trophoblastic disease II: classification and management of gestational trophoblastic neoplasia. *Am J Obstet Gynecol*. 2011;204(1):11–18. http://dx.doi.org/10.1016/j.ajog.2010.06.072.
190. Seckl MJ, Sebire NJ, Berkowitz RS. Gestational trophoblastic disease. *Lancet*. 2010;376(9742):717–729.
191. Bolze P-A, Riedl C, Massardier J, et al. Mortality of gestational trophoblastic neoplasia with a FIGO score of 13 and higher. *Am J Obstet Gynecol*. 2016;214(3):390. http://dx.doi.org/10.1016/j.ajog.2015.09.083.
192. Bolze P-A, Attia J, Massardier J, et al. Formalised consensus of the European Organisation for treatment of trophoblastic diseases on management of gestational trophoblastic diseases. *Eur J Cancer*. 2015;51(13):1725–1731. http://dx.doi.org/10.1016/j.ejca.2015.05.026.
193. Judson H, Hayward BE, Sheridan E, Bonthron DT. A global disorder of imprinting in the human female germ line. *Nature*. 2002;416(6880):539–542. http://dx.doi.org/10.1038/416539a.
194. Oppelt P, Strick R, Strissel PL, Winzierl K, Beckmann MW, Renner SP. Expression of the human endogenous retroviruse-W envelope gene syncytin in endometriosis lesions. *Gynecol Endocrinol*. 2009;25(11):741–747. http://dx.doi.org/10.3109/09513590903184142.
195. Zhou H, Li J, Podratz KC, et al. Hypomethylation and activation of syncytin-1 gene in endometriotic tissue. *Curr Pharm Des*. 2014;20(11):1786–1795.
196. Hu L, Hornung D, Kurek R, et al. Expression of human endogenous gammaretroviral sequences in endometriosis and ovarian cancer. *AIDS Res Hum Retrovirus*. 2006;22(6):551–557. http://dx.doi.org/10.1089/aid.2006.22.551.
197. Wang-Johanning F, Liu J, Rycaj K, et al. Expression of multiple human endogenous retrovirus surface envelope proteins in ovarian cancer. *Int J Cancer*. 2006;120(1):81–90. http://dx.doi.org/10.1002/ijc.22256.

198. Arimi MM, Nyachieo A, Langat DK, Abdi AM, Mwenda JM. Evidence for expression of endogenous retroviral sequences on primate reproductive tissues and detection of cross-reactive ERVS antigens in the baboon ovary: a review. *East Afr Med J.* 2006;83(2):106–112.
199. Köhler G, Milstein C. Continuous cultures of fused cells secreting antibody of predefined specificity. *Nature.* 1975;256(5517):495–497.
200. Pawelek JM. Tumour-cell fusion as a source of myeloid traits in cancer. *Lancet Oncol.* 2005;6(12):988–993. http://dx.doi.org/10.1016/S1470-2045(05)70466-6.
201. Larizza L, Schirrmacher V, Graf L, Pflüger E, Peres-Martinez M, Stöhr M. Suggestive evidence that the highly metastatic variant ESb of the T-cell lymphoma Eb is derived from spontaneous fusion with a host macrophage. *Int J Cancer.* 1984;34(5):699–707.
202. Munzarova M, Lauerova L, Capkova J. Are advanced malignant melanoma cells hybrids between melanocytes and macrophages? *Melanoma Res.* 1992;2(2):127–129.
203. Larsson L-I, Holck S, Christensen IJ. Prognostic role of syncytin expression in breast cancer. *Hum Pathol.* 2007;38(5):726–731. http://dx.doi.org/10.1016/j.humpath.2006.10.018.
204. Schiavetti F, Thonnard J, Colau D, Boon T, Coulie PG. A human endogenous retroviral sequence encoding an antigen recognized on melanoma by cytolytic T lymphocytes. *Cancer Res.* 2002;62(19):5510–5516.
205. Grow EJ, Flynn RA, Chavez SL, et al. Intrinsic retroviral reactivation in human preimplantation embryos and pluripotent cells. *Nature.* 2015;522(7555):221–225. http://dx.doi.org/10.1038/nature14308.
206. Santoni FA, Guerra J, Luban J. HERV-H RNA is abundant in human embryonic stem cells and a precise marker for pluripotency. *Retrovirology.* 2012;9(1):111. http://dx.doi.org/10.1186/1742-4690-9-111.
207. Zeng M, Hu Z, Shi X, et al. MAVS, cGAS, and endogenous retroviruses in T-independent B cell responses. *Science.* 2014;346(6216):1486–1492. http://dx.doi.org/10.1126/science.346.6216.1486.
208. Strick R, Strissel PL, Baylin SB, Chiappinelli KB. Unraveling the molecular pathways of DNA-methylation inhibitors: human endogenous retroviruses induce the innate immune response in tumors. *Oncoimmunology.* 2016;5,e1122160. http://dx.doi.org/10.1080/2162402X.2015.1122160.
209. Kassiotis G, Stoye JP. Immune responses to endogenous retroelements: taking the bad with the good. *Nat Publ Group.* 2016;16(4):207–219. http://dx.doi.org/10.1038/nri.2016.27.
210. Armando RG, Gomez DM, Gomez DE. AZT exerts its antitumoral effect by telomeric and non-telomeric effects in a mammary adenocarcinoma model. *Oncol Rep.* 2016;36(5):2731–2736. http://dx.doi.org/10.3892/or.2016.5094.

CHAPTER SIX

Role of Exosomes in Placental Homeostasis and Pregnancy Disorders☆

C. Salomon*,†,1, **G.E. Rice***,†

*Exosome Biology Laboratory, Centre for Clinical Diagnostics, UQ Centre for Clinical Research, Royal Brisbane and Women's Hospital, The University of Queensland, Brisbane, QLD, Australia
†Ochsner Baptist Hospital, New Orleans, LA, United States
[1]Corresponding author: e-mail address: c.salomongallo@uq.edu.au

Contents

Abstract

The human placenta is a unique organ that performs the function of the majority of fetal organs across gestation. How the placenta communicates with maternal tissues to prepare them for pregnancy is not fully understood. Recently, it has been established that placental cells can communicate with maternal tissues to regulate their biological function via extracellular vesicles (EVs). EVs are subclassified into exosomes or microvesicles (MVs) according to their size, cell or tissue of origin, functions, and physical features. Exosomes are a specific type of EVs from an endocytic origin, while MVs are released via budding from the plasma membrane. With regards to pregnancy, the role of EVs has been described in several functions such as immune responses and maternal metabolic adaptation to gestation. Interestingly, EVs of placental origin can be detected in a variety of body fluids including urine and blood, and have been identified in the maternal circulation at as early as 6 weeks of gestation. Moreover, the number of exosomes across gestation is higher in complications of pregnancies such as pre-eclampsia and gestational diabetes mellitus compared to normal pregnancies. Circulating exosomes contains proteins and RNAs that are representative of the cell of origin, including surface and cytoplasmic protein, messenger RNA, and micro-RNAs. Finally,

☆*Disclosure statement*: The authors report no conflict of interest.

Progress in Molecular Biology and Translational Science, Volume 145
ISSN 1877-1173
http://dx.doi.org/10.1016/bs.pmbts.2016.12.006

exosomes are capable of transferring their contents to other cells and regulating the biological function of the target cell. In this review, we will discuss the effect of the maternal microenvironment on secretion and content of placenta-derived EVs, and how this may lead to complications of pregnancies with a special emphasis on exosomes.

1. INTRODUCTION

Extracellular vesicles (EVs) are classified into at least three subpopulations according to various features, including size, cell, or tissue of origin and function. These populations are nanovesicles and exosomes, microvesicles (MVs), and apoptotic bodies.[1] These vesicles contribute to cell-to-cell communication by transporting signaling molecules that are encapsulated within the vesicle, embedded in their bilipid vesicle membrane, and/or bound to the surface of the vesicle. Moreover, the release of these vesicles is regulated by physiological and pathological conditions.[2]

Exosomes are lipid bilayer nanovesicles. They are distinguished by their size range of 40–120 nm, buoyant density of 1.12–1.19 g/mL. Also unique to exosomes are their endosomal biogenesis, which involves inward budding of multivesicular bodies. Consistent with their biogenesis, exosomes express specific late endosomal markers, such as CD63. Exosomes contain an array of signaling molecules, such as proteins, lipids, and RNAs including micro-RNAs and mRNAs. Exosomes are released via exocytosis into biofluid compartments. Their secretion is regulated by local environmental factors (e.g., oxygen tension, glucose, and free fatty acid concentration).[1,3] These vesicles regulate the activity of both proximal and distal target cells by various interactions, including: modification of the extracellular milieu of the cellular target, activating cell membrane receptors, or endocytosis by the target cell where cell contents are released intracellularly.[4] Exosomes have been reported to regulate a diverse range of cellular activities in target cells,[5] such as translational activity, angiogenesis, proliferation, metabolism, and apoptosis.

The trophoblast cells are one of the most important cell type in the human placenta[6] and can be differentiated into syncytiotrophoblast (ST) and extravillous trophoblast (EVT). The ST cells are in direct contact with the maternal circulation. Meanwhile, EVT is an invasive cell type that migrates to the maternal tissue to remodel the uterine spiral arteries.[7] Placental cells (i.e., both ST and EVT) release exosomes into maternal circulation throughout gestation. Placental exosomes engage in cell-to-cell communication between the placenta and contiguous maternal organs, as well as more distal interactions such as preparing remote tissues for gestational metabolic changes. Given that the content of exosomes is cell type specific, we hypothesize that exosomes may provide a unique "signature" of pregnancy progression and metabolic state of the human placenta. This "signature" may be of clinical utility in the early identification of women at risk of developing pregnancy complications and the development of more efficacious management. As potential biomarkers, exosomes are highly stable and can be isolated from several biological fluids, such as blood. Given these qualities, exosomes may potentially develop as a noninvasive biopsy of the placenta. Thus, this commentary aims to review the current knowledge regarding exosomes and others EVs during gestation. It will focus on how exosomes and other EVs participate in pregnancy progression, and thus how they may lead to early diagnosis and improved clinical management of pregnancy complications.

2. INFLUENCE OF OXYGEN TENSION ON PLACENTAL EXOSOMES RELEASE AND CONTENT

Oxygen tension is a regulator of placentation, as key events, from implantation to maternal perfusion of placenta, are affected by site-specific oxygen tensions.[8] EVT invasion into the maternal tissue is a critical process in placentation. Intrauterine oxygen tension at the time of embryo implantation is ~3%,[9] while the decidua and myometrium oxygen tension is ~8%–12%.[10] As the invasive capacity of EVTs is inversely related to oxygen tension, the uterine oxygen gradient may promote and direct the invasion of EVT cells into the decidua and myometrium. EVTs are involved in the remodeling of spiral arterioles.[11,12] A low-oxygen environment is required for normal early fetal and placental development. This environment is brought about by the occlusion of spiral arterioles by intraluminal EVTs. Low resistance and high-capacity flow is reestablished toward the end of the first trimester by the perfusion of the placental intravillous space with maternal blood. Consequently, an effective maternofetal exchange interface is established. It is thought that pregnancy complications are clinical manifestations of a common developmental lesion (i.e., inadequate invasion by EVT cells), leading to failed remodeling of maternal uterine spiral arteries. The severity of this lesion may determine the extent of problems in pregnancy. While mild dysfunction may lead to intrauterine growth restriction, preeclampsia, and preterm birth, severe dysfunction may lead to miscarriage.

The understanding of cell-to-cell communication has undergone a paradigm shift over the past decade with increased understanding of how exosomes mediate intercellular signaling.[13] Exosomes have been identified in plasma under both normal and pathological conditions[14] and their concentration increases with disease severity and/or progression[15] and in response to oxidative stress.[16] Similarly, exosomes are involved in placental changes during gestation. Recently, we demonstrated that exosomes are released from first-trimester placental mesenchymal stem cells (pMSC). These exosomes subsequently increase endothelial cell migration and vascular tube formation.[17] In addition, the release of exosomes from pMSC was increased under low-oxygen tension. Similar effects of oxygen tension on exosome release and protein content have been observed for cytotrophoblast (CT) cells.[18] In this study, using a bioinformatics approach (ingenuity pathway analysis, IPA), changes in HIFα and IL-8 signaling pathways depending on the oxygen tension were identified in exosomal proteins. Additionally, incubation under low oxygen tension meant that the exosomal protein profile was largely associated with activation of particular pathways, namely, MMP-9, TGF-β, MAPK, VEGF, p38MAPK, TIMP1, and ERK1/2. It remains to be established whether or not specific changes in the exosomal protein profile are associated with alterations in EVT invasion and proliferation. Moreover, the mechanisms by which EVT invasion is orchestrated are also unclear. However, within the current body of literature, there is a consensus that uterine wall invasion involves upregulation of particular proteins, including MMPs, integrins (α5β1 and α1β1), and VE-cadherin. The pathway analysis of our study suggests that CT-derived exosomes promote EVT invasion by promoting activation of MMPs, MAPK, and associated invasiveness pathways. The understanding of the how placental-derived exosomes influence early pregnancy events by shaping the activity of key cell types involved in maternal–fetal vascular communication is still in the preliminary stages. The molecular mechanisms that govern the secretion and bioactivity of exosomes under hypoxic conditions also require further study. The release of CT-derived exosomes under hypoxic conditions within the placenta may be an adaptive response to promote proliferation and invasion of EVT cells.

These data are consistent with the hypothesis that placental cells release exosomes in response to changes in environmental conditions, such as oxygen tension, and these exosomes modify the phenotype of recipient cells.

Previously, we have investigated how extracellular oxygen tension and glucose concentrations interact to regulate release and bioactivity of first-trimester trophoblast-derived exosomes.[19] As mentioned, early events in placentation are influenced by oxygen tension.[20,21] Previous studies suggest that oxygen tension itself may be influenced by fluctuations in substrate concentration, including glucose.[22,23] The two factors were shown to coregulate cell function in the ACH-3P cell line.[24] In addition, a study by Frohlich *et al.* demonstrated that glucose (5.5 and 25 mM) and oxygen tension (2.5%, 8%, and 21%) together affected cell proliferation and notably, the formation of reactive oxygen species (ROS).[25] While glucose concentrations at oxygen tensions below 8% did not affect cell numbers, superoxic conditions (>21%) led to a 65% reduction in cell numbers. Glucose also induced formation of ROS, although this was independent of oxygen tension. It is yet to be investigated how glucose-induced ROS formation affects exosome secretion.

Using a well-characterized in vitro primary human cell culture model (first-trimester trophoblast cells),[17,26,27] we examined exosomal signaling across a range of oxygen tensions and extracellular glucose concentrations. The data obtained in this study confirm that high D-glucose concentrations and oxygen tension interact synergistically to regulate the release and bioactivity of exosomes derived from first-trimester trophoblast cells. Interaction of low oxygen tension (1%) and high glucose concentration (25 mM) elicited maximal exosome secretion. These results are notable in the clinical context particularly with regards to maternal insulin resistance and subsequent hyperglycemia, as well as preeclampsia (PE), which is underpinned by hypoxia and placental insufficiency.

Oxygen tension-dependent effects on placentation may also be affected by HIF signaling. A study by King *et al.* found that transfection of cells with HIF-1α siRNA led to inhibition of hypoxia-induced exosome secretion from breast cancer cell lines.[28] The effect of other components of exosomal content, such as miRNA, on the interaction between glucose and oxygen tension as mediators of placental exosome secretion requires further study. Previous research suggests that glucose can modify exosomal miRNA cargo in exosomes. Interestingly, improving the glycemic control of diabetic patients was able to partially reverse the dysregulation of exosomal miRNA.[29]

3. EFFECTS OF PLACENTAL EXOSOMES ON TARGET CELLS

Glucose enhances the specific bioactivity of exosomes released from first-trimester trophoblast cells. By assessing exosome-induced cytokine release from human umbilical vein endothelial cells (HUVEC), it was found that culturing first-trimester trophoblast cells under high glucose concentration was associated with a threefold increase in bioactivity of exosomes cultured under normoglycemic conditions.[30] This effect was dependent on exosome integrity and heat stability, and was observed across all tested oxygen tensions. Hence, these results suggest that exposure of trophoblast cells to high glucose concentrations shapes the concentration of bioactivity in and associated with exosomes.

Recent studies have highlighted that exosome internalization induces recipient cells to release of proinflammatory cytokines. Bretz *et al.* isolated from ascites, amniotic fluid, serum, and urine promote the release of cytokines (including IL-6 and TNF-α) from THP-1 cells. Furthermore, exosomal induction of cytokine was release was shown to be dependent on rapid activation of NFκb (and latent induction of STAT3) mediate by Toll-like receptor (TLR) signaling.[31]

Mouse models of obesity have demonstrated that adipose tissue-derived exosomes lead to differentiation of monocytes into activated macrophages. These exosomes have also been shown to promote insulin resistance in an obese mouse model.[32] Moreover, exosomes isolated from mouse insulinoma were found to induce the secretion of inflammatory cytokines, such as IL-6 and TNF-α, in splenocytes cultured from nonobese diabetic mice.[33] Using exosomes from microvascular endothelial cells (HMEC-1), it was additionally found that environmental stress (such as hypoxia, glucose, and TNF-α release) can alter the protein and mRNA content of exosomes.[34]

Interestingly, the chromosome 19 miRNA cluster (C19MC) is the biggest miRNA gene cluster and is mainly expressed in the human placenta, and these miRNAs can be transfer to nonplacental cells via exosomes.[35] The miRNA profile in placental cells (e.g., BeWo and JEG-3) showed that hypoxia did not affect the expression of C19MC miRNAs with the exception of the miR-520-3p. However, other study has shown that, in response to metabolic challenges (i.e., oxygen tension and glucose concentration), trophoblast cells specifically package C19MC miRNA into the exosomes they release.[36] The expression pattern of C19MC miRNAs between EVT and villous trophoblast cells (VTs) has also been established.[37] C19MC miRNAs lessen cell migration with miR-519d targeting specific proteins that could play a role in the invasive phenotype. As such, this would suggest that C19MC miRNAs can contribute to the phenotypic differences between the VTs and the more invasive EVT.

Further, it has been establish that placental exosomes play a role in viral infection during pregnancy. Viral infections can be harmful during pregnancy when maternal to fetal transmission could have serious pathological outcomes, such as fetal infection, growth restriction, birth defects, and/or fetal death. As trophoblasts form the protective barrier between the two interfaces, it is a critical physical and immunological barrier to help restrict the spread of pathogens into the fetal microenvironment. Interestingly, the expression of C19MC miRNAs in trophoblast cells protect them from viral infection.[38] Moreover, trophoblast cells can transfer the resistance capacity to viral infection to nonplacental cells via exosomes.[39]

4. CIRCULATING EXOSOMES ACROSS NORMAL GESTATION

The placental mass is an important factor in dictating placenta-derived exosomal concentration. This is illustrated by the finding that in normal pregnancies, during the third trimester, the concentration of placenta-derived exosomes is positively correlated with placental weight.[40] It has also been found that in gestational diabetes mellitus (GDM), placental mass markedly increases compared to normal pregnancies.[41] Interestingly, total exosome concentration was negatively correlated with placental weight at delivery at late gestation (32–33 weeks), while placenta-derived exosome concentration was positively correlated. Both placental and nonplacental originating exosomes are elevated in GDM pregnancies, although the potential role of these nanovesicles during GDM pregnancies is yet to be fully understood.

Several reports have illustrated the effect of placenta-derived exosomes on maternal immune modulation during pregnancy. Exosomes were shown to activate the NK cell receptor NKG2D[42] and express the proapoptotic molecules Fas and TRAIL.[43] The role of exosomes in cell signaling remains formative given the array of signaling molecules transported by these vesicles. Recently, dysregulation of exosomal miRNA in diabetes patients has been reported. Changes in the glycemic control were associated with exosomal miRNA profiling involving the regulation of adiponectin pathway.[29] Despite these interesting results, a commercial kit was used to isolate extracellular vesicles. Exosomes are a specific population of EVs and this study cannot distinguish between exosomes and other EVs.

Recently, we have completed a study analyzing the effect of maternal BMI on the exosomal concentration across gestation.[44] Using quantum dots coupled with CD63 or PLAP antibodies, we have established that the number of placental and nonplacental exosomes present in material circulation across gestation are different in lean, overweight, and obese women who did not develop pregnancy complications. Interestingly, the proportion of placental exosomes compared to nonplacental exosomes increases at mid-gestation (i.e., ~20 weeks) and this is maintained until end of pregnancy (Fig. 1).

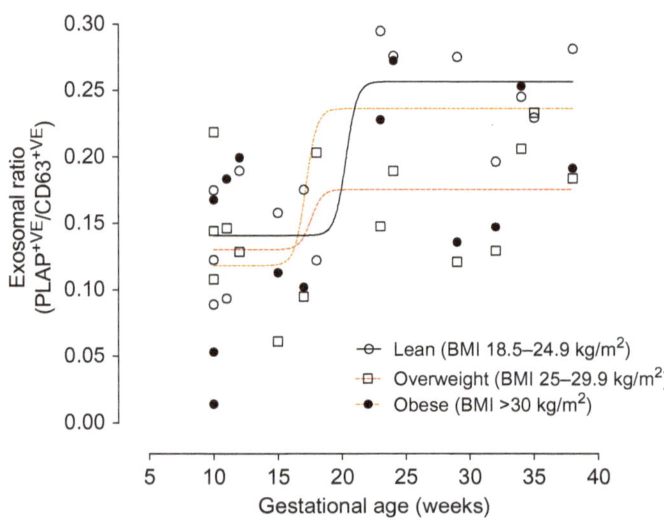

Fig. 1 Contribution of placental-derived exosomes in the maternal circulation across gestation—effect of maternal BMI. Enriched exosome populations were quantified using nanoparticle tracking analysis in fluorescence mode in peripheral plasma of lean, overweight (OW), and obese women across gestation. Nonlinear regression analysis of the ratio of placental exosomes ($PLAP^{+ve}$) and nonplacental exosomes ($CD63^{+ve}$ and $PLAP^{-ve}$). *Figure modified from Elfeky et al., Placenta, 2017,*[44] *http://dx.doi.org/10.1016/j.placenta.2016.12.020.*

5. EXOSOMAL PROFILE IN GDM PREGNANCIES

The release of diabetogenic autocoids from the placenta is a normal process during pregnancy, leading to the induction of insulin resistance and hyperinsulinemia.[45] It is when insulin release cannot compensate for this heightened insulin resistance that GDM develops. The progressive physiological changes that occur during pregnancy are crucial for the support and protection of the developing fetus. Moreover, these changes are essential for preparing for parturition.[46] In humans, these physiological changes in both the maternal physiology and the fetal development are regulated by the placenta. While pregnancy-associated insulin resistance has been attributed to the release of placental hormones, changes in the placental hormone profile do not directly correlate with changes in maternal insulin resistance.[45] Tissue-specific nanovesicles (i.e., exosomes) are a key area of research for the diagnosis of disease onset and treatment monitoring for GDM.[40,47]

As mentioned previously, exosomes are released in response to different oxygen tensions from primary placental (trophoblast) cells in vitro, the most abundant human placental cell type. In turn, the fusion of trophoblast cells gives rise to the syncytiotrophoblasts that sense and regulate oxygen and nutritional exchange at the maternofetal interface throughout gestation.[7,48] Exchange at this interface may be adversely affected by hypoxia, hyperinsulinemia, and hyperglycemia, which are significant risk factors for GDM. We conducted a longitudinal study to characterize changes in the exosomal profile, including concentration and bioactivity, in GDM pregnancies. We reported that there was an increase in the maternal plasma concentration of exosomes throughout both normal and GDM pregnancies; however, this effect was enhanced in GDM pregnancies that experienced a twofold increase in exosomal concentration. The results suggest that it may be possible to distinguish presymptomatic women who proceed to develop frank GDM (diagnosed between 24 and 28 weeks) earlier in pregnancy (11–14 weeks) using the plasma exosome profile. It was also observed that the contribution of placental exosomes to the total exosome concentration, as identified by using placental alkaline phosphatase (PLAP) as a placental marker, was comparatively lower in GDM pregnancies. Even though total exosome number was higher in GDM, the ratio of exosomal PLAP/total exosomes was lower in GDM compared to normal pregnancies. This may be indicative of alterations in placental exosome release or increased release of nonplacental-derived exosomes, although further study is required to determine which scenario is more likely.[49] Furthermore, the bioactivity of placenta-derived exosomes may exacerbate the proinflammatory state that is normally associated with pregnancy. In the early stages, the metabolic and immune status of the mother may alter the metabolism and function of the placenta. Accordingly, maternal inflammation, as identified by leucocyte number at the beginning of gestation, has been associated with increased risk of GDM.[50] This suggests that the inflammatory environment modulates the maternal glucose metabolism. This phenomenon may be associated with changes in the bioactivity of exosomes of both placental and nonplacental origin. The influence of proinflammatory cytokines on placental-derived exosome release has not been established. Recently, we established that first-trimester trophoblast cells increases proinflammatory cytokine release from endothelial cells in response to high glucose concentration.[30] Further study into the effect of placental exosomes on the release of exosomes from nonplacental sources is also required as this may be an important factor in the pathogenesis of GDM.

Hyperglycemia-induced oxidative stress makes an important contribution to the etiology of GDM, with consequences for both mother and baby.[51] In support of an etiological role of hypoglycemia and attendant oxidative stress in poor pregnancy outcome, the HAPO study reported a strong and continuous association between maternal glucose concentrations and pregnancy outcome; it confirmed a relationship between birth weight and maternal hyperglycemia.[52] Normal pregnancy is in a proinflammatory state associated with high concentrations of proinflammatory cytokines, a phenomenon that is even higher in GDM. Exosomes isolated from normal and GDM across gestation are bioactive with the capacity to be internalized by endothelial cells and increase the cytokine release.[49] Others studies have provided evidence that high glucose significantly increase the concentrations of TNF-α and IL-6 in the culture supernatants of HUVEC[53] and induced the adhesion of monocyte to endothelial cells.[54] Interestingly, release of the antiinflammatory cytokine IL-4 from endothelial cells was also increased in the presence of exosomes, suggesting a dual effect of exosomes in the regulation of the proinflammatory response. Moreover, imbalance between circulating pro- and antiinflammatory cytokines in patients with GDM has been previously reported.[55] While the mechanisms by which exosomes modulate cytokines released from endothelial cells remain to be elucidated, these nanovesicles with a payload of receptors, proteins, and/or oligonucleotides that have been specifically preconditioned by the GDM placenta to be delivered to maternal response systems. The extent and impact of placenta-derived exosomes on maternal physiology, however, remains to be elucidated.

6. EXOSOMAL PROFILE IN PE

PE is defined as the new onset of hypertension during the second half of pregnancy. PE and associated hypertensive disorders affect 10 million women worldwide and claims the lives of 76,000 women each year. In addition, 500,000 babies die due to PE each year.[56] Evidently PE and its related diseases is one of the most serious complications of pregnancy and is responsible for nearly 40% of premature births delivered before 35 weeks of gestation.[57]

Release of membranous material into the maternal circulation by the trophoblast cells is a feature of normal pregnancy and is increased in PE.[58] Recent research highlights the potential use of EVs in the diagnosis of PE.[59–61] Placental-derived exosomes (i.e., $PLAP^{+}$) are present in maternal circulation and their concentration increases during the course of a pregnancy and in association with other complications of pregnancy (e.g., GDM). The observation that oxygen tension regulates placental-derived exosomes release[17,27,62] is interesting in the context of PE. This is because PE is often underpinned by placental hypoxia leading to deficient spiral artery remodeling in the first 20 weeks of pregnancy.[63] These observations may indicate the possibility of identifying women at risk of PE during early pregnancy by assessing placental-derived exosomal profile. However, little is currently known about exosomes profile during PE pregnancies.

Various immunological and metabolic functions have been proposed to explain the role of exosomes during gestation, which is a relatively new area of research.[1,3] Through a payload of ribonucleic acids, proteins, and bioactive lipids, exosomes are essential in the adaptations of the maternal system to pregnancy-related physiological changes.[64] Among the protein content of exosomes is syncytin, which may contribute to the formation of the syncytiotrophoblast.[65,66] Interestingly, exosomal syncytin-2 concentration is significantly lower circulation in women with PE, compared to normal pregnancies compared to normal pregnancies.[67] The concentration of placenta-derived exosomes is higher in early onset-PE and late onset-PE compared to normal pregnancies matched by gestational age (i.e., third trimester).[68] Interestingly, a longitudinal study using samples at first trimester of pregnancies obtained from women who develop PE later during gestation has showed that the total number of exosomes and placental exosomes present in maternal circulation is higher in women who develop PE compared to normal.[69] These data illustrate the potential for placental exosomes in maternal circulation to be used as a biomarker for PE. Importantly, they provide an important source of information regarding the function and metabolic status of the placenta. Hence, placental exosomes may be a candidate for a noninvasive "biopsy" of the placental mass.

7. CONCLUSIONS AND PERSPECTIVES

Based on the available data, we propose that first-trimester trophoblast cells act as environmental sensors (Fig. 2). As such, these cells may respond to environmental flux, such as changes in the glucose and oxygen

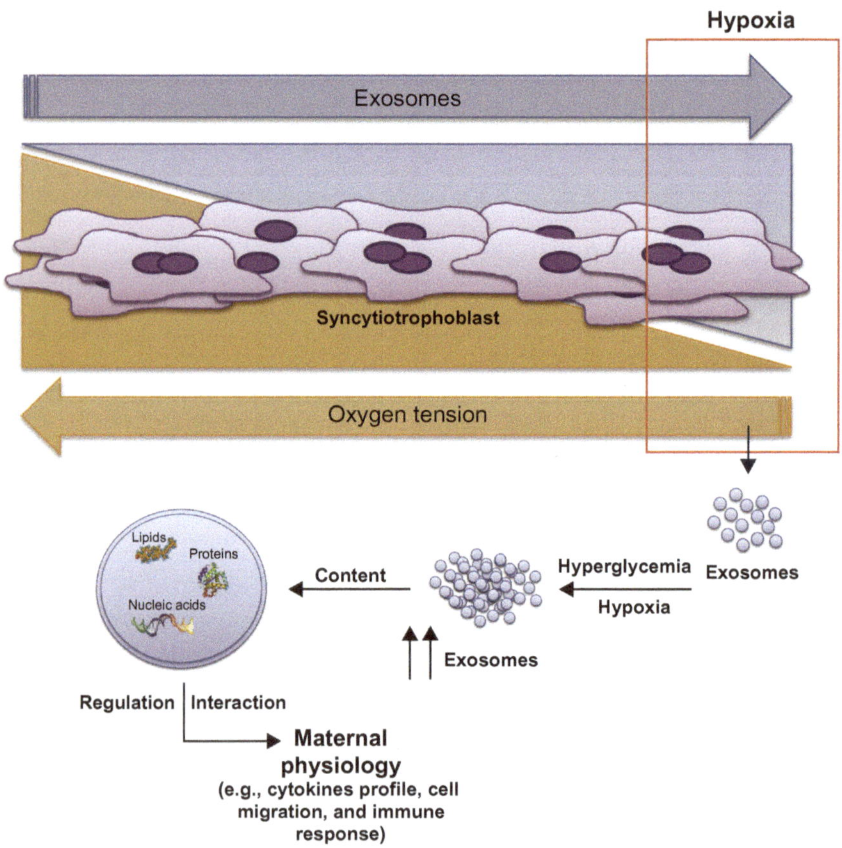

Fig. 2 Placental cells sense the maternal environment and change their exosomal bioactivity. Syncytiotrophoblast in response to changes in oxygen tension modifies the bioactivity of exosomes, thereby, regulating maternal phenotype. Exosomes released from syncytiotrophoblasts in response to low oxygen tension and/or hyperglycemia may alter maternal physiology via the process of exosomal placentomaternal transfection. This process involves a "payload" of proteins and/or miRNA that have been specifically preconditioned by the placenta to be delivered to and taken up by target cells.

concentrations, by synthesizing and releasing exosomes based on the particular environmental conditions. In turn, exosomes may go on to shape the phenotype of proximal cells (e.g., decidua and myometrial vascular cells), and distal cells by way of the maternal circulation.[70–71] Trophoblast-derived exosomes have multiple effects on target cells, including induction of cell migration, apoptosis, and proinflammatory cytokine release. Exosomal signaling, involving both the rate of release and bioactivity of exosomes may be altered in response to perturbations in homeostasis. A promising area for further investigation is defining the regulation of exosome internalization, and in particular, the role of the TLR family and subsequent activation of the NFKB signaling cascade. Increased exosome release from trophoblast cells in response to environmentally challenging conditions including elevated glucose concentrations and low oxygen tension may disrupt the T_h1/T_h2 T_h17/Teg-2 cytokine balance.[72] Notably, this balance is a requisite for normal implantation, placentation, and successful pregnancy outcome. The ability to identify aberrant trophoblast exosome release during early pregnancy (i.e., in maternal blood) and the identification of interventions to ameliorate their effect (e.g., inhibition of release or use of decoy exosomes) may be of clinical utility in successful management of such conditions.

REFERENCES

1. Mitchell MD, Peiris HN, Kobayashi M, et al. Placental exosomes in normal and complicated pregnancy. *Am J Obstet Gynecol.* 2015;213(4 suppl):S173–S181.
2. Tannetta D, Dragovic R, Alyahyaei Z, Southcombe J. Extracellular vesicles and reproduction-promotion of successful pregnancy. *Cell Mol Immunol.* 2014;11(6):548–563.
3. Mincheva-Nilsson L, Baranov V. Placenta-derived exosomes and syncytiotrophoblast microparticles and their role in human reproduction: immune modulation for pregnancy success. *Am J Reprod Immunol.* 2014;72(5):440–457.
4. Pegtel DM, van de Garde MD, Middeldorp JM. Viral miRNAs exploiting the endosomal-exosomal pathway for intercellular cross-talk and immune evasion. *Biochim Biophys Acta.* 2011;1809(11–12):715–721.
5. Kobayashi M, Salomon C, Tapia J, Illanes SE, Mitchell MD, Rice GE. Ovarian cancer cell invasiveness is associated with discordant exosomal sequestration of Let-7 miRNA and miR-200. *J Transl Med.* 2014;12:4.
6. Gude NM, Roberts CT, Kalionis B, King RG. Growth and function of the normal human placenta. *Thromb Res.* 2004;114(5–6):397–407.
7. Cartwright JE, Fraser R, Leslie K, Wallace AE, James JL. Remodelling at the maternal-fetal interface: relevance to human pregnancy disorders. *Reproduction.* 2010;140(6):803–813.
8. Burton GJ, Jauniaux E, Charnock-Jones DS. The influence of the intrauterine environment on human placental development. *Int J Dev Biol.* 2010;54(2–3):303–312.
9. Rodesch F, Simon P, Donner C, Jauniaux E. Oxygen measurements in endometrial and trophoblastic tissues during early pregnancy. *Obstet Gynecol.* 1992;80(2):283–285.

10. Jauniaux E, Gulbis B, Burton GJ. Physiological implications of the materno-fetal oxygen gradient in human early pregnancy. *Reprod Biomed Online*. 2003;7(2):250–253.
11. Kaufmann P, Black S, Huppertz B. Endovascular trophoblast invasion: implications for the pathogenesis of intrauterine growth retardation and preeclampsia. *Biol Reprod*. 2003;69(1):1–7.
12. Lyall F. Mechanisms regulating cytotrophoblast invasion in normal pregnancy and pre-eclampsia. *Aust N Z J Obstet Gynaecol*. 2006;46(4):266–273.
13. Simpson RJ, Jensen SS, Lim JW. Proteomic profiling of exosomes: current perspectives. *Proteomics*. 2008;8(19):4083–4099.
14. Mincheva-Nilsson L, Baranov V. The role of placental exosomes in reproduction. *Am J Reprod Immunol*. 2010;63(6):520–533.
15. Redman CW, Sargent IL. Circulating microparticles in normal pregnancy and pre-eclampsia. *Placenta*. 2008;29(suppl A):S73–S77.
16. Hedlund M, Nagaeva O, Kargl D, Baranov V, Mincheva-Nilsson L. Thermal- and oxidative stress causes enhanced release of NKG2D ligand-bearing immunosuppressive exosomes in leukemia/lymphoma T and B cells. *PLoS One*. 2011;6(2):e16899.
17. Salomon C, Ryan J, Sobrevia L, et al. Exosomal signaling during hypoxia mediates microvascular endothelial cell migration and vasculogenesis. *PLoS One*. 2013;8(7): e68451.
18. Salomon C, Kobayashi M, Ashman K, Sobrevia L, Mitchell MD, Rice GE. Hypoxia-induced changes in the bioactivity of cytotrophoblast-derived exosomes. *PLoS One*. 2013;8(11):e79636. http://dx.doi.org/10.1371/journal.pone.0079636. eCollection 2013.
19. Rice GE, Scholz-Romero K, Sweeney E, et al. The effect of glucose on the release and bioactivity of exosomes from first trimester trophoblast cells. *J Clin Endocrinol Metabol*. 2015;100(10):E1280–E1288.
20. Jauniaux E, Hempstock J, Greenwold N, Burton GJ. Trophoblastic oxidative stress in relation to temporal and regional differences in maternal placental blood flow in normal and abnormal early pregnancies. *Am J Pathol*. 2003;162(1):115–125.
21. Jauniaux E, Van Oppenraaij RH, Burton GJ. Obstetric outcome after early placental complications. *Curr Opin Obstet Gynecol*. 2010;22(6):452–457.
22. Hulme CH, Westwood M, Myers JE, Heazell AE. A high-throughput colorimetric-assay for monitoring glucose consumption by cultured trophoblast cells and placental tissue. *Placenta*. 2012;33(11):949–951.
23. Aalberts M, van Dissel-Emiliani FMF, van Adrichem NPH, et al. Identification of distinct populations of proteasomes that differentially express prostate stem cell antigen, annexin A1, and GLIPR2 in humans. *Biol Reprod*. 2012;86(3):82. http://dx.doi.org/10.1095/biolreprod.111.095760.
24. Hiden U, Wadsack C, Prutsch N, et al. The first trimester human trophoblast cell line ACH-3P: a novel tool to study autocrine/paracrine regulatory loops of human trophoblast subpopulations—TNF-alpha stimulates MMP15 expression. *BMC Dev Biol*. 2007;7:137.
25. Frohlich JD, Huppertz B, Abuja PM, Konig J, Desoye G. Oxygen modulates the response of first-trimester trophoblasts to hyperglycemia. *Am J Pathol*. 2012;180(1): 153–164.
26. Peiris HN, Salomon C, Payton D, et al. Myostatin is localized in extravillous trophoblast and up-regulates migration. *J Clin Endocrinol Metab*. 2014;99(11):E2288–E2297.
27. Salomon C, Kobayashi M, Ashman K, Sobrevia L, Mitchell MD, Rice GE. Hypoxia-induced changes in the bioactivity of cytotrophoblast-derived exosomes. *PLoS One*. 2013;8(11):e79636.
28. King HW, Michael MZ, Gleadle JM. Hypoxic enhancement of exosome release by breast cancer cells. *BMC Cancer*. 2012;12:421.

29. Santovito D, De Nardis V, Marcantonio P, et al. Plasma exosome microRNA profiling unravels a new potential modulator of adiponectin pathway in diabetes: effect of glycemic control. *J Clin Endocrinol Metab*. 2014;99(9):E1681–E1685.
30. Rice GE, Scholz-Romero K, Sweeney E, et al. The effect of glucose on the release and bioactivity of exosomes from first trimester trophoblast cells. *J Clin Endocrinol Metab*. 2015;100(10):E1280–E1288.
31. Bretz NP, Ridinger J, Rupp AK, et al. Body fluid exosomes promote secretion of inflammatory cytokines in monocytic cells via toll-like receptor signaling. *J Biol Chem*. 2013;288(51):36691–36702.
32. Deng ZB, Poliakov A, Hardy RW, et al. Adipose tissue exosome-like vesicles mediate activation of macrophage-induced insulin resistance. *Diabetes*. 2009;58(11):2498–2505.
33. Sheng H, Hassanali S, Nugent C, et al. Insulinoma-released exosomes or microparticles are immunostimulatory and can activate autoreactive T cells spontaneously developed in nonobese diabetic mice. *J Immunol*. 2011;187(4):1591–1600.
34. de Jong OG, Verhaar MC, Chen Y, et al. Cellular stress conditions are reflected in the protein and RNA content of endothelial cell-derived exosomes. *J Extracell Vesicles*. 2012:1. http://dx.doi.org/10.3402/jev.v1i0.18396. eCollection 2012.
35. Donker RB, Mouillet JF, Chu T, et al. The expression profile of C19MC microRNAs in primary human trophoblast cells and exosomes. *Mol Hum Reprod*. 2012;18(8): 417–424.
36. Almohammadi D, Scholz-Romero K, Duncombe G, Rice GE, Salomon C. Oxygen, glucose and insulin modulate the expression of miRNAs from chromosome 19 cluster in trophoblast cells and exosomes. *Reprod Sci*. 2016;23:156a.
37. Xie L, Mouillet JF, Chu T, et al. C19MC microRNAs regulate the migration of human trophoblasts. *Endocrinology*. 2014;155(12):4975–4985.
38. Delorme-Axford E, Bayer A, Sadovsky Y, Coyne CB. Autophagy as a mechanism of antiviral defense at the maternal-fetal interface. *Autophagy*. 2013;9(12):2173–2174.
39. Delorme-Axford E, Donker RB, Mouillet JF, et al. Human placental trophoblasts confer viral resistance to recipient cells. *Proc Natl Acad Sci USA*. 2013;110(29): 12048–12053.
40. Salomon C, Torres MJ, Kobayashi M, et al. A gestational profile of placental exosomes in maternal plasma and their effects on endothelial cell migration. *PLoS One*. 2014;9(6): e98667. http://dx.doi.org/10.1371/journal.pone.0098667.
41. Pala HG, Artunc-Ulkumen B, Koyuncu FM, Bulbul-Baytur Y. Three-dimensional ultrasonographic placental volume in gestational diabetes mellitus. *J Matern Fetal Neonatal Med*. 2016;29(4):610–614. http://dx.doi.org/10.3109/14767058.2015.1012066.
42. Mincheva-Nilsson L, Nagaeva O, Chen T, et al. Placenta-derived soluble MHC class I, chain-related molecules down-regulate NKG2D receptor on peripheral blood mononuclear cells during human pregnancy: A possible novel immune escape mechanism for fetal survival. *J Immunol*. 2006;176(6):3585–3592.
43. Taylor DD, Akyol S, Gercel-Taylor C. Pregnancy-associated exosomes and their modulation of T cell signaling. *J Immunol*. 2006;176(3):1534–1542.
44. Elfeky O, Longo S, Lai A, Rice GE, Salomon C. Influence of maternal BMI on the exosomal profile during gestation and their role on maternal systemic inflammation. *Placenta*. 2017;50:60–69. http://dx.doi.org/10.1016/j.placenta. 2016.12.020.
45. Barbour LA, McCurdy CE, Hernandez TL, Kirwan JP, Catalano PM, Friedman JE. Cellular mechanisms for insulin resistance in normal pregnancy and gestational diabetes. *Diabetes Care*. 2007;30(suppl 2):S112–S119.
46. Hoile SP, Lillycrop KA, Thomas NA, Hanson MA, Burdge GC. Dietary protein restriction during F_0 pregnancy in rats induces transgenerational changes in the hepatic transcriptome in female offspring. *PLoS One*. 2011;6(7)e21668.

47. Sarker S, Scholz-Romero K, Perez A, et al. Placenta-derived exosomes continuously increase in maternal circulation over the first trimester of pregnancy. *J Transl Med.* 2014;12:204.
48. Costa SL, Proctor L, Dodd JM, et al. Screening for placental insufficiency in high-risk pregnancies: is earlier better? *Placenta.* 2008;29(12):1034–1040.
49. Salomon C, Scholz-Romero K, Sarker S, et al. Gestational diabetes mellitus is associated with changes in the concentration and bioactivity of placenta-derived exosomes in maternal circulation across gestation. *Diabetes.* 2015;65(3):598–609. http://dx.doi.org/10.2337/db15-0966.
50. Wolf M, Sauk J, Shah A, et al. Inflammation and glucose intolerance: a prospective study of gestational diabetes mellitus. *Diabetes Care.* 2004;27(1):21–27.
51. Lappas M, Hiden U, Desoye G, Froehlich J, Hauguel-de Mouzon S, Jawerbaum A. The role of oxidative stress in the pathophysiology of gestational diabetes mellitus. *Antioxid Redox Signal.* 2011;15(12):3061–3100.
52. HSCR Group, Metzger BE, Lowe LP, et al. Hyperglycemia and adverse pregnancy outcomes. *N Engl J Med.* 2008;358(19):1991–2002.
53. Chen YY, Chen J, Hu JW, Yang ZL, Shen YL. Enhancement of lipopolysaccharide-induced toll-like receptor 2 expression and inflammatory cytokine secretion in HUVECs under high glucose conditions. *Life Sci.* 2013;92(10):582–588.
54. Shanmugam N, Reddy MA, Guha M, Natarajan R. High glucose-induced expression of proinflammatory cytokine and chemokine genes in monocytic cells. *Diabetes.* 2003;52(5):1256–1264.
55. Kuzmicki M, Telejko B, Zonenberg A, et al. Circulating pro- and anti-inflammatory cytokines in Polish women with gestational diabetes. *Horm Metab Res.* 2008;40(8):556–560.
56. Kuklina EV, Ayala C, Callaghan WM. Hypertensive disorders and severe obstetric morbidity in the United States. *Obstet Gynecol.* 2009;113(6):1299–1306.
57. Ananth CV, Keyes KM, Wapner RJ. Pre-eclampsia rates in the United States, 1980–2010: age-period-cohort analysis. *BMJ.* 2013;347:f6564. http://dx.doi.org/10.1136/bmj.f6564.
58. Burton GJ, Yung HW, Cindrova-Davies T, Charnock-Jones DS. Placental endoplasmic reticulum stress and oxidative stress in the pathophysiology of unexplained intrauterine growth restriction and early onset preeclampsia. *Placenta.* 2009;30(suppl A):S43–S48.
59. Xiao DY, Ohlendorf J, Chen YL, et al. Identifying mRNA, MicroRNA and protein profiles of melanoma exosomes. *PLoS One.* 2012;7(10):e46874. http://dx.doi.org/10.1371/journal.pone.0046874.
60. Rabinowits G, Gercel-Taylor C, Day JM, Taylor DD, Kloecker GH. Exosomal MicroRNA: a diagnostic marker for lung cancer. *Clin Lung Cancer.* 2009;10(1):42–46.
61. Tannetta DS, Dragovic RA, Gardiner C, Redman CW, Sargent IL. Characterisation of syncytiotrophoblast vesicles in normal pregnancy and pre-eclampsia: expression of Flt-1 and endoglin. *PLoS One.* 2013;8(2):e56754.
62. Salomon C, Scholz-Romero K, Kobayashi M, et al. Oxygen tension regulates glucose-induced biogenesis and release of different subpopulations of exosome vesicles from trophoblast cells: a gestational age profile of placental exosomes in maternal plasma with gestational diabetes mellitus. *Placenta.* 2015;36(4):488.
63. Cartwright JE, Keogh RJ, Tissot van Patot MC. Hypoxia and placental remodelling. *Adv Exp Med Biol.* 2007;618:113–126.
64. Mincheva-Nilsson L, Baranov V. The role of placental exosomes in reproduction. *Am J Reprod Immunol.* 2010;63(6):520–533.
65. Record M. Intercellular communication by exosomes in placenta: a possible role in cell fusion? *Placenta.* 2014;35(5):297–302.

66. Vargas A, Zhou S, Et´hier-Chiasson M, et al. Syncytin proteins incorporated in placenta exosomes are important for cell uptake and show variation in abundance in serum exosomes from patients with preeclampsia. *FASEB J.* 2014;28(8):3703–3719.
67. Vargas A, Zhou S, Ethier-Chiasson M, et al. Syncytin proteins incorporated in placenta exosomes are important for cell uptake and show variation in abundance in serum exosomes from patients with preeclampsia. *FASEB J.* 2014;28(8):3703–3719.
68. Pillay P, Maharaj N, Moodley J, Mackraj I. Placental exosomes and pre-eclampsia: maternal circulating levels in normal pregnancies, and early and late onset pre-eclamptic pregnancies. *Placenta.* 2016;46:18–25.
69. Salomon C, Yee S, Sarker S, et al. Placenta-derived exosomes promote trophoblast invasion and spiral arterial remodeling—a possible role in the physiopathology of preeclampsia. *Reprod Sci.* 2015;22:290a.
70. Salomon C, Torres MJ, Kobayashi M, et al. A gestational profile of placental exosomes in maternal plasma and their effects on endothelial cell migration. *PLoS One.* 2014;9(6): e98667.
71. Sarker S, Scholz-Romero K, Perez A, et al. Placenta-derived exosomes continuously increase in maternal circulation over the first trimester of pregnancy. *J Transl Med.* 2014;12(1):204.
72. Toldi G, Molvarec A, Stenczer B, et al. Peripheral T(h)1/T(h)2/T(h)17/regulatory T-cell balance in asthmatic pregnancy. *Int Immunol.* 2011;23(11):669–677.

CHAPTER SEVEN

Novel Regulators of Hemodynamics in the Pregnant Uterus

N.C. Clark, C.A. Pru, J.K. Pru[1]
Center for Reproductive Biology, Washington State University, Pullman, WA, United States
[1]Corresponding author e-mail address: jpru@wsu.edu

Contents

Abstract

The uterus is a highly dynamic organ, undergoing dramatic physiological changes during normal cyclicity and pregnancy. Many of these changes involve remodeling of the uterine vasculature in order to provide oxygen and nutrients to the developing embryo/fetus. Vasculogenesis, angiogenesis, vasodilation/vasoconstriction, and vascular permeability are coordinated by a vast network of autocrine, paracrine, and endocrine-signaling factors that derive from a number of cellular sources at the maternal:fetal interface, as well as from tissue outside the uterus. In this chapter, the dynamic changes that occur in uterine vasculature during pregnancy are described, and some of the

Progress in Molecular Biology and Translational Science, Volume 145
ISSN 1877-1173
http://dx.doi.org/10.1016/bs.pmbts.2016.12.007

hemodynamic regulatory factors are reviewed. These include uterine natural killer cells, sex steroid hormones, the calcitonin gene-related peptide family, angiopoietins, sphingolipids, and the renin–angiotensin system. Aberrancies in these factors are associated with disorders of uterine vascular remodeling, leading to conditions such as early pregnancy loss, preeclampsia, uterine hemorrhage, and intrauterine growth restriction. In addition, we introduce the role of the mas-related gene family in angiotensin signaling and endothelial function during pregnancy. Finally, this chapter introduces the novel concept that in addition to remodeling the vasculature to bring oxygenated maternal blood to the embryo, the gravid uterus synthesizes its own hemoglobin. Overall, this chapter provides an overview of the regulators of uterine vascular remodeling and hemodynamics during pregnancy and pregnancy-associated pathologies.

1. INTRODUCTION

Properly regulated uterine vascular remodeling and expansion are essential for successful pregnancy. Unless associated with a pathophysiological condition, the vascular system of the pregnancy female remains in relative stasis outside of subtle changes in vasodilation and vasoconstriction. However, this is not the case in the uterus. Changes in the endometrial vasculature during the menstrual/estrous cycle are restricted primarily to the small arterioles and capillaries (i.e., microvasculature). In contrast, pregnancy initiates dramatic changes in the uterine vasculature with physiological implications for both the mother and the developing offspring. As such, uterine hemodynamic events that occur during pregnancy mark a unique physiological occurrence that is seen nowhere to this extent outside of prenatal and early postnatal development in mammals.

Abnormalities in uterine vascular remodeling have adverse outcomes for both the mother and offspring during the pregnancy, often resulting in early pregnancy loss, preeclampsia, preterm delivery, and/or fetal growth restriction. Such abnormalities that occur during pregnancy negatively impact the health and quality of life of both the mother and offspring following parturition. Preeclamptic pregnancies are commonly associated with later cardiovascular complications in women long after childbirth. A history of preeclampsia is associated with a 2-fold increased risk for developing cardiovascular complications later in life, as well as a 5- to 12-fold increased risk of end-stage renal disease.[1] Furthermore, it is now well established that adult metabolic and cardiovascular diseases, among others, can be traced back to disruption in the uteroplacental vascular network during prenatal life, a phenomenon now commonly referred to as the "Barker Hypothesis."[2]

The uterine vascular remodeling that accompanies pregnancy is a subset of what occurs systemically in the female as she adapts to the nutrient and gas demands of the fetus. Vascular resistance drops systemically in pregnant women with a concomitant increase in cardiac output, blood volume, ventilation, and basal metabolic rate. Coincident with these systemic changes is the dramatic decline in local uteroplacental vascular resistance that accommodates an estimated 20% of the total maternal cardiac output by the end of gestation.[3] While much of the research focus has been placed on remodeling of uterine arteries, particularly the spiral arteries, uterine veins also become enlarged during pregnancy in humans and rodents.[4–6] These uterine vascular changes during pregnancy are coordinated by a combination of maternal and fetal sex steroids, the invading trophectoderm, uterine natural killer cells (uNKs), endocrine products, and local paracrine factors. The combination of these cues alters arterial tone, diameter, and length. The objective of this review is to provide a general overview of the hemodynamic events that occur in the gravid uterus, which result in increased uteroplacental blood flow required for fetal expansion, and to discuss some of the cues that initiate these uterine vascular events.

2. AN OVERVIEW OF VASCULAR REMODELING DURING PREGNANCY

2.1 Uterine Vascular Anatomy

The uterus is comprised of three tissue layers that include the outer perimetrium, smooth muscle of the myometrium, and inner endometrium. The endometrium of most mammals including humans can be regionally partitioned into the deeper *basalis* and the more superficial (i.e., luminal) *functionalis*. The endometrium is composed of endothelial, smooth muscle, immune, epithelial, and fibroblast-like stromal cells. Unlike most vascular networks that enter an organ from a single arterial source, the uterus receives blood from both the superiorly positioned ovarian artery originating from the descending aorta and the more caudal uterine artery, which branches from the iliac artery in most mammalian species.[7] Because both the ovarian and uterine arteries conjoin, blood enters the uterus through two different routes. This evolutionary adaptation in most mammalian species ensures that the implanted embryo or fetus has an uncompromised flow of blood perfusing the uterus. This is particularly important in litter-bearing species. The amalgamated utero-ovarian artery branches into arcuate arteries that encircle the uterus. Radial arteries then branch from the arcuate artery and penetrate deeply into the myometrium. At the myometrial–endometrial border, the radial arteries bifurcate into basal arteries that feed the endometrial *basalis* and spiral arteries that extend toward the uterine lumen. In the *basalis*, the arteries are composed of endothelial cells surrounded by vascular smooth muscle cells (VSMCs),[8] but as the arteries extend toward the luminal surface the vessel walls become thin and are composed exclusively of endothelial cells.[9] The vessels form a subepithelial capillary plexus that is then coupled to a venous drainage system that juxtaposes the arterial network.[7]

2.2 Vascular Changes During the Menstrual Cycle

Vasculogenesis is the formation of new blood vessels by differentiation of hemangiogenic progenitor cells into endothelial cells.[10] Angiogenesis is the development of microvessels from preexisting vessels, and this includes branching processes such as sprouting angiogenesis or splitting angiogenesis (i.e., intussusception), as well as nonbranching processes like vessel elongation and the incorporation of circulating endothelial precursor cells into existing vessels. While vasculogenesis and angiogenesis generally do not occur in the adult except under pathological and wound healing conditions, these processes occur under physiological conditions in the adult uterus.[10] The human uterus is a dynamic organ, undergoing a monthly cycle governed by the ovarian sex steroid hormones estrogen and progesterone. The first 14 days of the menstrual cycle is the proliferative phase, during which estrogen drives the build-up of the *functionalis* by proliferation of the epithelial cells, angiogenesis, and edema. Following ovulation, the next 14 days of the cycle is termed the secretory phase, during which progesterone coordinates decidualization of the stromal cells into epithelioid cells and spiraling of the uterine arties.[7] If pregnancy occurs, the embryo implants and further endometrial vascular changes take place that precede the development of the placenta. If pregnancy does not occur, the *functionalis* is shed and the *basalis* regenerates the endometrium in preparation for an ensuing cycle.

Vascular changes occur in three distinct phases of the normal menstrual cycle. During the proliferative phase, angiogenesis via vessel elongation and intussusception occurs in the *functionalis* as this layer expands in response to estrogen.[10] Proliferation of endothelial and VSMCs occurs in the *functionalis*, but not the *basalis*.[11,12] During the secretory phase, the subepithelial capillary plexus matures and more peripherally located spiral arteries grow by lengthening and coiling in response to progesterone. Finally, following shedding of the *functionalis* during menstruation, angiogenesis occurs in the *basalis* as the *functionalis* is reconstituted.[10] Overall, most of the vascular changes that occur in the nongravid uterus take place at the level of the microvasculature.

2.3 Vascular Changes During Early Pregnancy: Implantation and Decidualization

Hemochorial implanting species undergo an essential process of decidualization, in which uterine stromal cells differentiate into epithelioid cells. Decidualization, which occurs spontaneously in humans in response to progesterone during the secretory phase, is an essential step in the establishment and maintenance of pregnancy.[13] Decidualized stromal cells serve several functions including: (1) providing nutrients for the embryo until the placenta develops; (2) regulating trophoblast expansion; (3) modulating the maternal immune system to provide an immune privileged environment for the semi-allogenic embryo; (4) sensing or distinguishing between high and low quality embryos; and (5) regulating embryonic access to maternal vasculature under tight spatial and temporal control.[14,15] Women with recurrent pregnancy loss show a severely disrupted decidualization program that leads to increased expression of proinflammatory cytokines.[16] Decidual cells from women with recurrent pregnancy loss also lose their sensing capacity and fail to distinguish between high and low quality embryos.[17,18] As an embryo implants, the uterine vascular network changes in order to provide adequate nutrient supply for the embryo. These changes include vasodilation, increased vessel permeability, and vasculogenesis and angiogenesis.[19]

Compared to the secretory endometrium, decidual tissue has fewer but larger blood vessels. The larger vessel diameter facilitates greater blood supply demanded by the embryo and fetus.[20] In addition, the spiral arteries undergo changes from high-resistance vessels to a low-resistance vasodilated vascular network, once again providing the increased blood supply needed during pregnancy.[21] Remodeling of the spiral arteries occurs through a coordinated network of different cells and signaling factors. For instance, leukocytes including uNKs and macrophages secrete factors to disrupt the VSMCs underlying the endothelial cells of the spiral arteries.[22] Embryonic trophoblast cells begin invading into the uterus during implantation and develop into two layers: cytotrophoblasts (mononuclear) and syncytiotrophoblasts (multinucleated).[23] Extravillous trophoblast cells, which invade through the uterine interstitium toward the decidua and myometrium, begin to penetrate and replace the endothelial and VSMCs of the spiral arteries.[22] As the cells penetrate the endothelium, they accumulate in the vessel lumen, plugging the spiral arteries.[24] These plugs block maternal blood flow toward the embryonic tissues during the first trimester, generating a lower oxygen tension environment to presumably protect the developing embryo from oxidative stress. These plugs disintegrate toward the end of the first trimester, allowing contact between maternal and fetal blood within a functioning placenta.[25]

2.4 Vascular Changes During Placentation

Humans develop a discoid, hemochorial placenta, by classification the same as that developed in most rodents. Hemochorial placentation involves angiogenesis to form new blood vessels, dilation of these vessels, and increased permeability such that maternal blood enters into embryonic trophoblast cell-lined sinuses.[26] Both maternal and embryonic factors regulate these maternal vascular changes. However, there are key differences between human and rodent placentation. Of note, decidualization begins prior to embryo implantation in humans, and consequently the placenta develops earlier during gestation than it does in rodents. The main structural unit of the human placenta is the chorionic villus, which is apparent by day 21 after ovulation,[27] whereas rodents have a labyrinth structure that is not functional until halfway through gestation.[26,28]

While rodent and primate[29–31] models can offer some insight into human placentation, generally the uniqueness of the human placenta demands that definitive studies of vascular networks in the placenta must be done in humans. The evaluation of placental angiogenesis in women has utilized three methodologies: (1) Doppler sonography during pregnancy to evaluate blood flow;[32] (2) perfusion casting of delivered placentas followed by scanning electron microscopy to evaluate blood vessel branching;[33,34] and (3) histological evaluation of hysterectomy specimens obtained during gestation, which, while rare, have offered valuable insight into early vascular changes during gestation.[35] Such studies have revealed that the maternal (decidual) and fetal (placental) vascular systems connect late in the first trimester, establishing the functional placenta. As mentioned earlier, trophoblast cells invade the maternal blood vessels, forming maternal blood-filled sinuses; this establishes the architecture necessary for the development of the chorionic villi, through which fetal blood circulates. By 4 weeks gestation, all features of the mature placenta are present. The mature human placenta is comprised of fetal circulation that ends in capillary loops in the chorionic villi, which penetrate the maternal blood-filled space that is supplied by the spiral arteries and drained by uterine veins.[23]

Multiple changes in vascular structure occur in order to develop the uteroplacental unit. Vasculogenesis and branching angiogenesis contribute to the development of the placenta during the first trimester; branching angiogenesis contributes during the second; and finally nonbranching angiogenesis occurs during the third.[20] The vasculogenesis in the placenta begins with hemangioblastic cell cords, which begin to form endothelial tubes by 21 days gestation. However, these cords have no connection to the embryonic circulation until 4 weeks gestation, at which point the villous capillaries fuse with each other and with the allantoic vessels. Branching angiogenesis enables these capillaries to form networks during the remainder of the first trimester. During the second trimester, some of the peripheral capillaries regress, and central stem vessels are formed. Stem villi, which make up the mature form of the placenta, are characterized by a fibrosed stromal core. Finally, during the third trimester, nonbranching angiogenesis lengthens and expands the tips of capillaries, forming loops.[36]

2.5 Vascular Changes During Parturition and Involution

Many vascular changes occur during the 6-week puerperal period during which the uterus changes from carrying a full-term fetus and fully developed placenta to being at prepregnancy size and functionality. Prior to and during labor, the blood vessels in the uterus are filled with fibrin and platelets, endothelial cells, amniotic cells, and mucus.[37] During birth, the fetus is delivered, and the placenta separates from the wall of the uterus. After birth, the uterus weighs approximately 1000 g and must return to its prepregnancy state of 50–100 g.[38] Part of this involution process includes the discharge of fluid, debris, and red blood cells from the uterus. This discharge, termed lochia, is present for several weeks.[39] In addition, the remaining deciduum is eliminated by necrosis and neutrophil-mediated phagocytosis.[40]

The uterus undergoes changes in vasculature during involution. The walls of the endometrial arteries thicken within 5 h postdelivery but return to normal within 6 weeks. The walls of the serosal veins undergo fibrous hardening, which persists beyond 6 weeks.[41] During the first 5 days following parturition, there is a steady increase in vascular resistance. This increase in vascular resistance does not occur in women whose uterus fails to return to normal size after parturition,[42] indicating that the increase in vascular resistance is a critical component of converting the uterus from a condition of large vascular sinuses and permeability back to its prepregnancy condition of smaller diameter, thick-walled, high-resistance spiral arteries lined with

VSMCs. Within a few weeks, the vascular channels become less prominent by Doppler sonography.[40] Involution involves not only a reduction of uterine size and removal of the deciduum, placental bed, and pregnancy-related vascular changes, but also regeneration of the normal uterine lining. After the superficial (*functionalis*) layer of the uterus is shed in the lochia, the *basalis* contains residual endometrial glands and regenerates the endometrium as during normal menstrual cycling. This regeneration occurs within 16 days postparturition.[43] The placental site takes longer to regenerate, as new endometrium must be generated from the sides of the site as well as glands that remained in the lower portion of the decidua.[44] This regeneration occurs within 6 weeks postparturition.[43]

3. REGULATORS OF UTERINE HEMODYNAMICS

There are many factors involved in regulating the complex changes that occur in uterine vasculature during pregnancy. Many are beyond the scope of this review, and we refer the reader to other reviews describing the actions of vascular endothelial growth factor (VEGF);[45,46] bradykinin and kallikrein;[46,47] nitric oxide (NO);[46,48] growth factors including IGF, PGF, and FGF;[49–52] matrix metalloproteinases;[6] human chorionic gonadotropin (hCG);[19] interleukin-6 and other decidual cytokines;[53,54] relaxin;[55,56] and the Wnt pathway.[57]

3.1 Uterine Natural Killer Cells

During early pregnancy, the spiral arteries undergo extensive remodeling. This remodeling is driven by extravillous trophoblast cells as described earlier, as well as by uNKs.[58] uNKs are evolutionarily conserved, having been shown to accumulate in the uterus during pregnancy in species including rodent, pig, and human.[59] uNK cells are recruited to the uterus via local chemokine actions. For instance, trophoblasts secrete CXCL12 and endometrial cells secrete CXCL10 and CXCL11 to attract uNK cells. uNK cells bind these chemokines with receptors including CXCR3 and CXCR4.[60] Recruitment of uNK cells also involves the chemokine CXCL14, as evidenced by the fact that *Cxcl14*$^{-/-}$ mice have significantly decreased uNK cells in the uterus during pregnancy.[61]

uNK cells can cause trophoblast-independent spiral artery remodeling and facilitate the invasion of extravillous trophoblasts. uNK cells play a role in spiral artery remodeling by secreting factors such as interferon-gamma (IFNG).[62] IFNG finely regulates the migration of extravillous trophoblasts.[63] IFNG also regulates gene expression in human uterine microvasculature endothelial cells, inducing genes involved in angiogenesis and uNK cell recruitment.[64] In addition, estrogen induces uNK cells to produce the chemokine CCL2,[65] which is another known angiogenic factor.[66] In addition to spiral artery remodeling, uNK cells induce angiogenesis and increase microvasculature density in part via expression of VEGFA.[67]

While uNK cells are critical for remodeling spiral arteries and inducing angiogenesis during early pregnancy, these cells must be closely regulated to prevent embryonic death. Increased uNK cell numbers are associated with recurrent reproductive failure, possibly due to early onset of maternal blood flow leading to fetal oxidative stress.[68] Similarly, uNK cell numbers are dysregulated during the menstrual cycle of women with heavy menstrual bleeding.[69]

3.2 Sex Steroids

The sex steroids estradiol (E2) and progesterone (P4) play a pivotal role in the establishment and maintenance of pregnancy by modulating uterine blood flow following remodeling of uterine arteries, as well as placental angiogenesis and vasculogenesis. These events ensure that an appropriate exchange apparatus is established to optimize the transfer of oxygen, nutrients, and waste between maternal and fetal circulation. E2 and P4 mainly function by inducing the expression of growth factors and other paracrine-signaling molecules. P4 is critical for uterine vasodilation during the first trimester.[70] P4 enhances nitric oxide (NO) signaling, prostaglandin production, and calcium signaling, leading to vasodilation.[71] P4 is also the hormone that drives decidualization of the stromal compartment. Decidualized stromal cells produce factors that regulate vascular tone during trophoblast invasion. Aberrant levels of or response to P4 during decidualization can lead to pregnancy loss and hemorrhage,[72] highlighting the importance of P4 signaling in regulation of uterine vascular dynamics.

E2 is the more profound angiogenic hormone. Angiogenesis occurs in the cycling uterus under the control of high E2. In addition, E2 induces endothelial proliferation and migration[73] as well as vascular permeability.[74] E2 also mediates angiogenesis by inducing VEGF expression,[75] regulating NO signaling,[76] and recruiting proinflammatory/proangiogenic cells such as macrophages to the uterus.[77] These changes are mediated via both ER-α and ER-β, which are both expressed in the uterus. However, these receptors have structural differences, suggesting that differential receptor activation is involved in modulating E2 actions.[78] Notably, while E2 is generally considered proangiogenic, it has also been found to be antiangiogenic under some circumstances.[79]

3.3 Calcitonin Gene-Related Peptide Family

As mentioned earlier, blood flow to the uterus increases during pregnancy to provide for the growing embryo. One molecule mediating the necessary increases in uterine artery size is the vasodilator calcitonin gene-related peptide (CGRP).[80] Receptors for this peptide are expressed at a higher level in the pregnant uterus,[81] and CGRP has a potent vasodilatory effect on uterine arteries of pregnant women.[82] CGRP signals through two receptors: (1) CGRP-A receptors, which comprise the calcitonin receptor-like receptor (CRLR) and its accessory receptor activity modifying protein 1 (RAMP1); and (2) the CGRP-B receptor.[83]

Another protein that signals through CRLR/RAMP2 is adrenomedullin (ADM). Similarly to CGRP, ADM decreases vascular resistance via vasodilation.[84] ADM is synthesized by placental syncytiotrophoblast and fetal membranes, thus increasing the amniotic fluid ADM concentration above that of systemic plasma.[85] CGRP and ADM relax VSMCs in the uterus during pregnancy.[86] Similarly, adrenomedullin 2 (ADM2), otherwise known as intermedin (IMD), is a vasodilator that signals through the CRLR/RAMP receptor complex.[87] ADM2 facilitates trophoblast invasion into the decidua and maternal vasculature during the first trimester by regulating MMP2 and MUC1 expression.[88]

At a molecular level, these peptides signal in an autocrine/paracrine manner. For instance, ADM has been shown to be secreted by vascular endothelial and smooth muscle cells and also stimulates the proliferation of VSMCs.[89] This allows for vasodilation, angiogenesis, and smooth muscle relaxation.[90] The receptor for these peptides, CRLR, is trafficked to the membrane by RAMP proteins, which confers ligand specificity.[89] Upon peptide ligand binding to the G-protein-coupled CRLR, different cell signaling pathways are activated. These include MAP kinase pathways,[89] cAMP,[91] and calcium signaling to activate potassium channels that in turn cause smooth muscle relaxation.[92]

While each of the CGRP peptides (CGRP, ADM, ADM2) signal through different combinations of the CRLR and RAMP proteins, all ligands and receptors in the family are expressed in decidual, placental, and fetal tissues. Plasma levels of CGRP, ADM, and ADM2 are regulated during gestation and influence placental development.[93] Aberrancies in these signaling pathways are associated with disorders of uteroplacental vasculature. CGRP has been suggested to facilitate the compensatory vasodilatory changes that occur in response to hypertension during pregnancy.[94] However, CGRP and RAMP1 levels are decreased in fetoplacental vessels during preeclampsia, such that CGRP-induced relaxation is deficient.[95] Similarly, both ADM and CGRP levels are decreased in placental villi in women with preeclampsia.[96] Dysregulation of ADM and RAMP2 mRNA levels occurs during pregnancy-induced hypertension.[97] Furthermore, ADM2 is decreased in women with preeclampsia.[98] Because changes in the expression of CGRP, ADM, and ADM2 (and their receptors) in either the plasma or the uterine/placental tissues are associated with uterine vascular disorders during pregnancy, these signaling pathways offer great potential as diagnostic or therapeutic targets. For instance, magnesium sulfate has been shown to correct the aberrant levels of ADM and CGRP in maternal circulation of women with pregnancy-induced hypertension and preeclampsia.[99]

3.4 Angiopoietins

Angiopoietins are one of the secreted paracrine factors involved in signaling that promotes angiogenesis in the uterus. Angiopoietins are glycosylated proteins that bind to TIE (Tyr kinase with Ig and epidermal growth factor homology domains) receptors.[100] While the VEGF system initiates uterine angiogenesis, angiopoietins act later in vessel maturation, specifically in regulating VSMC recruitment and proliferation and maintaining the delicate balance between vessel development and quiescence.[101] The ANG–TIE system achieves this capacity for complex differential actions through a variety of mechanisms. First, there are multiple angiopoietin ligands including ANG1–ANG4 and angiopoietin-like ANGPTL1–ANGPTL8.[102] Second, there are different angiopoietin receptors, TIE1 and TIE2 (also known as TEK).[103] Third, the different ligands have different actions upon the receptors, some acting as agonists and others acting as antagonists.[101,104] Finally, the cellular signaling milieu affects ANG–TIE functioning. For instance, the presence or absence of VEGF and other factors affects the biological outcome of ANG:TIE signaling.[101]

ANG1 is upregulated in the uterus during the secretory phase of the menstrual cycle and in response to progesterone.[105] Notably, ANG1 is highly expressed in the VSMCs of the spiral arteries, and its expression correlates with VSMC proliferation.[106] Studies of early pregnancy in mice reveal that ANG1 is expressed in stromal cells on day of pregnancy (DOP) 2 and in epithelial cells on DOP4. Overall, ANG1 expression increases during gestation.[107] Studies of placentation in primates establish that ANG1 is expressed in the syncytiotrophoblast.[108]

ANG2 expression does not change significantly throughout the menstrual cycle.[105,109] Studies in pregnant mice show that ANG2 is not expressed until DOP6, with expression confined to the primary zone of decidualization in the uterus. Interestingly, in contrast to ANG1, ANG2 expression is induced by estrogen and decreased by progesterone.[110] Studies in pregnant primates and humans show that ANG2 is expressed exclusively in endothelial cells of maternal blood vessels.[108,109] In addition, ANG2 levels increase dramatically in the serum of women during pregnancy. This ANG2 is produced by decidual endothelial cells.[111]

The ANG1 and ANG2 expression patterns are quite distinct during the menstrual cycle and during pregnancy. This is in accordance with their molecular actions: the two ligands compete with each other for TIE2 binding and have opposing actions on TIE2 activity.[112] While both ligands bind TIE2, ANG1:TIE2 binding is affected by TIE1 such that suppression of TIE1 increases ANG1:TIE2 binding. In contrast, ANG2:TIE2 binding is not affected by TIE1.[113] While canonical research shows that ANG1 is a TIE2 agonist while ANG2 is a TIE2 antagonist, recent research has shown that ANG2 actions are actually much more nuanced. When ANG1 is absent, ANG2 is a TIE2 agonist, though to a weaker extent than ANG1. When ANG1 is present, ANG2 is a TIE2 antagonist.[114] When ANG1 activates the TIE2 receptor, it promotes maturation of vessels. When ANG2 antagonizes activation of the TIE2 receptor, it promotes destabilization of vessels and initiates neovascularization.[115] However, this neovascularization is dependent on VEGF expression. VEGFA enables ANG2 to promote endothelial cell migration and proliferation, but lack of VEGFA causes ANG2 to initiate endothelial cell death.[101,115] Thus, a balance of ANG1 and ANG2 actions along with other factors is involved in uterine vascular development and regulation.

Mechanistically, TIE2 is a tyrosine kinase receptor. TIE2 is expressed in endothelial cells. Upon ligand binding, TIE2 is phosphorylated and the ligand:receptor complex is internalized and degraded.[116] TIE2 activation leads to PI3K/AKT pathway activation, increasing angiogenesis through the NO pathway.[117] In addition, phosphorylated TIE2 is bound by ShcA adaptor proteins, linking to Ras/MAPK pathways involved in migration and sprouting and three-dimensional organization of vascular networks.[118]

While less studied, the other angiopoietins are also functionally important for regulation of uterine vascular remodeling during pregnancy. ANG3 is the mouse counterpart of ANG4. While derived from the homologous genetic locus, the two proteins have different expression distributions and functions.[104] ANG3 expression has been studied in the mouse uterus. ANG3 expression begins on DOP6, increasing through DOP8 in the primary decidua. ANG3 expression is induced by estrogen and progesterone.[119] In general, ANG3 has been shown to be a TIE2 antagonist, while ANG4 has been identified as a TIE2 agonist.[104,120] In human umbilical vein endothelial cells, ANG4 causes TIE2 phosphorylation and subsequently AKT phosphorylation, promoting cell survival and migration. However, in an in vivo assay of mouse lung microvascular endothelial cells, the effects were reversed, with ANG3 causing TIE2 and AKT phosphorylation to a greater extent than ANG4. Both proteins induce strong angiogenesis in vivo.[120] However, another study using human ANG4 in Matrigel implants in nude mice revealed that ANG4 inhibits migration of human umbilical vein endothelial cells and inhibits angiogenesis.[121] Such studies show that even though they are interspecies orthologs, ANG3 and ANG4 can have differing effects on TIE2 in their respective or the opposite species. This also reveals that agonizing or antagonizing TIE2 can have effects on vascular remodeling and angiogenesis.

Like the angiopoietins, the angiopoietin-like proteins have diverse effects on angiogenesis and cooperate to regulate vascular remodeling. ANGPTL1, also known as angioarrestin, is a key antiangiogenic factor that inhibits proliferation, migration, tube formation, and adhesion of endothelial cells.[122] However, ANGPTL1 is also antiapoptotic, promoting endothelial cell survival via phosphorylation of ERK1/2 and AKT.[123] This highlights the complex pathways that must intersect to regulate cell proliferation and survival. Even though AKT phosphorylation causes angiogenesis when induced by ANG1, AKT phosphorylation inhibits angiogenesis when induced by ANGPTL1. Further research is merited to elucidate the other factors involved in angiogenic regulation via the angiopoietin system.

The angiopoietin-like proteins play multiple roles in regulation of vascular remodeling. ANGPTL1 inhibits angiogenesis, ANGPTL2 induces angiogenesis and endothelial cell migration, ANGPTL4 has been implicated in angiogenesis, and ANGPTL6 has been implicated in endothelial dysfunction during pregnancy-induced hypertension.[102] However, the angiopoietin-like proteins can play additional roles beyond regulation of angiogenesis. ANGPTL3, ANGPTL4, and ANGPTL8 inhibit lipoprotein lipase. ANGPTL5 is involved in the expansion of hematopoietic stem cells in human cord blood. ANPTL7 is involved in extracellular matrix remodeling in glaucoma.[102] However, these proteins may all play roles in the uterus during implantation, as they are all expressed in the mouse uterus during implantation. In particular, ANGPTL4 is induced in implantation sites, with expression in both endothelial cells and stromal cells of the maternal aspect as well as the ectoplacental cone and trophoblast cells of the embryonic aspect. This suggests that ANGPTL4 may be involved in regulation of lipid metabolism and angiogenesis during pregnancy.[124]

The importance of the angiopoietin signaling network in regulation of uterine vascular dynamics is highlighted by the fact that mutations and dysregulated expression of these proteins is found in many disorders of uterine vascularization. Abnormal angiogenesis and vascular dynamics can contribute to conditions including abnormal bleeding during menstrual cycling, endometrial cancer, and endometriosis. During pregnancy, aberrancies in angiogenesis and placental vascular remodeling can cause gestational trophoblastic disease, placenta accreta, preeclampsia, and intrauterine growth restriction. Angiopoietins are implicated in all of these disorders as described later.

In the cycling uterus, lack of blood vessel stability can lead to abnormally heavy menstrual bleeding, termed menorrhagia. Distinct studies have demonstrated aberrancies in the ANG-TIE system in idiopathic menorrhagia. For instance, one study found dysregulation of ANG1 expression in the secretory phase, right before menstruation.[125] Another study found increases in ANG2 and TIE2 and TIE1 expression in menorrhagic tissues.[126] The combination of these changes leads to a 50% decrease in the ANG1:ANG2 ratio. Loss of normal ANG1 expression and increase in ANG2 likely destabilize endometrial vasculature, leading to the heavy bleeding.[115] A similar decrease in the ANG1:ANG2 ratio is seen in endometrial cancer, which could be involved in the angiogenesis associated with the progression of the disease.[127]

Endometriosis is a condition in which endometrial tissue implants ectopically in the peritoneal cavity. Growth of endometrial lesions is supported by angiogenesis.[128] Both ANG1 and ANG2 have been found to have higher expression in the eutopic and ectopic endometrium of women with endometriosis compared to healthy endometrium.[129,130] This suggests that angiopoietins could drive the excessive angiogenesis seen in endometriosis.

During pregnancy, many changes occur in the uterine vasculature. If placental development does not occur correctly, the trophoblasts may develop into tumors, generating gestational trophoblastic disease. Gestational trophoblastic disease includes molar pregnancies, placental site trophoblastic tumors, choriocarcinoma, and other conditions. ANG2 expression is lower and more heterogeneous in molar pregnancies compared to normal placenta.[131] However, such changes in angiopoietin expression are specific to the type of gestational trophoblastic disease, as ANG1 and ANG2 expression in placental site trophoblastic tumors is higher than ANG1 and ANG2 expression in choriocarcinoma.[132] Another disorder in placentation is placenta accreta, in which the placenta invades too deeply into the uterine lining due to abnormal uteroplacental neovascularization. This can cause massive maternal peripartum hemorrhage. A study has found downregulation of TIE2 in syncytiotrophoblasts and upregulation of ANG2 in placental lysates in women with placenta accreta.[133]

As pregnancies progress, insufficient placental angiogenesis can lead to increased vascular resistance and maternal hypertension, causing pregnancy-induced hypertension and preeclampsia. Insufficient placental angiogenesis can lead to intrauterine growth restriction and small for gestational age neonates. There is a polymorphism of ANG1 associated with increased plasma levels and a reduced risk for preeclampsia and small for gestational age births. Increased ANG1 at the maternal–fetal interface promotes angiogenesis, thereby preventing disorders due to insufficient angiogenesis.[134] Conversely, elevated plasma levels of ANGPTL6 are associated with pregnancy-induced hypertension.[135]

These studies highlight the importance of checks and balances to regulate uterine angiogenesis—too much or too little development of blood vessels during menstrual cycling or during pregnancy can be problematic. The angiopoietin system, with its plethora of pro- and antiangiogenic ligands and receptors that function in a context-dependent manner, is a critical component of this delicate and complex regulation of uterine vascular dynamics.

3.5 Sphingolipids

Sphingolipids are a component of eukaryotic cell membranes and play roles in membrane structure and signaling. Sphingolipids in the membranes of vascular cells have been shown to regulate vascular tone, permeability, proliferation, and migration.[136] In particular, sphingosine-1-phosphate (S1P) has been shown in in vitro and in vivo models to induce endothelial cell migration and proliferation, recruitment of smooth muscle cells to developing vessels, and vessel morphogenesis.[137] In the uterus, the enzymes catalyzing the interconversion of sphingomyelin, ceramide, sphingosine, and S1P are upregulated during embryo implantation and decidualization.[138] S1P is abundant in the gravid uterus of pregnant ewes.[139] In addition, the cognate G-protein-coupled receptors (GPCRs) that bind S1P are expressed in the decidual microvasculature of mice.[140]

The fact that the proangiogenic S1P molecule and its receptors are present in the uteri of pregnant ewes and mice suggests that S1P may play a role in angiogenesis during pregnancy. Indeed, mice deficient in sphingosine kinases, which are responsible for the generation of S1P, are infertile due to endometrial hemorrhage and decidual dysfunction.[141] Other lipids such as lysophosphatidic acid (LPA) are also implicated in pregnancy. For instance, targeted deletion of an LPA receptor in mice leads to delayed implantation and embryonic development as well as hypertrophic placentas and embryonic death.[142] Both S1P and LPA are implicated in decidual angiogenesis, lymphocyte trafficking, trophoblast cell function, and regulation of prostaglandin production and signaling.[143] The importance of sphingolipids in uteroplacental vasculature is highlighted by the fact that women with adverse pregnancies, such as preeclampsia and recurrent miscarriage, have altered placental sphingolipid composition.[144] Further research is warranted to understand the roles of sphingolipids in uterine vascular remodeling during pregnancy and pregnancy disorders.

3.6 Renin–Angiotensin System and MAS-Related Gene Family

During pregnancy, maternal blood volume increases in order to enhance perfusion to the growing fetus. This increase in volume is accomplished through increased sodium and water retention by the kidney.[145] The renin–angiotensin system (RAS) is responsible for regulation of systemic blood volume. In short, a decrease in blood volume or blood pressure induces the kidney to release renin. As shown in Fig. 1, activation of the classical RAS pathway begins with the pre/prohormone angiotensinogen that is generated by the liver and secreted into general circulation in its inactive form. Renin proteolytically cleaves angiotensinogen to form angiotensin I (Ang I). Angiotensin-converting enzymes (ACE or ACE1) cleave Ang I to form the active hormone, angiotensin II (Ang II). The actions of Ang II are mediated by angiotensin receptors type 1 (AT1) and AT2. Ang II stimulates arterial vasoconstriction to increase blood pressure. More recently, Ang II has been shown to be processed further through proteolysis to

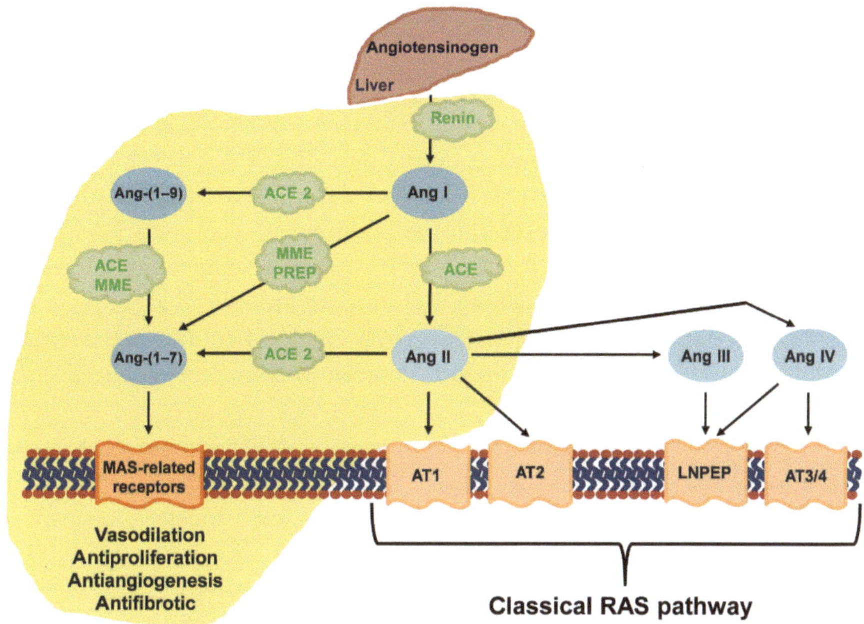

Fig. 1 The classical and nonclassical pathways within the renin–angiotensin system. In the classical renin–angiotensin system (*right side*), serum-derived angiotensinogen produced by the liver is proteolytically converted to angiotensin I (Ang I) by the enzyme renin. Angiotensin-converting enzyme (ACE or ACE1) then converts the Ang I prohormone hormone to the biologically active peptide Ang II. Angiotensin receptor type 1 (AT1) is the prototypical receptor for Ang II. This ligand:receptor interaction causes vasodilation. Ang II can also bind and activate AT2. Finally, Ang II can be converted by further proteolytic cleavage to Ang III and Ang IV. These smaller peptide-signaling molecules have been shown to bind and activate AT3 and/or AT4. Alternatively, Ang III and Ang IV can bind and activate the peptidase and receptor leucyl/cystinyl aminopeptidase (LNPEP, also called placental leucine aminopeptidase or insulin-regulated aminopeptidase). As shown in the *yellow-shaded area*, Ang I and II can also become processed by ACE 2, metalloendopeptidase (MME), and prolyl endopeptidase (PREP) enzymatic activity to Ang-(1–9) and Ang-(1–7). Ang-(1–7) serves as a novel ligand for at least some members of the MAS-related G-protein-coupled receptor superfamily. Activation of the prototypical MAS receptor MAS1 has been shown to cause vasodilation. Activated MAS1 also has antiproliferative, antiangiogenic, and antifibrotic functions in this nonclassical arm of the RAS.

Ang III and Ang IV. Less is known about the biological actions of these signaling peptides, but their actions are mediated by AT3, AT4, and leucyl/cystinyl aminopeptidase (LNPEP).[146]

In addition to signaling through the classical RAS, Ang I and Ang II can be converted to smaller peptide-signaling molecules in the nonclassical arm of the RAS (Fig. 1). Beyond being converted into Ang II, Ang I can also be converted into Ang-(1–7) by the enzymes neprilysin[147] and angiotensin-converting enzyme 2 (ACE2, Fig. 1).[148] Contrary to the vasoconstrictor actions of Ang II, Ang-(1–7) is a vasodilator.[149]

A secondary response to angiotensin signaling is the production and release of aldosterone by the adrenal gland. Aldosterone in turn initiates a series of events that allows for retention of sodium and water which increases blood volume.[150] Activity of the RAS increases during pregnancy, and this is thought to contribute to an increase in blood volume. However, pregnant women undergo a simultaneous increase in both blood volume and decrease in blood pressure due to vasodilation, as well as resistance to the pressor effects of Ang II.[151]

Despite being originally evaluated in the context of renal physiology and regulation of systemic blood volume, the RAS has been found more recently to function at a local level to regulate hemodynamic processes, and this is particularly true for the gravid uterus. Components of the RAS from ligands to receptors to proteolytic enzymes that activate angiotensinogen are also expressed and dynamically regulated locally in the uterine tissues. Renin is present in all components of the human uteroplacental unit.[152] Angiotensinogen is present in the human placenta and decidua,[153] and while some of this protein may be sequestered from the high angiotensinogen levels found in the plasma of pregnant women,[154] some is produced locally as evidenced by the presence of angiotensinogen mRNA in the placenta and decidua.[155] The human uterus, placenta, and fetal tissues have high local activity of ACE1.[156] The AT1 receptor is predominant in the placenta,[157] and AT2 is abundantly expressed in fetal tissues and the female reproductive tract.[158] ACE2 and Ang-(1–7) are expressed in the decidua and uterine epithelium during implantation in rats.[159] Ang-(1–7) acts through the MAS receptor, which is expressed in the human endometrium during the menstrual cycle[160] with increased expression during early pregnancy.[161]

During pregnancy, estrogen increases the activity of both systemic and local uteroplacental RAS by increasing angiotensinogen and renin expression.[162] Regulation of these angiotensin-signaling peptides is also likely controlled by the embryo.[159] As shown in Fig. 2, we have characterized the uterine expression of the metabolic machinery that converts angiotensinogen to the various active angiotensin-signaling products, and these include renin (*Ren1*), *Ace*, prolyl endopeptidase (*Prep*), metalloendopeptidase (*Mme*), and renin-binding protein (*Renbp*) using qPCR analysis on DOP 4, 7, and 9 in the mouse. Of note, with the exception of *Prep*, each of the genes is

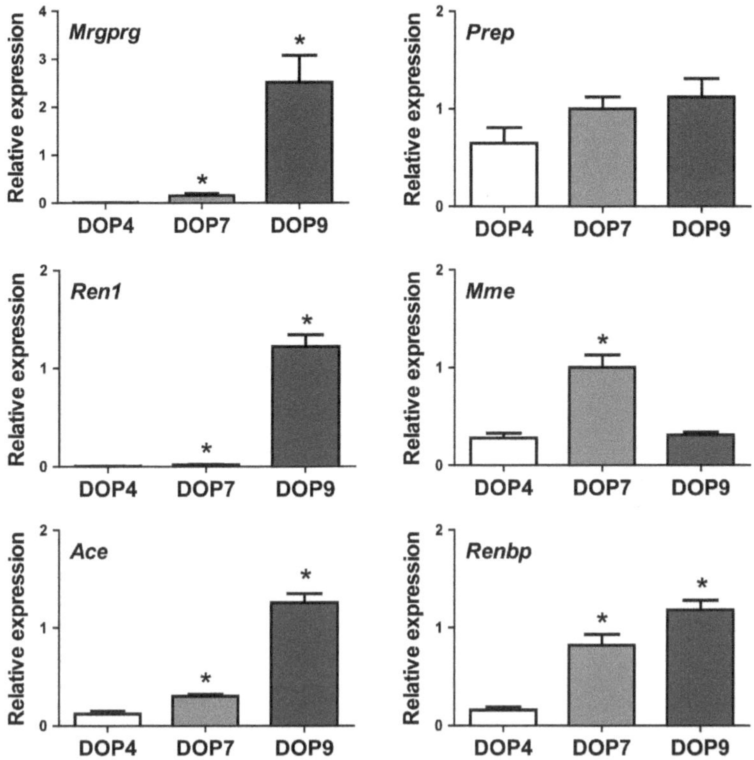

Fig. 2 qPCR analysis was used to evaluate the expression of various components of the RAS on day of pregnancy (DOP) 4, 7, and 9. While *mas-related G-protein receptor G (Mrgprg)*, *renin* (*Ren1*), *angiotensin-converting enzyme* (*Ace*), and *renin-binding protein* (*Renbp*) all increased significantly at each time point (*$P<0.05$), *membrane metalloendopeptidase* (*Mme*) increased significantly only from DOP 4 to 7. *Prolyl endopeptidase* (*Prep*) showed no change in expression from DOP 4 to 9, $n=3$.

significantly upregulated in uterine tissue as pregnancy progresses, suggesting an important role for local RAS signaling as the embryo begins to develop its placenta during the transition to fetal life.

Ang II has multiple effects in uterine vascular remodeling, and these include stimulation of decidualization, regulation of uteroplacental blood flow, stimulation of hormone secretion, and promotion of angiogenesis and vascular permeability.[158] These changes occur through activation of the G-protein-coupled AT1 and AT2 receptors and subsequent signaling events including calcium mobilization, protein kinase activation, and phosphatase action.[163] Conversely, Ang-(1–7) may inhibit angiogenesis. This occurs through activation of the G-protein-coupled MAS receptors, described in greater detail below. Activation of the MAS receptor by Ang-(1–7) leads to arachidonic acid production and activation of NO synthase.[132] Because Ang II and Ang-(1–7) have opposing actions on angiogenesis, the RAS likely plays a role in the delicate balance between angiogenic and anti-angiogenic activities in the gravid uterus. The Ang II:Ang-(1–7) ratio may be important for correct uterine vascular remodeling as pregnancy is established.

GPCRs constitute the largest and most functionally varied family of transmembrane receptor proteins. GPCRs respond to diverse stimuli that include light, peptide and polypeptide hormones, amines, neuropeptides, bioactive lipids and steroids. Among the greater than 800 GPCRs, approximately 100 remain "orphaned," with no identified ligand. The MAS-related gene (MRG) cluster was first described in 2001 as a group of approximately 50 GPCRs that retain sequence and structural similarities with the protooncogene MAS1.[164] MRGs were initially described as being expressed in specific subsets of nociceptive sensory neurons in dorsal (spinal) root ganglia.[164] MRGs retain high sequence identity with MAS1, the prototypical member of this family. Members of the murine MRGA, MRGB, and MRGC subgroups have 22, 13, and 14 members, respectively. Other family members include MRGD, MRGE, MRGF, and MRGG (also called MRGPRG).

Functional and mechanistic data on MRG family members are limited. Several studies have established that MRGs are expressed in neurons of dorsal root ganglia and vasculature. Interestingly, MAS1 has been shown in a number of species to mediate the antidiuretic actions of Ang-(1–7) in water-loaded animals, suggesting a vasodilatory function for MAS1.[165] MAS1 and MAS-like receptors are now known to mediate vasodilatory, antiproliferative, antiangiogenic, and antifibrotic actions (Fig. 1). The effects of Ang-(1–7) on vasodilation have been tested physiologically in mutant mice in which MAS1 deficiency completely ablates the vasodilatory effects of Ang-(1–7). *Mas1*$^{-/-}$ mice also display with endothelial dysfunction and high blood pressure compared to WT counterparts.[165–167] The cumulative findings therefore indicate that MAS1 acts to reduce blood pressure, to attenuate hemostasis, and to maintain normal endothelial cell function possibly through upregulation of endothelial NO synthase gene expression.[168–170]

Kashiwagi *et al.* found nearly 1500 genes to be differentially expressed in the uterine decidua of pregnant mice, when compared to artificially stimulated deciduomal (pseudopregnant) tissue.[171] This general approach was used to identify genes that were regulated in the uterus of early pregnancy by the embryo. One of the most significantly upregulated genes was found to encode the previously uncharacterized GPCR called *Mas-related g-protein receptor G* (*Mrgprg*).

Our lab has begun characterizing *Mrgprg* and other members of the MAS-related receptor family. Fig. 2 shows the dramatic upregulation of *Mrgprg* from DOP 4–9 of murine pregnancy. Furthermore, *Mrgprg* expression in the uterus is regulated by the embryo. As presented in Fig. 3A, *Mrgprg* is not expressed in the nongravid uterus. In female mice stimulated to undergo artificial decidualization by injection of sesame oil into the uterine lumen on day 4 of pseudopregnancy, *Mrgprg* is weakly upregulated in deciduomal tissue, but not in the unstimulated contralateral horn, indicating that *Mrgprg* expression occurs only in decidualized endometrium. In contrast, *Mrgprg* expression increases over 12-fold in decidual tissue on DOP7 compared with oil-induced deciduomal tissue, providing evidence of embryonic regulation. Other members of the MRG family expressed

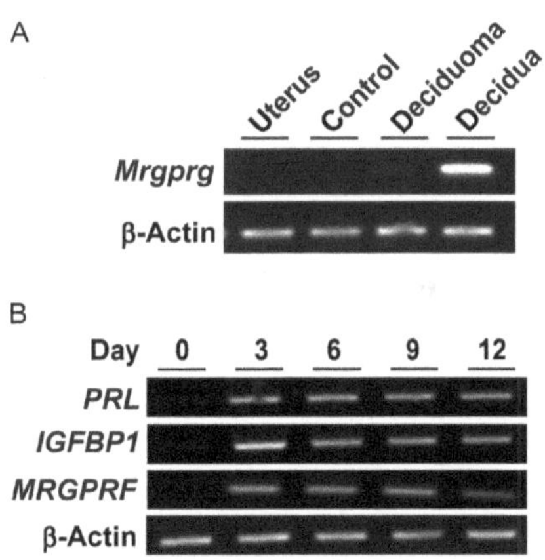

Fig. 3 Expression of MAS-related receptors in uterine cells during decidualization. (A) Evaluation of murine *Mrgprg* in nongravid uterus (uterus), undecidualized (control), and artificially decidualized uterus (deciduoma) on day 7 of pseudopregnancy and in decidual tissue (decidua) on day of pregnancy 7. β-Actin was used as an internal reference gene. (B) Human endometrial stromal cells (HESC) were treated with estrogen, progesterone, and cAMP to induce decidualization in vitro, and collected at time 0 as well as days 3, 6, 9, and 12 after stimulation. As demonstrated by conventional RT-PCR, prolactin (*PRL*) and insulin-like growth factor-binding protein 1 (*IGFBP1*) were not expressed until day 3 after providing induction medium. The MAS-related receptor *MRGPRF* was shown to be induced in response to in vitro decidualization. β-Actin was used as an internal reference gene, $n = 3$.

in the decidualizing endometrium of murine pregnancy include *Mrgpra2*, *Mrgpre*, and *Mprgprf* (Pru, unpublished result). While MRGPRG is not expressed in artificially decidualized human endometrial stromal cells (HESC), Fig. 3B shows that MRGPRF becomes expressed in response to HESC decidualization in vitro similarly to the classical markers of HESC decidualization prolactin (PRL) and insulin-like growth factor-binding proteins 1 (IGFBP1). These initial expression studies of components of the RAS and members of the MAS-related receptor family have been used to develop a hypothetical model for how Ang-(1–7) might contribute to establishing a functional uteroplacental vascular network (Fig. 4).

In this hypothetical model, embryonic paracrine-signaling factors induce or upregulate the expression of MRGPRG and renin. The presence of these two proteins, when combined with the metabolic machinery that are expressed in the endometrium in response to sex steroid hormones during decidualization, promotes the production of Ang-(1–7) and its signaling activation of MRGPRG (and/or other members of the MAS-related

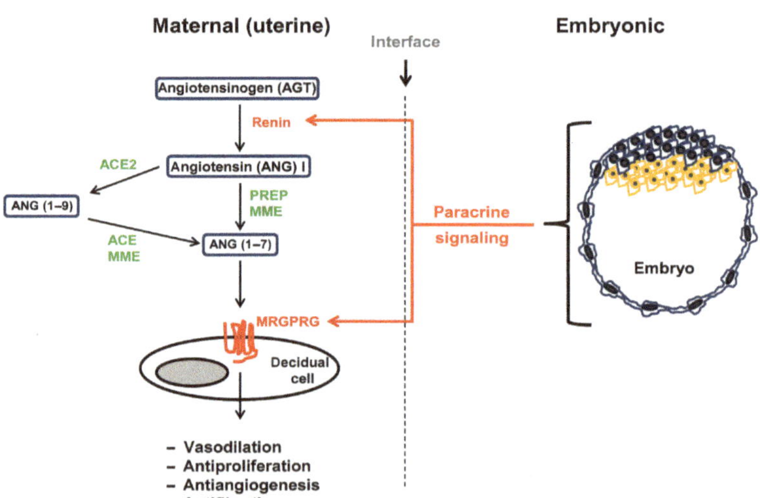

Fig. 4 Hypothetical model of how the nonclassical pathway of the renin–angiotensin system could contribute to the establishment of a uterine environment that is permissive to embryonic/fetal growth and development. Embryonic paracrine-signaling factors induce or upregulate uterine expression of MRGPRG and renin. The presence of these two proteins, when combined with the metabolic machinery that are expressed in the endometrium in response to sex steroid hormones during decidualization, promotes the production of Ang-(1–7) and its signaling activation of MRGPRG (and/or other members of the MAS-related receptor family). This in turn regulates vascular remodeling including vasodilation.

receptor family). This in turn helps establish a uterine environment that is permissive to embryonic/fetal growth and development, principally by reducing vascular resistance through vasodilation. Clearly many elements of this model need to be validated, but this provides a starting framework with which to begin experimentation.

The necessity of the coordinated interaction between systemic and local RASs during pregnancy is highlighted by the fact that aberrancies in these systems are associated with obstetrical complications stemming from vascular remodeling and blood pressure changes. For example, disruption of the RAS is thought to be causally linked to the development of preeclampsia in human pregnancy, as circulating levels of Ang-(1–7) are diminished in preeclamptic women.[172] Ang-(1–7) also regulates sFlt1 release in normal but not preeclamptic women.[173] While the mechanistic links between RAS dysregulation and preeclampsia are not fully elucidated, one potential mechanism has been studied. Preeclamptic women have autoantibodies that agonize AT1 (AT1-AA).[174,175] AT1-AA induces NADPH oxidase and MAPK/ERK pathway, leading to increase in factors like sFlt1 and endothelin-1 that are increased in preeclamptic women.[176] In addition, AT1-AA promotes platelet aggregation,[177] vasoconstriction, and endothelial cell injury,[178] all of which could be causally related to the increased blood pressure found in preeclampsia. The AT1-AA has effects beyond the uterus, contributing to the vascular injuries involved in retinopathy[179] and renal dysfunction[180] sometimes observed in preeclamptic women.

4. UTERINE HEMOGLOBIN BIOSYNTHESIS DURING GESTATION

It is interesting to note that in most mammalian species, the embryo exists inside the female reproductive tract without direct access to the maternal vasculature until 25%–30% of pregnancy has transpired, the time when the placenta becomes functionally competent. How then does the embryo obtain oxygen, particularly after implantation? There is clear evidence that like most stem cell niches, embryonic stem cells that will give rise to extraembryonic and embryonic structures are heavily influenced by oxygen gradients. This was first demonstrated by direct measurements of oxygen tension in endometrial and embryonic tissues during early pregnancy.[181,182] Oxygen diffusion within tissues is limited to a distance of about 150 μm.[183,184] Without an established uteroplacental network, delivery of oxygen to the embryo is thus dependent on the simple property of diffusion.

This places the developing embryo in a relatively hypoxic environment. The successful artificial manipulation of embryonic and adult stem cells in culture has been greatly advanced the by growth of such stem cells under hypoxic conditions.[185–187] In fact, it is now recognized that careful maintenance of low oxygen tension is as critical to the growth, maintenance, expansion, and differentiation of stem cells as the various cytokines that also drive these processes. Collectively, the combination of paracrine factors, extracellular scaffolding, biophysical properties, and oxygen tension all contributes significantly to what was originally coined by Schofield as the "stem cell niche."[188] However, despite the hypoxic environment of the implantation site, the embryo still requires oxygen for ATP synthesis, particularly as it expands during organogenesis prior to development of a functional placenta.

Another aspect of providing the embryo/fetus with ample oxygen for growth and expansion that has been almost completely overlooked, or at least not well studied, is the local production of hemoglobin by the gravid uterus. Several tissues produce globin molecules to sequester oxygen. For example, cardiac and skeletal muscle have myoglobin and neural tissues have neuroglobin.[189–192] The liver, cornea, and ovarian follicle all produce various isoforms of hemoglobin.

Interestingly, like the preimplantation blastocyst or implanted embryo, the cornea and ovarian follicle exist without a direct vascular supply. The ovarian follicle was recently shown to express hemoglobin.[193] The function of hemoglobin in granulosa cells is not presently known, but it could serve as an oxygen repository to ensure that at least some aerobic respiration occurs in this highly metabolic tissue. Likewise, the alpha and beta isoforms of hemoglobin are expressed in both murine and human decidual cells during early gestation.[194,195] Efforts in our lab have confirmed the expression of β-hemoglobin in both the epithelial and stromal compartments of the endometrium on DOP 3 and 7 (Fig. 5). Furthermore, we evaluated the expression of delta-aminolevulinate synthase (ALAS2) during early murine gestation. ALAS2 is the rate-limiting enzyme in an eight-step metabolic pathway that functions to generate heme. ALAS2 is reported to have an expression pattern restricted exclusively to erythroid cells. As shown in Fig. 5, ALAS2, like β-hemoglobin, is robustly expressed in epithelial and stromal cells throughout the endometrium during early gestation. While the exact function of hemoglobin in the uterus is not presently known, it could play a role similar to myoglobin in muscle, neuroglobin in the brain, or hemoglobin in the ovarian follicle, and that is to serve as an oxygen buffer to ensure that the embryo has a ready supply of oxygen regardless of

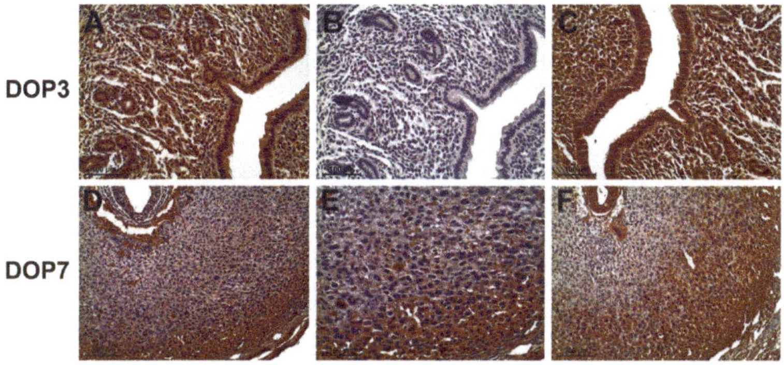

Fig. 5 Immunohistochemical staining of β-hemoglobin and delta-aminolevulinate synthase 2 (ALAS2) in gravid murine uteri on day of pregnancy (DOP) 3 and 7. β-Hemoglobin (A) and ALAS2 (C) are abundantly expressed on DOP3 throughout the preimplanted uterus. Panel B shows a negative control in which primary antibody was omitted. β-Hemoglobin (D, E) and ALAS2 (F) are abundantly expressed in undecidualized stromal tissue near the myometrial border with much reduced expression in decidualized cells immediately surrounding the embryo, $n=4$.

metabolic demands placed on the mother. This intriguing finding provides yet another level of complexity to understanding hemodynamic events during early pregnancy.

5. CONCLUDING REMARKS

Vascular remodeling is an essential process for a successful pregnancy. Blood flow must increase to the uterus to provide nutrients and oxygen for the fetus. This occurs via vasculogenesis and angiogenesis, increase in diameter and length of spiral arteries, and ultimately invasion of maternal blood vessels by embryonic cells. Lack of remodeling can lead to diseases such as intrauterine growth restriction. However, this vascular remodeling must be regulated, as too much blood flow can lead to oxidative damage to the embryo or to hemorrhage. This finely tuned homeostatic regulation of vascular remodeling is accomplished via a variety of signaling pathways that have capacity for pro- and/or antiangiogenic actions. While much research has been done on this topic, there are still many questions remaining, and novel regulators are still being identified. In this chapter, the MAS-related gene family and uterine hemoglobin biosynthesis were introduced as additional components involved in regulating vascular dynamics and oxygen/nutrient provision to the developing embryo. Understanding the role of the various regulators of vascular dynamics will provide further insight into the etiologies and treatment of miscarriage, preeclampsia, intrauterine growth restriction, and other disorders of uterine vascular remodeling.

ACKNOWLEDGMENTS

Supported by NIH R21OD010488 and R21OD016564.

Disclosure Statement: The authors have no relevant financial or nonfinancial relationships to disclose.

REFERENCES

[1] Paauw ND, Luijken K, Franx A, Verhaar MC, Lely AT. Long-term renal and cardiovascular risk after preeclampsia: towards screening and prevention. *Clin Sci (Lond)*. 2016;130(4):239–246.

[2] Barker DJ. The fetal and infant origins of adult disease. *BMJ*. 1990;301(6761):1111.

[3] Palmer SK, Zamudio S, Coffin C, Parker S, Stamm E, Moore LG. Quantitative estimation of human uterine artery blood flow and pelvic blood flow redistribution in pregnancy. *Obstet Gynecol*. 1992;80(6):1000–1006.

[4] Forbes TR, Taku E. Vein size in intact and hysterectomized mice during the estrous cycle and pregnancy. *Anat Rec*. 1975;182(1):61–65.

[5] Page KL, Celia G, Leddy G, Taatjes DJ, Osol G. Structural remodeling of rat uterine veins in pregnancy. *Am J Obstet Gynecol*. 2002;187(6):1647–1652.

[6] Osol G, Mandala M. Maternal uterine vascular remodeling during pregnancy. *Physiology (Bethesda)*. 2009;24:58–71.

[7] Smith SK. Regulation of angiogenesis in the endometrium. *Trends Endocrinol Metab*. 2001;12(4):147–151.

[8] Abberton KM, Healy DL, Rogers PA. Smooth muscle alpha actin and myosin heavy chain expression in the vascular smooth muscle cells surrounding human endometrial arterioles. *Hum Reprod*. 1999;14(12):3095–3100.

[9] Kohnen G, Campbell S, Jeffers MD, Cameron IT. Spatially regulated differentiation of endometrial vascular smooth muscle cells. *Hum Reprod*. 2000;15(2):284–292.

[10] Demir R, Yaba A, Huppertz B. Vasculogenesis and angiogenesis in the endometrium during menstrual cycle and implantation. *Acta Histochem*. 2010;112(3):203–214.

[11] Rogers PA, Lederman F, Taylor N. Endometrial microvascular growth in normal and dysfunctional states. *Hum Reprod Update*. 1998;4(5):503–508.

[12] Abberton KM, Taylor NH, Healy DL, Rogers PA. Vascular smooth muscle cell proliferation in arterioles of the human endometrium. *Hum Reprod*. 1999;14(4): 1072–1079.

[13] Wetendorf M, DeMayo FJ. The progesterone receptor regulates implantation, decidualization, and glandular development via a complex paracrine signaling network. *Mol Cell Endocrinol*. 2012;357(1–2):108–118.

[14] Gellersen B, Brosens JJ. Cyclic decidualization of the human endometrium in reproductive health and failure. *Endocr Rev*. 2014;35(6):851–905.

[15] Cha J, Sun X, Dey SK. Mechanisms of implantation: strategies for successful pregnancy. *Nat Med*. 2012;18(12):1754–1767.

[16] Salker MS, Nautiyal J, Steel JH, et al. Disordered IL-33/ST2 activation in decidualizing stromal cells prolongs uterine receptivity in women with recurrent pregnancy loss. *PLoS One*. 2012;7(12):e52252.

[17] Weimar CH, Kavelaars A, Brosens JJ, et al. Endometrial stromal cells of women with recurrent miscarriage fail to discriminate between high- and low-quality human embryos. *PLoS One*. 2012;7(7):e41424.

[18] Weimar CH, Macklon NS, Post Uiterweer ED, Brosens JJ, Gellersen B. The motile and invasive capacity of human endometrial stromal cells: implications for normal and impaired reproductive function. *Hum Reprod Update*. 2013;19(5):542–557.
[19] Zygmunt M, Herr F, Munstedt K, Lang U, Liang OD. Angiogenesis and vasculogenesis in pregnancy. *Eur J Obstet Gynecol Reprod Biol*. 2003;110(suppl 1):S10–S18.
[20] Plaisier M. Decidualisation and angiogenesis. *Best Pract Res Clin Obstet Gynaecol*. 2011;25(3):259–271.
[21] Craven CM, Morgan T, Ward K. Decidual spiral artery remodelling begins before cellular interaction with cytotrophoblasts. *Placenta*. 1998;19(4):241–252.
[22] Harris LK. IFPA Gabor than award lecture: transformation of the spiral arteries in human pregnancy: key events in the remodelling timeline. *Placenta*. 2011;32(suppl 2):S154–S158.
[23] Kliman HJ. The placenta revealed. *Am J Pathol*. 1993;143:332–336.
[24] Weiss G, Sundl M, Glasner A, Huppertz B, Moser G. The trophoblast plug during early pregnancy: a deeper insight. *Histochem Cell Biol*. 2016;146:749–756.
[25] Jauniaux E, Watson AL, Hempstock J, Bao YP, Skepper JN, Burton GJ. Onset of maternal arterial blood flow and placental oxidative stress. A possible factor in human early pregnancy failure. *Am J Pathol*. 2000;157(6):2111–2122.
[26] Cross JC, Hemberger M, Lu Y, et al. Trophoblast functions, angiogenesis and remodeling of the maternal vasculature in the placenta. *Mol Cell Endocrinol*. 2002;187(1-2):207–212.
[27] Malassine A, Frendo JL, Evain-Brion D. A comparison of placental development and endocrine functions between the human and mouse model. *Hum Reprod Update*. 2003;9(6):531–539.
[28] Maltepe E, Bakardjiev AI, Fisher SJ. The placenta: transcriptional, epigenetic, and physiological integration during development. *J Clin Invest*. 2010;120(4):1016–1025.
[29] Blankenship TN, Enders AC. Trophoblast cell-mediated modifications to uterine spiral arteries during early gestation in the macaque. *Acta Anat (Basel)*. 1997;158(4): 227–236.
[30] Enders AC, Blankenship TN, Fazleabas AT, Jones CJ. Structure of anchoring villi and the trophoblastic shell in the human, baboon and macaque placenta. *Placenta*. 2001;22(4):284–303.
[31] Enders AC, Blankenship TN. Modification of endometrial arteries during invasion by cytotrophoblast cells in the pregnant macaque. *Acta Anat (Basel)*. 1997;159(4): 169–193.
[32] Pretorius DH, Nelson TR, Baergen RN, Pai E, Cantrell C. Imaging of placental vasculature using three-dimensional ultrasound and color power Doppler: a preliminary study. *Ultrasound Obstet Gynecol*. 1998;12(1):45–49.
[33] Gordon Z, Elad D, Almog R, Hazan Y, Jaffa AJ, Eytan O. Anthropometry of fetal vasculature in the chorionic plate. *J Anat*. 2007;211(6):698–706.
[34] Lee MM, Yeh MN. Fetal circulation of the placenta: a comparative study of human and baboon placenta by scanning electron microscopy of vascular casts. *Placenta*. 1983; 4(Spec No):515–526.
[35] Pijnenborg R, Bland JM, Robertson WB, Brosens I. Uteroplacental arterial changes related to interstitial trophoblast migration in early human pregnancy. *Placenta*. 1983;4(4):397–413.
[36] Kaufmann P, Mayhew TM, Charnock-Jones DS. Aspects of human fetoplacental vasculogenesis and angiogenesis. II. Changes during normal pregnancy. *Placenta*. 2004;25(2-3):114–126.
[37] Leong AS, Norman JE, Smith R. Vascular and myometrial changes in the human uterus at term. *Reprod Sci*. 2008;15(1):59–65.

[38] Verguts J, Ameye L, Bourne T, Timmerman D. Normal data for uterine size according to age and gravidity and possible role fo the classical golden ratio. *Ultrasound Obstet Gynecol.* 2013;42:713–717.
[39] Sherman D, Lurie S, Frenkel E, Kurzweil Y, Bukovsky I, Arieli S. Characteristics of normal lochia. *Am J Perinatol.* 1999;16(8):399–402.
[40] Kurjak A, Chervenak FA. *Donald School Textbook of Ultrasound in Obstetrics and Gynecology.* Panama City: Jaypee Brothers Medical Publishers; 2011.
[41] Wakasa T, Wakasa K, Nakayama M, et al. Change in morphology and oxytocin receptor expression in the uterine blood vessels during the involution process. *Gynecol Obstet Invest.* 2009;67(2):137–144.
[42] Sohn C, Fendel H, Kesternich P. Involution-induced changes in arterial uterine blood flow. *Z Geburtshilfe Perinatol.* 1988;192(5):203–209.
[43] Blackburn ST. *Maternal, Fetal, and Neonatal Physiology: A Clinical Perspective.* 3rd ed. St. Louis: Saunders; 2007.
[44] Cunningham FG, Leveno KJ, Bloom SL, Hauth JC, Gilstrap LC, Wenstrom KD. *Williams Obstetrics.* 22nd ed. New York: McGraw-Hill; 2005.
[45] Chennazhi KP, Nayak NR. Regulation of angiogenesis in the primate endometrium: vascular endothelial growth factor. *Semin Reprod Med.* 2009;27(1):80–89.
[46] Valdes G, Corthorn J. Review: the angiogenic and vasodilatory utero-placental network. *Placenta.* 2011;32(suppl 2):S170–S175.
[47] Valdes G, Corthorn J, Figueroa CD. The kallikrein-kinin system along the different stages of gestation: experimental and clinical findings. *Biol Res.* 1998;31(3): 235–241.
[48] Boeldt DS, Yi FX, Bird IM. eNOS activation and NO function: pregnancy adaptive programming of capacitative entry responses alters nitric oxide (NO) output in vascular endothelium—new insights into eNOS regulation through adaptive cell signaling. *J Endocrinol.* 2011;210(3):243–258.
[49] Klammt J, Pfaffle R, Werner H, Kiess W. IGF signaling defects as causes of growth failure and IUGR. *Trends Endocrinol Metab.* 2008;19(6):197–205.
[50] Smith SK. Prostaglandins and growth factors in the endometrium. *Baillieres Clin Obstet Gynaecol.* 1989;3(2):249–270.
[51] Vrachnis N, Kalampokas E, Sifakis S, et al. Placental growth factor (PlGF): a key to optimizing fetal growth. *J Matern Fetal Neonatal Med.* 2013;26(10):995–1002.
[52] Smith SK. Angiogenesis, vascular endothelial growth factor and the endometrium. *Hum Reprod Update.* 1998;4(5):509–519.
[53] Motro B, Itin A, Sachs L, Keshet E. Pattern of interleukin 6 gene expression in vivo suggests a role for this cytokine in angiogenesis. *Proc Natl Acad Sci U S A.* 1990;87(8): 3092–3096.
[54] Okada H, Tsuzuki T, Shindoh H, Nishigaki A, Yasuda K, Kanzaki H. Regulation of decidualization and angiogenesis in the human endometrium: mini review. *J Obstet Gynaecol Res.* 2014;40(5):1180–1187.
[55] Marshall SA, Senadheera SN, Parry LJ, Girling JE. The role of Relaxin in normal and abnormal uterine function during the menstrual cycle and early pregnancy. *Reprod Sci.* 2016, Epub ahead of print.
[56] Goldsmith LT, Weiss G. Relaxin in human pregnancy. *Ann N Y Acad Sci.* 2009; 1160:130–135.
[57] Kiewisz J, Wasniewski T, Kmiec Z. Participation of WNT and beta-catenin in physiological and pathological endometrial changes: association with angiogenesis. *Biomed Res Int.* 2015;2015:854056.
[58] Tessier DR, Yockell-Lelievre J, Gruslin A. Uterine spiral artery remodeling: the role of uterine natural killer cells and extravillous trophoblasts in normal and high-risk human pregnancies. *Am J Reprod Immunol.* 2015;74(1):1–11.

[59] Leonard S, Murrant C, Tayade C, van den Heuvel M, Watering R, Croy BA. Mechanisms regulating immune cell contributions to spiral artery modification—facts and hypotheses—a review. *Placenta*. 2006;27(suppl A):S40–S46.
[60] Du MR, Wang SC, Li DJ. The integrative roles of chemokines at the maternal-fetal interface in early pregnancy. *Cell Mol Immunol*. 2014;11(5):438–448.
[61] Cao Q, Chen H, Deng Z, et al. Genetic deletion of Cxcl14 in mice alters uterine NK cells. *Biochem Biophys Res Commun*. 2013;435(4):664–670.
[62] Ratsep MT, Felker AM, Kay VR, Tolusso L, Hofmann AP, Croy BA. Uterine natural killer cells: supervisors of vasculature construction in early decidua basalis. *Reproduction*. 2015;149(2):R91–R102.
[63] Hu Y, Tan R, MacCalman CD, et al. IFN-gamma-mediated extravillous trophoblast outgrowth inhibition in first trimester explant culture: a role for insulin-like growth factors. *Mol Hum Reprod*. 2008;14(5):281–289.
[64] Kitaya K, Yasuo T, Yamaguchi T, Fushiki S, Honjo H. Genes regulated by interferon-gamma in human uterine microvascular endothelial cells. *Int J Mol Med*. 2007;20(5):689–697.
[65] Gibson DA, Greaves E, Critchley HO, Saunders PT. Estrogen-dependent regulation of human uterine natural killer cells promotes vascular remodelling via secretion of CCL2. *Hum Reprod*. 2015;30(6):1290–1301.
[66] Steiner JL, Murphy EA. Importance of chemokine (CC-motif) ligand 2 in breast cancer. *Int J Biol Markers*. 2012;27(3):e179–e185.
[67] Li G, Huang W, Xia Q, et al. Role of uterine natural killer cells in angiogenesis of human decidua of the first-trimester pregnancy. *Sci China C Life Sci*. 2008;51(2):111–119.
[68] Quenby S, Nik H, Innes B, et al. Uterine natural killer cells and angiogenesis in recurrent reproductive failure. *Hum Reprod*. 2009;24(1):45–54.
[69] Biswas Shivhare S, Bulmer JN, Innes BA, Hapangama DK, Lash GE. Menstrual cycle distribution of uterine natural killer cells is altered in heavy menstrual bleeding. *J Reprod Immunol*. 2015;112:88–94.
[70] Dickey RP, Hower JF. Relationship of estradiol and progesterone levels to uterine blood flow during early pregnancy. *Early Pregnancy*. 1996;2(2):113–120.
[71] Maliqueo M, Echiburu B, Crisosto N. Sex steroids modulate uterine-placental vasculature: implications for obstetrics and neonatal outcomes. *Front Physiol*. 2016;7:152.
[72] Schatz F, Guzeloglu-Kayisli O, Arlier S, Kayisli UA, Lockwood CJ. The role of decidual cells in uterine hemostasis, menstruation, inflammation, adverse pregnancy outcomes and abnormal uterine bleeding. *Hum Reprod Update*. 2016;22(4):497–515.
[73] Morales DE, McGowan KA, Grant DS, et al. Estrogen promotes angiogenic activity in human umbilical vein endothelial cells in vitro and in a murine model. *Circulation*. 1995;91(3):755–763.
[74] Hyder SM, Stancel GM. Regulation of angiogenic growth factors in the female reproductive tract by estrogens and progestins. *Mol Endocrinol*. 1999;13(6):806–811.
[75] Shifren JL, Tseng JF, Zaloudek CJ, et al. Ovarian steroid regulation of vascular endothelial growth factor in the human endometrium: implications for angiogenesis during the menstrual cycle and in the pathogenesis of endometriosis. *J Clin Endocrinol Metab*. 1996;81(8):3112–3118.
[76] Hishikawa K, Nakaki T, Marumo T, Suzuki H, Kato R, Saruta T. Up-regulation of nitric oxide synthase by estradiol in human aortic endothelial cells. *FEBS Lett*. 1995;360(3):291–293.
[77] Losordo DW, Isner JM. Estrogen and angiogenesis: a review. *Arterioscler Thromb Vasc Biol*. 2001;21(1):6–12.

[78] Pastore MB, Jobe SO, Ramadoss J, Magness RR. Estrogen receptor-alpha and estrogen receptor-beta in the uterine vascular endothelium during pregnancy: functional implications for regulating uterine blood flow. *Semin Reprod Med.* 2012;30(1):46–61.
[79] Rogers PA, Donoghue JF, Walter LM, Girling JE. Endometrial angiogenesis, vascular maturation, and lymphangiogenesis. *Reprod Sci.* 2009;16(2):147–151.
[80] Gangula PR, Thota C, Wimalawansa SJ, Bukoski RD, Yallampalli C. Mechanisms involved in calcitonin gene-related Peptide-induced relaxation in pregnant rat uterine artery. *Biol Reprod.* 2003;69(5):1635–1641.
[81] Yallampalli C, Gangula PR, Kondapaka S, Fang L, Wimalawansa S. Regulation of calcitonin gene-related peptide receptors in the rat uterus during pregnancy and labor and by progesterone. *Biol Reprod.* 1999;61(4):1023–1030.
[82] Nelson SH, Steinsland OS, Suresh MS. Possible physiologic role of calcitonin gene-related peptide in the human uterine artery. *Am J Obstet Gynecol.* 1993;168(2): 605–611.
[83] Yallampalli C, Chauhan M, Thota CS, Kondapaka S, Wimalawansa SJ. Calcitonin gene-related peptide in pregnancy and its emerging receptor heterogeneity. *Trends Endocrinol Metab.* 2002;13(6):263–269.
[84] Beltowski J, Jamroz A. Adrenomedullin—what do we know 10 years since its discovery? *Pol J Pharmacol.* 2004;56(1):5–27.
[85] Marinoni E, Di Iorio R, Letizia C, Villaccio B, Scucchi L, Cosmi EV. Immunoreactive adrenomedullin in human fetoplacental tissues. *Am J Obstet Gynecol.* 1998;179(3 pt 1):784–787.
[86] Thota C, Yallampalli C. Progesterone upregulates calcitonin gene-related peptide and adrenomedullin receptor components and cyclic adenosine 3′5′-monophosphate generation in Eker rat uterine smooth muscle cell line. *Biol Reprod.* 2005;72(2): 416–422.
[87] Roh J, Chang CL, Bhalla A, Klein C, Hsu SY. Intermedin is a calcitonin/calcitonin gene-related peptide family peptide acting through the calcitonin receptor-like receptor/receptor activity-modifying protein receptor complexes. *J Biol Chem.* 2004;279(8):7264–7274.
[88] Chauhan M, Balakrishnan M, Chan R, Yallampalli C. Adrenomedullin 2 (ADM2) regulates mucin 1 at the maternal-fetal interface in human pregnancy. *Biol Reprod.* 2015;93(6):136.
[89] Shichiri M, Fukai N, Ozawa N, Iwasaki H, Hirata Y. Adrenomedullin is an autocrine/paracrine growth factor for rat vascular smooth muscle cells. *Regul Pept.* 2003;112(1-3):167–173.
[90] Nikitenko LL, Brown NS, Smith DM, MacKenzie IZ, Bicknell R, Rees MC. Differential and cell-specific expression of calcitonin receptor-like receptor and receptor activity modifying proteins in the human uterus. *Mol Hum Reprod.* 2001;7(7):655–664.
[91] Hay DL, Smith DM. Adrenomedullin receptors: molecular identity and function. *Peptides.* 2001;22(11):1753–1763.
[92] Dong Y, Betancourt A, Chauhan M, et al. Pregnancy increases relaxation in human omental arteries to the CGRP family of peptides. *Biol Reprod.* 2015;93(6):134.
[93] Yallampalli C, Chauhan M, Endsley J, Sathishkumar K. Calcitonin gene related family peptides: importance in normal placental and fetal development. *Adv Exp Med Biol.* 2014;814:229–240.
[94] Yallampalli C, Wimalawansa SJ. Calcitonin gene-related peptide (CGRP) is a mediator of vascular adaptations during hypertension in pregnancy. *Trends Endocrinol Metab.* 1998;9(3):113–117.
[95] Dong YL, Green KE, Vegiragu S, et al. Evidence for decreased calcitonin gene-related peptide (CGRP) receptors and compromised responsiveness to CGRP of

fetoplacental vessels in preeclamptic pregnancies. *J Clin Endocrinol Metab.* 2005;90(4):2336–2343.

[96] Knerr I, Dachert C, Beinder E, et al. Adrenomedullin, calcitonin gene-related peptide and their receptors: evidence for a decreased placental mRNA content in preeclampsia and HELLP syndrome. *Eur J Obstet Gynecol Reprod Biol.* 2002;101(1): 47–53.

[97] Makino Y, Shibata K, Makino I, Kangawa K, Kawarabayashi T. Alteration of the adrenomedullin receptor components gene expression associated with the blood pressure in pregnancy-induced hypertension. *J Clin Endocrinol Metab.* 2001;86(10): 5079–5082.

[98] Chauhan M, Balakrishnan M, Vidaeff A, et al. Adrenomedullin2 (ADM2)/Intermedin (IMD): a potential role in the pathophysiology of Preeclampsia. *J Clin Endocrinol Metab.* 2016;101:4478–4488. jc20161333.

[99] Fei X, Hongxiang Z, Qi C, Daozhen C. Maternal plasma levels of endothelial dysfunction mediators including AM, CGRP, sICAM-1 and tHcy in pre-eclampsia. *Adv Clin Exp Med.* 2012;21(5):573–579.

[100] Krikun G, Schatz F, Lockwood CJ. Endometrial angiogenesis: from physiology to pathology. *Ann N Y Acad Sci.* 2004;1034:27–35.

[101] Girling JE, Rogers PA. Regulation of endometrial vascular remodelling: role of the vascular endothelial growth factor family and the angiopoietin-TIE signalling system. *Reproduction.* 2009;138(6):883–893.

[102] Santulli G. Angiopoietin-like proteins: a comprehensive look. *Front Endocrinol (Lausanne).* 2014;5:4.

[103] Augustin HG, Koh GY, Thurston G, Alitalo K. Control of vascular morphogenesis and homeostasis through the angiopoietin-Tie system. *Nat Rev Mol Cell Biol.* 2009;10(3):165–177.

[104] Valenzuela DM, Griffiths JA, Rojas J, et al. Angiopoietins 3 and 4: diverging gene counterparts in mice and humans. *Proc Natl Acad Sci U S A.* 1999;96(5): 1904–1909.

[105] Hirchenhain J, Huse I, Hess A, Bielfeld P, De Bruyne F, Krussel JS. Differential expression of angiopoietins 1 and 2 and their receptor Tie-2 in human endometrium. *Mol Hum Reprod.* 2003;9(11):663–669.

[106] Nayak NR, Kuo CJ, Desai TA, et al. Expression, localization and hormonal control of angiopoietin-1 in the rhesus macaque endometrium: potential role in spiral artery growth. *Mol Hum Reprod.* 2005;11(11):791–799.

[107] Ma H, Zhu G. Expression of angiopoietin-1/-2 in the process of mouse embryo implantation. *J Huazhong Univ Sci Technolog Med Sci.* 2007;27(2):200–202.

[108] Wulff C, Wilson H, Dickson SE, Wiegand SJ, Fraser HM. Hemochorial placentation in the primate: expression of vascular endothelial growth factor, angiopoietins, and their receptors throughout pregnancy. *Biol Reprod.* 2002;66(3): 802–812.

[109] Krikun G, Schatz F, Finlay T, et al. Expression of angiopoietin-2 by human endometrial endothelial cells: regulation by hypoxia and inflammation. *Biochem Biophys Res Commun.* 2000;275(1):159–163.

[110] Guo B, Wang W, Li SJ, et al. Differential expression and regulation of angiopoietin-2 in mouse uterus during preimplantation period. *Anat Rec (Hoboken).* 2012;295(2): 338–346.

[111] Woolnough C, Wang Y, Kan CY, Morris JM, Tasevski V, Ashton AW. Source of angiopoietin-2 in the sera of women during pregnancy. *Microvasc Res.* 2012;84(3): 367–374.

[112] Kim M, Allen B, Korhonen EA, et al. Opposing actions of angiopoietin-2 on Tie2 signaling and FOXO1 activation. *J Clin Invest.* 2016;126(9):3511–3525.

[113] Hansen TM, Singh H, Tahir TA, Brindle NP. Effects of angiopoietins-1 and -2 on the receptor tyrosine kinase Tie2 are differentially regulated at the endothelial cell surface. *Cell Signal.* 2010;22(3):527–532.

[114] Yuan HT, Khankin EV, Karumanchi SA, Parikh SM. Angiopoietin 2 is a partial agonist/antagonist of Tie2 signaling in the endothelium. *Mol Cell Biol.* 2009;29(8):2011–2022.

[115] Hewett P, Nijjar S, Shams M, Morgan S, Gupta J, Ahmed A. Down-regulation of angiopoietin-1 expression in menorrhagia. *Am J Pathol.* 2002;160(3):773–780.

[116] Bogdanovic E, Nguyen VP, Dumont DJ. Activation of Tie2 by angiopoietin-1 and angiopoietin-2 results in their release and receptor internalization. *J Cell Sci.* 2006;119(pt 17):3551–3560.

[117] Karar J, Maity A. PI3K/AKT/mTOR pathway in angiogenesis. *Front Mol Neurosci.* 2011;4:51.

[118] Audero E, Cascone I, Maniero F, et al. Adaptor ShcA protein binds tyrosine kinase Tie2 receptor and regulates migration and sprouting but not survival of endothelial cells. *J Biol Chem.* 2004;279(13):13224–13233.

[119] Guo B, Zhang XM, Li SJ, et al. Differential expression and regulation of angiopoietin-3 in mouse uterus during preimplantation period. *J Exp Zool B Mol Dev Evol.* 2012;318(4):316–324.

[120] Lee HJ, Cho CH, Hwang SJ, et al. Biological characterization of angiopoietin-3 and angiopoietin-4. *FASEB J.* 2004;18(11):1200–1208.

[121] Olsen MW, Ley CD, Junker N, Hansen AJ, Lund EL, Kristjansen PE. Angiopoietin-4 inhibits angiogenesis and reduces interstitial fluid pressure. *Neoplasia.* 2006;8(5):364–372.

[122] Dhanabal M, LaRochelle WJ, Jeffers M, et al. Angioarrestin: an antiangiogenic protein with tumor-inhibiting properties. *Cancer Res.* 2002;62(13):3834–3841.

[123] Kubota Y, Oike Y, Satoh S, et al. Cooperative interaction of Angiopoietin-like proteins 1 and 2 in zebrafish vascular development. *Proc Natl Acad Sci U S A.* 2005;102(38):13502–13507.

[124] Scott CA, van Huyen D, Bany BM. Angiopoietin-like gene expression in the mouse uterus during implantation and in response to steroids. *Cell Tissue Res.* 2012;348(1):199–211.

[125] Mints M, Blomgren B, Palmblad J. Expression of angiopoietins 1, 2 and their common receptor tie-2 in relation to the size of endothelial lining gaps and expression of VEGF and VEGF receptors in idiopathic menorrhagia. *Fertil Steril.* 2010;94(2):701–707.

[126] Blumenthal RD, Taylor AP, Goldman L, Brown G, Goldenberg DM. Abnormal expression of the angiopoietins and Tie receptors in menorrhagic endometrium. *Fertil Steril.* 2002;78(6):1294–1300.

[127] Saito M, Watanabe J, Fujisawa T, et al. Angiopoietin-1, 2 and Tie2 expressions in endometrial adenocarcinoma—the Ang2 dominant balance up-regulates tumor angiogenesis in the presence of VEGF. *Eur J Gynaecol Oncol.* 2006;27(2):129–134.

[128] Rocha AL, Reis FM, Taylor RN. Angiogenesis and endometriosis. *Obstet Gynecol Int.* 2013;2013:859619.

[129] Jingting C, Yangde Z, Yi Z, et al. Expression of heparanase and angiopoietin-2 in patients with endometriosis. *Eur J Obstet Gynecol Reprod Biol.* 2008;136(2):199–209.

[130] Hur SE, Lee JY, Moon HS, Chung HW. Angiopoietin-1, angiopoietin-2 and Tie-2 expression in eutopic endometrium in advanced endometriosis. *Mol Hum Reprod.* 2006;12(7):421–426.

[131] Nagymanyoki Z, Growdon WB, Sarno J, et al. Vascularization and expression of angiogenic factors in partial and complete molar pregnancies. *J Reprod Med.* 2008;53(8):589–594.

[132] Singh M, Kindelberger D, Nagymanyoki Z, et al. Vascular endothelial growth factors and their receptors and regulators in gestational trophoblastic diseases and normal placenta. *J Reprod Med.* 2012;57(5-6):197–203.
[133] Tseng JJ, Chou MM. Differential expression of growth-, angiogenesis- and invasion-related factors in the development of placenta accreta. *Taiwan J Obstet Gynecol* 2006;45(2):100–106.
[134] Andraweera PH, Dekker GA, Thompson SD, et al. A functional variant in ANGPT1 and the risk of pregnancies with hypertensive disorders and small-for-gestational-age infants. *Mol Hum Reprod.* 2012;18(6):325–332.
[135] Tuuri AL, Jauhiainen MS, Ehnholm CP, Tikkanen MJ, Nicholls MG, Kaaja RJ. Elevated serum angiopoietin-like protein 6 in women with subsequent pregnancy-induced hypertension: a preliminary study. *Hypertens Pregnancy.* 2013;32(3):203–213.
[136] Bao JX, Su YT, Cheng YP, Zhang HJ, Xie XP, Chang YM. Vascular sphingolipids in physiological and pathological adaptation. *Front Biosci (Landmark Ed).* 2016;21: 1168–1186.
[137] English D, Brindley DN, Spiegel S, Garcia JG. Lipid mediators of angiogenesis and the signalling pathways they initiate. *Biochim Biophys Acta.* 2002;1582(1-3):228–239.
[138] Kaneko-Tarui T, Zhang L, Austin KJ, et al. Maternal and embryonic control of uterine sphingolipid-metabolizing enzymes during murine embryo implantation. *Biol Reprod.* 2007;77(4):658–665.
[139] Dunlap KA, Kwak HI, Burghardt RC, et al. The sphingosine 1-phosphate (S1P) signaling pathway is regulated during pregnancy in sheep. *Biol Reprod.* 2010;82(5):876–887.
[140] Skaznik-Wikiel ME, Kaneko-Tarui T, Kashiwagi A, Pru JK. Sphingosine-1-phosphate receptor expression and signaling correlate with uterine prostaglandin-endoperoxide synthase 2 expression and angiogenesis during early pregnancy. *Biol Reprod.* 2006;74(3):569–576.
[141] Mizugishi K, Li C, Olivera A, et al. Maternal disturbance in activated sphingolipid metabolism causes pregnancy loss in mice. *J Clin Invest.* 2007;117(10):2993–3006.
[142] Ye X, Hama K, Contos JJ, et al. LPA3-mediated lysophosphatidic acid signalling in embryo implantation and spacing. *Nature.* 2005;435(7038):104–108.
[143] Nagamatsu T, Iwasawa-Kawai Y, Ichikawa M, et al. Emerging roles for lysophospholipid mediators in pregnancy. *Am J Reprod Immunol.* 2014;72(2):182–191.
[144] Baig S, Lim JY, Fernandis AZ, et al. Lipidomic analysis of human placental syncytiotrophoblast microvesicles in adverse pregnancy outcomes. *Placenta.* 2013; 34(5):436–442.
[145] West CA, Sasser JM, Baylis C. The enigma of continual plasma volume expansion in pregnancy; critical role of the renin/angiotensin/aldosterone system. *Am J Physiol Renal Physiol.* 2016;311:F1125–F1134.
[146] Fyhrquist F, Saijonmaa O. Renin-angiotensin system revisited. *J Intern Med.* 2008;264(3):224–236.
[147] Allred AJ, Diz DI, Ferrario CM, Chappell MC. Pathways for angiotensin-(1-7) metabolism in pulmonary and renal tissues. *Am J Physiol Renal Physiol.* 2000;279(5): F841–F850.
[148] Guang C, Phillips RD, Jiang B, Milani F. Three key proteases—angiotensin-I-converting enzyme (ACE), ACE2 and renin—within and beyond the renin-angiotensin system. *Arch Cardiovasc Dis.* 2012;105(6-7):373–385.
[149] Neves LA, Chappell MC, Ferrario CM, Gallagher PE, Ganten D, Brosnihan KB. Effect of estrogen on neprilysin expression in uterus and kidney of Sprague-Dawley normotensive and heterozygous (mRen2)27-transgenic hypertensive rats. *Peptides.* 2006;27(11):2912–2918.

[150] Peach MJ. Renin-angiotensin system: biochemistry and mechanisms of action. *Physiol Rev*. 1977;57(2):313–370.
[151] Gant NF, Daley GL, Chand S, Whalley PJ, MacDonald PC. A study of angiotensin II pressor response throughout primigravid pregnancy. *J Clin Invest*. 1973;52(11): 2682–2689.
[152] Skinner SL, Lumbers ER, Symonds EM. Renin concentration in human fetal and maternal tissues. *Am J Obstet Gynecol*. 1968;101(4):529–533.
[153] Metsarinne KP, Pekonen F, Fyhrquist FY. Localization of renin substrate in fetoplacental tissue. *Acta Endocrinol (Copenh)*. 1990;123(1):113–117.
[154] Skinner SL, Lumbers ER, Symonds EM. Analysis of changes in the renin-angiotensin system during pregnancy. *Clin Sci*. 1972;42(4):479–488.
[155] Paul M, Wagner J, Dzau VJ. Gene expression of the renin-angiotensin system in human tissues. Quantitative analysis by the polymerase chain reaction. *J Clin Invest*. 1993;91(5):2058–2064.
[156] Kalenga MK, Thomas K, de Gasparo M, De Hertogh R. Determination of renin, angiotensin converting enzyme and angiotensin II levels in human placenta, chorion and amnion from women with pregnancy induced hypertension. *Clin Endocrinol (Oxf)*. 1996;44(4):429–433.
[157] Kalenga MK, De Hertogh R, Whitebread S, Vankrieken L, Thomas K, De Gasparo M. Distribution of the concentrations of angiotensin II (A II), A II receptors, hPL, prolactin, and steroids in human fetal membranes. *Rev Fr Gynecol Obstet*. 1991;86(10):585–591.
[158] Nielsen AH, Schauser KH, Poulsen K. Current topic: the uteroplacental renin-angiotensin system. *Placenta*. 2000;21(5-6):468–477.
[159] Neves LA, Stovall K, Joyner J, et al. ACE2 and ANG-(1-7) in the rat uterus during early and late gestation. *Am J Physiol Regul Integr Comp Physiol*. 2008;294(1): R151–R161.
[160] Vaz-Silva J, Carneiro MM, Ferreira MC, et al. The vasoactive peptide angiotensin-(1-7), its receptor Mas and the angiotensin-converting enzyme type 2 are expressed in the human endometrium. *Reprod Sci*. 2009;16(3):247–256.
[161] Yamaleyeva LM, Neves LA, Coveleskie K, Diz DI, Gallagher PE, Brosnihan KB. AT1, AT2, and AT(1-7) receptor expression in the uteroplacental unit of normotensive and hypertensive rats during early and late pregnancy. *Placenta*. 2013;34(6): 497–502.
[162] Anton L, Brosnihan KB. Systemic and uteroplacental renin-angiotensin system in normal and pre-eclamptic pregnancies. *Ther Adv Cardiovasc Dis*. 2008;2(5): 349–362.
[163] Poisner AM. The human placental renin-angiotensin system. *Front Neuroendocrinol*. 1998;19(3):232–252.
[164] Dong X, Han S, Zylka MJ, Simon MI, Anderson DJ. A diverse family of GPCRs expressed in specific subsets of nociceptive sensory neurons. *Cell*. 2001;106(5): 619–632.
[165] Santos RA, Simoes e Silva AC, Maric C, et al. Angiotensin-(1-7) is an endogenous ligand for the G protein-coupled receptor Mas. *Proc Natl Acad Sci U S A*. 2003; 100(14):8258–8263.
[166] Xu P, Costa-Goncalves AC, Todiras M, et al. Endothelial dysfunction and elevated blood pressure in MAS gene-deleted mice. *Hypertension*. 2008;51(2): 574–580.
[167] Xu P, Sriramula S, Lazartigues E. ACE2/ANG-(1-7)/Mas pathway in the brain: the axis of good. *Am J Physiol Regul Integr Comp Physiol*. 2011;300(4):R804–R817.
[168] Fraga-Silva RA, Pinheiro SV, Goncalves AC, Alenina N, Bader M, Santos RA. The antithrombotic effect of angiotensin-(1-7) involves mas-mediated NO release from platelets. *Mol Med*. 2008;14(1-2):28–35.

[169] Alenina N, Xu P, Rentzsch B, Patkin EL, Bader M. Genetically altered animal models for Mas and angiotensin-(1-7). *Exp Physiol.* 2008;93(5):528–537.
[170] Gembardt F, Grajewski S, Vahl M, Schultheiss HP, Walther T. Angiotensin metabolites can stimulate receptors of the Mas-related genes family. *Mol Cell Biochem.* 2008;319(1-2):115–123.
[171] Kashiwagi A, DiGirolamo CM, Kanda Y, et al. The postimplantation embryo differentially regulates endometrial gene expression and decidualization. *Endocrinology.* 2007;148(9):4173–4184.
[172] Merrill DC, Karoly M, Chen K, Ferrario CM, Brosnihan KB. Angiotensin-(1-7) in normal and preeclamptic pregnancy. *Endocrine.* 2002;18(3):239–245.
[173] Anton L, Merrill DC, Neves LA, Gruver C, Moorefield C, Brosnihan KB. Angiotensin II and angiotensin-(1-7) decrease sFlt1 release in normal but not preeclamptic chorionic villi: an in vitro study. *Reprod Biol Endocrinol.* 2010;8:135.
[174] Verdonk K, Visser W, Van Den Meiracker AH, Danser AH. The renin-angiotensin-aldosterone system in pre-eclampsia: the delicate balance between good and bad. *Clin Sci (Lond).* 2014;126(8):537–544.
[175] Seki H. The role of the renin-angiotensin system in the pathogenesis of preeclampsia—new insights into the renin-angiotensin system in preeclampsia. *Med Hypotheses.* 2014;82(3):362–367.
[176] Herse F, LaMarca B. Angiotensin II type 1 receptor autoantibody (AT1-AA)-mediated pregnancy hypertension. *Am J Reprod Immunol.* 2013;69(4):413–418.
[177] Bai K, Wang K, Li X, et al. Autoantibody against angiotensin AT1 receptor from preeclamptic patients enhances collagen-induced human platelet aggregation. *Acta Biochim Biophys Sin (Shanghai).* 2013;45(9):749–755.
[178] Yang X, Wang F, Lau WB, et al. Autoantibodies isolated from preeclamptic patients induce endothelial dysfunction via interaction with the angiotensin II AT1 receptor. *Cardiovasc Toxicol.* 2014;14(1):21–29.
[179] Liu F, Wang Y, Wang X, Zheng Y, Jin Z, Zhi J. Role of agonistic autoantibodies against type-1 angiotensin II receptor in the pathogenesis of retinopathy in preeclampsia. *Sci Rep.* 2016;6:29036.
[180] Cunningham Jr MW, Williams JM, Amaral L, et al. Agonistic autoantibodies to the Angiotensin II Type 1 receptor enhance Angiotensin II-induced renal vascular sensitivity and reduce renal function during pregnancy. *Hypertension.* 2016;68(5): 1308–1313.
[181] Mitchell JA, Yochim JM. Intrauterine oxygen tension during the estrous cycle in the rat: its relation to uterine respiration and vascular activity. *Endocrinology.* 1968;83(4):701–705.
[182] Rodesch F, Simon P, Donner C, Jauniaux E. Oxygen measurements in endometrial and trophoblastic tissues during early pregnancy. *Obstet Gynecol.* 1992;80(2):283–285.
[183] Folkman J, Hahnfeldt P, Hlatky L. Cancer: looking outside the genome. *Nat Rev Mol Cell Biol.* 2000;1(1):76–79.
[184] Gatenby RA, Gillies RJ. Why do cancers have high aerobic glycolysis? *Nat Rev Cancer.* 2004;4(11):891–899.
[185] Eliasson P, Jonsson JI. The hematopoietic stem cell niche: low in oxygen but a nice place to be. *J Cell Physiol.* 2010;222(1):17–22.
[186] Panchision DM. The role of oxygen in regulating neural stem cells in development and disease. *J Cell Physiol.* 2009;220(3):562–568.
[187] Silvan U, Diez-Torre A, Arluzea J, Andrade R, Silio M, Arechaga J. Hypoxia and pluripotency in embryonic and embryonal carcinoma stem cell biology. *Differentiation.* 2009;78(2-3):159–168.
[188] Schofield R. The relationship between the spleen colony-forming cell and the haemopoietic stem cell. *Blood Cells.* 1978;4(1-2):7–25.

[189] Ordway GA, Garry DJ. Myoglobin: an essential hemoprotein in striated muscle. *J Exp Biol.* 2004;207(pt 20):3441–3446.
[190] Hendgen-Cotta UB, Kelm M, Rassaf T. Myoglobin functions in the heart. *Free Radic Biol Med.* 2014;73:252–259.
[191] Baez E, Echeverria V, Cabezas R, Avila-Rodriguez M, Garcia-Segura LM, Barreto GE. Protection by neuroglobin expression in brain pathologies. *Front Neurol.* 2016;7:146.
[192] Burmester T, Hankeln T. Function and evolution of vertebrate globins. *Acta Physiol (Oxf).* 2014;211(3):501–514.
[193] Brown HM, Anastasi MR, Frank LA, et al. Hemoglobin: a gas transport molecule that is hormonally regulated in the ovarian follicle in mice and humans. *Biol Reprod.* 2015;92(1):26.
[194] Yasuda Y, Sasaki T, Takagawa M, et al. Erythropoietin contributes to implantation: ectopic hemoglobin synthesis in decidual cells of mice. *Congenit Anom (Kyoto).* 2007;47(1):22–33.
[195] Shiota M, Yasuda Y, Shimaoka M, et al. Erythropoietin is involved in hemoprotein syntheses in developing human decidua. *Congenit Anom (Kyoto).* 2013;53(1):18–26.

CHAPTER EIGHT

Regulation of Placental Amino Acid Transport and Fetal Growth

O.R. Vaughan[1], F.J. Rosario, T.L. Powell, T. Jansson
University of Colorado Anschutz Medical Campus, Aurora, CO, United States
[1]corresponding author: e-mail address: owen.vaughan@ucdenver.edu

Contents

Abstract

The fetus requires amino acids for the processes of protein synthesis, carbon accretion, oxidative metabolism, and biosynthesis, which ultimately determine growth rate in utero. The fetal supply of amino acids is critically dependent on the transport capacity of the placenta. System A amino acid transporters in the syncytiotrophoblast microvillous plasma membrane, directed toward maternal blood, actively accumulate amino acids, while system L exchangers mediate uptake of essential amino acids from the maternal circulation. The functional capacity and protein abundance of these transporters in the placenta are related to fetal growth in both humans and experimental animals. Maternal nutritional and endocrine signals including insulin, insulin-like growth factors, adipokines, and steroid hormones regulate placental amino acid transport, against the background of growth signals originating from the fetus. Anabolic signals of abundant maternal resource availability stimulate placental amino acid transport to optimize offspring fitness, whereas catabolic signals reduce placental amino acid transport in an attempt to ensure survival and long-term reproductive capacity of the mother

Progress in Molecular Biology and Translational Science, Volume 145
ISSN 1877-1173
http://dx.doi.org/10.1016/bs.pmbts.2016.12.008

when resources are scarce. These signals regulate placental amino acid transport by controlling transcription, translation, plasma membrane trafficking, and degradation of transporters. Adaptations in placental amino acid transport capacity may underlie either under- or overgrowth of the fetus when maternal nutrient and hormone levels are altered as a result of altered maternal nutrition or metabolic disease. Strategies to modulate placental amino acid transport may prove effective to normalize fetal growth in intrauterine growth restriction and fetal overgrowth.

1. INTRODUCTION

Fetal growth is a major determinant of pregnancy outcome and lifelong health. Both restricted and excessive growth in utero are associated with increased perinatal morbidity and mortality.[1,2] Moreover, birth weight predicts lifelong risk of cardiovascular and metabolic disease.[3] Fetal growth requires accretion of a net quantity of protein, synthesized entirely from the umbilical supply of amino acids.[4] To meet the needs of rapid fetal protein synthesis, particularly in skeletal muscle, liver, and gut, amino acids must be supplied at a rate estimated to be between 10 and 60 g/day per kg fetus.[5,6] Amino acids taken up by the fetus are also used for oxidative production of ATP, carbon accretion, interorgan nitrogen cycling and in the biosynthesis of other molecules including haem, porphyrins, nitric oxide, neurotransmitters, and nucleotides.[7,8] All of these requirements must be satisfied by placental transfer of amino acids from the mother to the fetus. Moreover, as in postnatal life, essential amino acids cannot be synthesized by the fetus and must therefore be derived directly from the maternal circulation. Hence, total placental amino acid delivery and fetal availability of specific amino acids are direct determinants of the rate of fetal growth. This is supported by studies showing plasma amino acid concentrations tend to be lower in human fetuses that are growth restricted,[9,10] and fetal uptake of essential amino acids has been reported to be reduced in growth-restricted fetuses both in humans and experimental animals.[11–14]

Several biophysical factors determine the net fetal umbilical uptake of each individual amino acid. The fetal concentration of some amino acids is correlated with maternal concentrations.[10] However, because uptake of amino acids from the uteroplacental circulation is carrier mediated, the rate of total amino acid uptake is influenced by the absolute size and surface area of the placenta, and the rate of uterine blood flow. Amino acids taken up by the syncytiotrophoblast from the maternal circulation may either be transferred directly to the fetus or utilized in oxidative or anabolic processes within the placenta. Therefore, the net umbilical uptake rate of each amino acid also depends upon its rate of metabolism within the placenta.[12,15] In some cases, placental amino acid metabolism may be an important mechanism by which the fetus obtains certain amino acids. For example, there is net placental uptake of glutamate from both the maternal and fetal circulations, and this glutamate is converted into glycine and released to the fetus.[16,17] The regulation of placental morphology and metabolism and their importance in determining fetal growth rates have been reviewed in detail recently.[18,19] This review will provide an overview of mechanisms of placental amino acid transport across the maternal blood/trophoblast interface and their importance for fetal growth. It will also examine the various endocrine and nutritional factors known to alter rates of placental amino acid transport and the molecular mechanisms that may underlie these processes, with particular emphasis on the haemochorial placentae of humans, nonhuman primates, and rodents.

2. PLACENTAL AMINO ACID TRANSPORT AND FETAL GROWTH

Concentrations of most amino acids are higher in fetal plasma than in maternal plasma, indicating that they are actively accumulated across the syncytiotrophoblast, the transporting and hormone-producing epithelium of the human placenta.[9,20] Such directional transfer requires the coordinated action of more than 20 different amino acid transporter proteins localized both to the maternal and fetal facing plasma membrane of the epithelium.[8,21,22] These proteins may be broadly classified as accumulative transporters, exchangers, or facilitated transporters (Fig. 1). Briefly, accumulative transporters mediate net uptake into the trophoblast from maternal or fetal blood, using either the inwardly directed electrochemical gradient for Na^+ or the transmembrane potential difference to drive active transport and thereby establish high intracellular concentrations of both neutral and charged amino acids. Exchangers use an antiport mechanism that effluxes accumulated amino acids from the trophoblast cytosol, allowing uptake of essential amino acids into the placenta. Finally, facilitated transporters mediate diffusion of amino acids down their concentration gradient, a process necessary to allow accumulated amino acids to efflux from the trophoblast cytosol into fetal circulation and thus net uptake of amino acids by the fetus.[23] The cellular localization of amino acid transporters on the

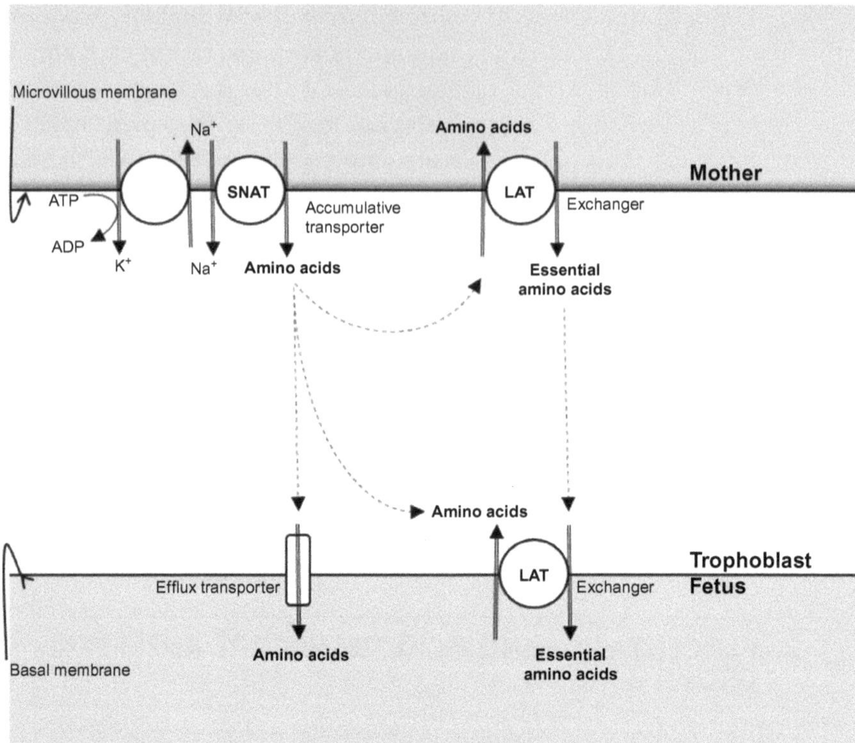

Fig. 1 Placental amino acid transport mechanisms. Schematic diagram showing mechanisms of net amino acid transport from mother to fetus and classification of amino acid transporters on the trophoblast plasma membrane. *LAT*, system L, sodium-independent neutral amino acid transporter; *SNAT*, system A, sodium-dependent neutral amino acid transporter.

syncytiotrophoblast epithelium is polarized, with differing complements in the maternal (microvillous) and fetal (basal) facing membranes.[8] Often, amino acid transporting systems are able to carry more than one type of amino acid, while each amino acid may be transported by several different systems, such that substrate competition exists at the level of the transporter protein.

Active uptake of neutral amino acids into the syncytiotrophoblast is mediated by the sodium-dependent system A family of accumulative amino acid transporters (SNAT), in parallel with the ASC, B^0, N, and Gly systems. System A activity in the placenta is attributable to the proteins SNAT1, 2, and 4, which are localized predominantly to the maternal facing membrane of the syncytiotrophoblast in humans, nonhuman primates, and rodents.[24–27] Expression and activity of the system A transporters increase with gestational age and increasing fetal size in all species studied to date.[24,28,29] The system L transporter proteins are exchangers responsible for the uptake of essential amino acids, such as leucine, in exchange for non-essential amino acids.[8] System L isoforms LAT1 and LAT2 have overlapping substrate specificity with the system A transporters and are located predominantly on the maternal face of the syncytiotrophoblast, with lower expression of LAT2 in the basal plasma membrane.[30] Recent evidence from a computational modeling approach suggests that system L transporters may also act as facilitated transporters in the microvillous membrane.[31–33] Thus, the combined activities of systems A and L are critical in determining the fetal supply of neutral, and in particular, essential amino acids for fetal biosynthesis and metabolism. In addition, the sodium-dependent system β transporter, TAUT, plays an important role in the transplacental delivery of the taurine to the fetus, where it is essential for antioxidant and osmoregulatory processes, as well as neurological development.[34,35] Transporters for cationic (systems y^+, y^+L, $b^{0,+}$) and anionic (system X^-_{AG}) amino acids are also present in the placentae of humans and experimental animals, although their precise importance in determining the rate of fetal growth is less well studied.[36–39]

Placental expression and activity of the system A amino acid transporters are strongly linked to fetal growth in humans and experimental animals. System A-dependent uptake of the nonmetabolizable amino acid analog, methylaminoisobutyric acid (MeAIB), along with SNAT2 gene expression and protein abundance are reduced in term microvillous membrane samples from women with intrauterine growth-restricted or small for gestational age fetuses.[40–43] Similarly, fetal growth restriction induced experimentally through surgical ligation of the uterine artery or genetic manipulation is associated with reduced syncytiotrophoblast membrane SNAT protein abundance and activity in mice, rats, and guinea pigs.[44–47] Indeed, when delivery of amino acids to the fetus by placental system A activity is inhibited by continuously infusing pregnant rats with nonmetabolizable MeAIB from mid-pregnancy, fetal weight is reduced near term.[48] System L-mediated leucine uptake capacity is reduced in microvillous and basal plasma membrane vesicles isolated from growth-restricted human placentae,[49] and system L activity may be further reduced in vivo due to diminished intracellular amino acid concentrations, secondary to lower system A activity, which inhibits the exchange function of these transporters. Intrauterine growth restriction is also associated with decreases in abundance and/or activity of the system ASC and system β amino acid transporters in vitro.[50,51] Studies clearly demonstrate that in vivo transplacental clearance of aminocyclopentane-1-carboxylic acid, a nonspecific, nonmetabolizable substrate for transport systems A, ASC, and L is reduced in growth-restricted sheep fetuses in late gestation.[52] In contrast, fetal overgrowth is associated with elevated system A and L transporter activity in isolated placental microvillous membranes from obese and diabetic women[53,54] and in vivo in experimentally overnourished rodents.[25,55] Taken together, these findings suggest that placental amino acid transport capacity is a major determinant of growth in utero and that alterations in placental amino acid transport may be a contributing factor to both restricted and excessive fetal weight. Alterations in in vivo amino acid transport capacity may reflect changes in total surface area of the syncytiotrophoblast epithelium, its membrane potential, and/or the abundance and activity of the amino acid transporters themselves.

3. PHYSIOLOGICAL REGULATORS OF PLACENTAL AMINO ACID TRANSPORT

Growth in utero is regulated by several endocrine and metabolic signals, including insulin, insulin-like growth factor (IGF)-I and -II, adipose-derived cytokines (adipokines), and glucocorticoids.[56–60] Abnormal fetal growth is often associated with perturbations in the levels of these signals in either the mother, the fetus, or both. Because fetal growth is ultimately dependent upon placental nutrient delivery and the placenta expresses receptors for the endocrine growth regulators,[61–65] it is a key target and effector of growth stimulatory and inhibitory signaling.[66] In addition, the placenta acts as a direct sensor of nutrient and oxygen availability and regulates both overall placental growth and nutrient transport capacity accordingly.[67] Consequently, restricted or excessive fetal growth may be secondary to up- or downregulation of placental amino acid transport in response to maternal and/or fetal signals, nutrient, and oxygen availability.

3.1 Oxygen and Nutrient Availability

Fetal weight at term is related to plasma glucose,[68] amino acid,[10,20] and fatty acid concentrations,[69,70] and blood pO_2[37,71] in both maternal and fetal circulations. Fetal growth is increased when maternal nutrient availability is increased in women with obesity or poorly controlled gestational diabetes[54] or by provision of a high calorie diet to experimental animals (Table 2). In part, these relationships may reflect the impact of maternal nutrient concentrations on placental amino acid transport capacity. Both system A and system L amino acid transport capacity are increased in microvillous membrane isolated from the placentae of women with type I or gestational diabetes, who tend to deliver overgrown fetuses.[53] Specific changes in the complement of system $y^{+}L$ transporter isoforms in the trophoblast may alter the transport capacity of the placenta for essential amino acids in gestational diabetes.[72] Maternal obesity also increases system A and system β amino acid transport in placental microvillous membrane vesicles and villous explants, in association with increased maternal protein, fat, and carbohydrate intake.[54,73,74] However, system A-mediated amino acid uptake has also been reported to be reduced in obese women and nonhuman primates, suggesting that it may be sensitive to the circulating levels of specific nutrients or hormones, rather than only maternal adiposity.[75,76] Incubation of primary human trophoblast cells in vitro with physiological levels of the monounsaturated fatty acid, oleic acid, stimulates uptake of the system A substrate MeAIB, whereas incubation with the long chain polyunsaturated fatty acid docosahexaenoic acid inhibits system A (Table 1). System L-mediated leucine uptake in primary human trophoblasts decreases in a dose-dependent manner with increasing glucose concentrations in the culture medium,[79] although system A-mediated MeAIB uptake does not change (Table 1). When pregnant mice are fed a diet high in fat and sugar from conception, fetal weight is initially reduced below control values.[93] Subsequently, placental system A amino acid transport is transiently increased in late gestation, concomitant with impaired maternal glucose tolerance[94] (Table 2). As a result, fetal growth rate is increased in late gestation and normal weight achieved at term, albeit via an alternate trajectory.[93] Provision of mice with a high calorie diet before pregnancy, such that dams are obese at conception, results in upregulation of placental system A transport capacity and SNAT2 expression, which is sustained up to term with overt fetal overgrowth (Table 2). System L transport may also be upregulated in the placentae of these overnourished mice, dependent on the degree of maternal prepregnancy obesity.[25] These observations suggest that an excess of glucose or fatty acids in the maternal circulation generally stimulates placental amino acid transport, favoring greater fetal growth.

Whereas maternal overnutrition can lead to accelerated fetal growth, birth weight is reduced in mothers subjected to protein–energy malnutrition during pregnancy.[109] Experimentally, restriction of maternal calorie or protein intake (Table 2), or chronic reductions in fetal oxygen, glucose, and amino acid availability reduce fetal growth.[110,111] Although little is known

Table 1 Effects of Incubation With Nutrients, Hormones, and Cytokines on System A-Mediated Amino Acid Transport in Human Trophoblast Cells

Manipulation		Cell Type	System A Amino Acid Transport Capacity	References
Oleic acid	400 μmol L^{-1}, 24 h	PHT	↑100%	77
Doxosahexanoic acid	25 μmol L^{-1}, 24 h	PHT	↓40%	78
	50 μmol L^{-1}, 24 h	PHT	↓60%	
Palmitic acid	100 μmol L^{-1}, 24 h	PHT	No Δ	
Glucose deprivation	0.5 vs 16.0 mmol L^{-1}, 24 h	PHT	No Δ	79
Amino acid deprivation	Nonessential amino acids, 6 h	BeWo	↑225%†	80
	All amino acids, 4 h	BeWo	↑600%†	81
Hypoxia	3% vs 20%, 24 h	PHT	↓37%	82
	1% vs 20%, 24 h	PHT	↓82%	
Insulin	0.6 ng mL^{-1}, 1 h	PVF	↑47%	83
	60 ng mL^{-1}, 24 h	PHT	↑25%	79
	300 ng mL^{-1}, 1 h	PVF	↑56%	83
	600 ng mL^{-1}, 4 h	PHT	↑70%	84
IGF1	300 ng mL^{-1}, 24 h	PHT	↑30%	79
Growth hormone	600 ng mL^{-1}, 1 h	PVF	No Δ	83
Leptin	500 ng mL^{-1}, 1 h	PVF	↑37%	83
	1000 ng mL^{-1}, 1 h	PVF	↑45%	85
TNFα	10 pg mL^{-1}, 24 h	PHT	↑60%	86
IL6	20 pg mL^{-1}, 24 h	PHT	↑100%	87
IL1β#	10 pg mL^{-1}, 24 h	PHT	↓27%	88
Adiponectin#	5 μg mL^{-1}, 20 h	PHT	↓58%	89
Cortisol	340 ng mL^{-1}, 1 h	PVF	No Δ	83
	1000 nmol L^{-1}, 24 h	BeWo	↑125%†	90
Dexamethasone	1000 nmol L^{-1}, 48 h	VE	↑28%	91
Angiotensin II	100 $mmol^{-1}$, 2 h	PVF	↑25%	85
	500 $nmol^{-1}$, 1 h	PVF	↓47%	92

BeWo, choriocarcinoma cell line; *PHT*, primary human trophoblast; *PVF*, primary villous fragments; *VE*, villous explant; ↑, increased, ↓, decreased, No Δ, no change relative to control, †, transcellular transport, #, insulin present (1 nmol L^{-1}).

Table 2 Effect of Maternal Experimental Manipulations on Placental System A Amino Acid Transport Capacity, Transport Abundance, and Fetal Weight Near Term in Rodents

			Placental System A Amino Acid Transport			
Manipulation		**Species**	**Capacity**	**Transporter Abundance**	**Fetal Weight**	**References**
High calorie diet	30% fat D0–19	Mouse	No Δ	No Δ	No Δ	93
	32% fat from before conception	Mouse	↑900%	↑SNAT2	↑42%	55
	41% fat from before conception	Mouse	↑425%	↑SNAT2	↑29%	25,95
Total calorie restriction	80% D3–D19	Mouse	↑66%	↑*Slc38a2*	↓13%	96,97
	50% D10–D19	Mouse	No Δ	↑*Slc38a1*, *Slc38a2*	↓48%	98
	50% D0–20	Rat	↓32%	?	↓15%	99
Low protein diet	18% vs 23% D3–19	Mouse	↓25%	↓*Slc38a4*	No Δ	100
	9% vs 23% D3–19	Mouse	No Δ	↓*Slc38a1*, *Slc38a4*	↓9%	
	4% vs 18% D2–21	Rat	↓36%	↓SNAT1, SNAT2	↓23%	101,102
	5% vs 20% D6–20	Rat	↓48%*	?	↓21%	39
	5% vs 21% D0–21	Rat	↓61%	?	↓28%	103
Hypoxia	13% vs 21% D14–19	Mouse	No Δ	↑*Slc38a1*	↓5%	104
	10% vs 21% D14–19	Mouse	↓39%	No Δ	↓9%	
Adiponectin	s.c. D15–19	Mouse	↓66%*	↓SNAT1, 2, 4	↓19%	105

Continued

Table 2 Effect of Maternal Experimental Manipulations on Placental System A Amino Acid Transport Capacity, Transport Abundance, and Fetal Weight Near Term in Rodents—cont'd

			Placental System A Amino Acid Transport			
Manipulation		**Species**	**Capacity**	**Transporter Abundance**	**Fetal Weight**	**References**
Testosterone	s.c. D15–19	Rat	↓30%	↓*Slc38a2*,↓SNAT2	↓7%	106
Corticosterone	Oral D11–16	Mouse	↑33%	↓*Slc38a2*	↓5%	107
	Oral D14–19	Mouse	↓46%	↑*Slc38a1*	↓16%	
Dexamethasone	Oral D11–16	Mouse	↑75%	↓*Slc38a2*	↓5%	108
	Oral D14–19	Mouse	No Δ	No Δ	↓5%	
	s.c. D14,15	Mouse	↓46%	No Δ	No Δ	29

s.c., subcutaneous; ↑, increased; ↓, decreased; No Δ, no change relative to control; *, measured ex vivo in trophoblast plasma membrane vesicles.

about placental amino acid transport in malnourished women, system A-mediated transport is decreased from the beginning of the third trimester through to term when baboons are fed 70% of normal daily food intake from early pregnancy, even though maternal plasma amino acid profiles are largely unchanged.[112] Downregulation of system A transport capacity precedes the reduction in fetal growth in these baboons.[26,113] Similarly, in rodents, restriction of dietary protein content from early pregnancy reduces near-term activity of the amino acid transport systems A, L, and y^+ on the maternal facing surface of the placental epithelium, in addition to system X_{AG}^- on the fetal facing surface (Table 2).[39] As in the calorie-restricted non-human primate,[26,113] downregulation of amino acid transport precedes fetal growth restriction,[101] and the degree of transport downregulation and fetal growth restriction are proportional to the severity and duration of protein restriction (Table 2). Global calorie restriction throughout pregnancy also appears to reduce system A amino acid transport in the rat (Table 2) and system L activity in the mouse,[98,114] further supporting a relationship between maternal nutritional state and placental function in determining fetal growth rates. However, culturing BeWo choriocarcinoma cells in medium depleted of amino acids for up to 48 h increases sodium-dependent uptake of MeAIB (Table 1), along with abundance and plasma membrane localization of SNAT2.[80] This response is known as adaptive regulation, although whether this type of regulation occurs also in the placenta in vivo remains to be fully established. It is however likely that a multitude of signals, sometimes with opposing effects, concurrently impinge on placental amino acid transporters, and the net effect on amino acids transport activity depends on the sum of all factors. Indeed, transplacental MeAIB clearance is increased when maternal plasma α-amino nitrogen is low in mice restricted to 80% of normal food intake.[96,97] Taken together, these observations may suggest that reduced amino acid availability induces adaptive upregulation of system A amino acid transport in trophoblast in vitro or in mice during mild global undernutrition. In vivo placental amino acid transport has been reported to be downregulated in response to maternal protein restriction, which is believed to be mediated by circulating maternal signals of negative energy balance, as reflected by the fact that amino acid availability in maternal blood is sustained by catabolism of maternal tissues.[115] Perturbed placental amino acid transport may also underlie reduced fetal growth as a result of low oxygen availability. This is observed both in pregnant women at high altitude and in experimental animals exposed to normobaric hypoxia.[116,117] System A amino acid transport has been clearly shown to be downregulated by reduced pO_2, both in trophoblast cells in vitro (Table 1) and in pregnant mice in vivo, depending on the degree of hypoxia and the associated alterations in maternal nutrition (Table 2). Collectively, these data indicate that the activity of system A amino acid transporters responds adaptively to signals of nutrient availability, but adaptation is also subject to regulation by general signals of maternal energy balance and oxygen delivery. Whether placental amino acid transport capacity per se is also directly responsive to fetal nutrient status is not clear, although there are alterations in net fetal amino acid uptake and metabolism during both chronic fetal hypo- and hyperglycaemia.[110,118–121]

3.2 Insulin and Insulin-Like Growth Factors

Insulin, derived from the fetal pancreas, is the predominant promoter of longitudinal growth in utero, although many of its effects are also mediated by IGFs acting at both local and systemic levels.[56] The placenta is exposed to circulating insulin and IGFs originating from both the mother and fetus, and since both *IGF1* and *IGF2* genes are expressed within the placenta, IGF1 and 2 may also act in a paracrine manner.[122,123] In humans, maternal concentrations of insulin and IGF1 are correlated with birth weight.[124] Often, fetal overgrowth occurs when maternal insulin levels are elevated gestational diabetes mellitus.[53] Moreover, experimental manipulation of maternal or fetal concentrations of insulin, IGF1 or IGF2, by direct infusion or transgenic overexpression increases fetal weight, whereas genetic deficiency of these hormones or their receptors results in fetal growth restriction.[56,125–129]

Elevated maternal insulin concentrations may, in part, underlie the increases in placental amino acid transport observed in diabetic and obese pregnant women.[53,54,74] Incubation of primary human trophoblast with either insulin or IGF1 increases both the rate of uptake and maximum accumulated concentration of nonmetabolizable MeAIB via system A transporters (Table 1).[83,84,130] This stimulatory effect of insulin in vitro is most apparent over the physiological range of concentrations (<1 ng/mL) (Fig. 2). Adenoviral transfection of BeWo cells with IGF1 enhances both system A- and system L-mediated amino acid uptake and transporter expression at the gene and protein levels.[131] Similarly, administration of IGF1 to guinea pigs in mid-pregnancy increases placental MeAIB transport and expression of the system A transporter gene, *Slc38a2*, concomitant with increased fetal growth.[132] Transgenic overexpression of bovine growth

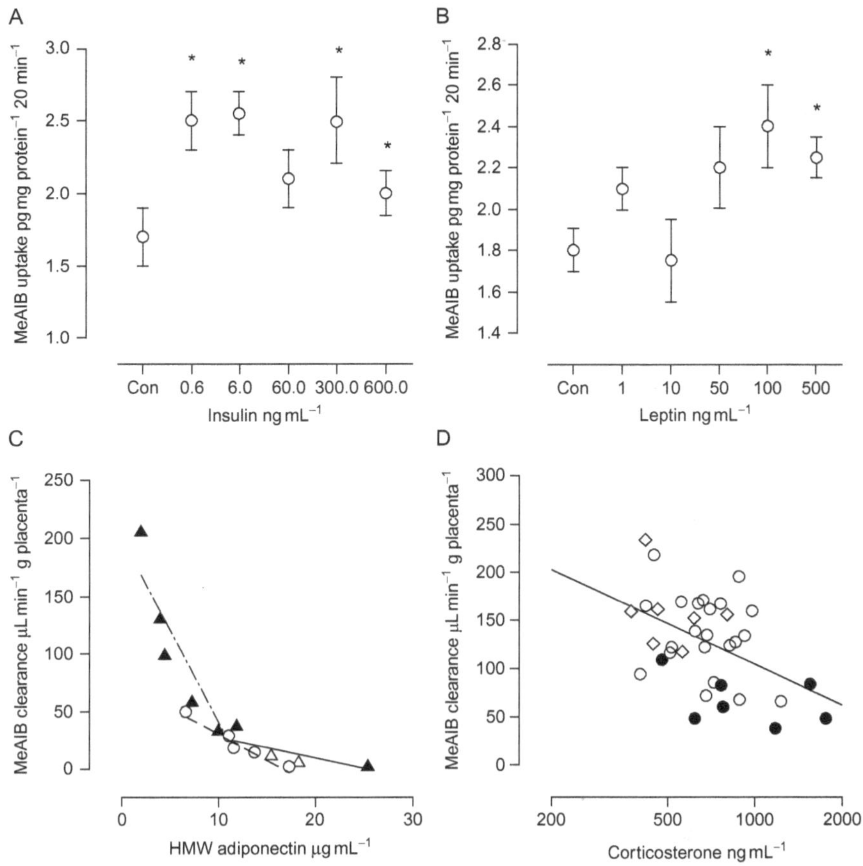

Fig. 2 Hormonal regulation of placental system A amino acid transport. Rates of system A-mediated uptake or transplacental clearance of radiolabeled methylaminoisobutyric acid (MeAIB) in response to different hormones, expressed per mg total protein in human placental villous fragments (A and B) or per gram placental weight in pregnant mice (C and D). Human placental villous fragments were incubated with insulin (A) or leptin (B) at varying concentrations for 24 h; *, significantly different from control according to original analysis.[83] Pregnant mice were infused subcutaneously with adiponectin (0.62 µg g^{-1} day^{-1}, 4 days, C) or given corticosterone orally (80 µg g^{-1} day^{-1}, 5 days, D). (C) *HMW*, high molecular weight; *open circles*, lean dams; *filled triangles*, obese dams infused with saline; *open triangles*, obese dams infused with adiponectin, individual linear regression lines for each experimental group, significantly nonzero slope in all cases. (D) *open circles*, control; *filled circles*, oral corticosterone day 14–19 of pregnancy; *open diamonds*, oral corticosterone day 11–16 of pregnancy, linear regression of all groups combined, significantly nonzero slope. *Data taken from Jansson N, Greenwood SL, Johansson BR, Powell TL, Jansson T. Leptin stimulates the activity of the system A amino acid transporter in human placental villous fragments.* J Clin Endocrinol Metab. *2003;88(3):1205–1211; Aye ILMH, Rosario FJ, Powell TL, Jansson T. Adiponectin supplementation in pregnant mice prevents the adverse effects of maternal obesity on placental function and fetal growth.* Proc Natl Acad Sci USA. *2015;112(41):12858–12863; Vaughan OR, Sferruzzi-Perri AN, Fowden AL. Maternal corticosterone regulates nutrient allocation to fetal growth in mice.* J Physiol. *2012;590:5529–5540.*

hormone, such that circulating IGF1 concentrations are elevated in the pregnant mouse, also increases abundance of system X^-_{AG} anionic amino acid transporters in the placenta,[133] although growth hormone itself has no effect on amino acid transport in vitro.[83] In contrast, rendering pregnant rats insulin deficient by pharmacological ablation of the maternal pancreas reduces placental transport of MeAIB, while this effect is rescued by administration of exogenous insulin.[134] Deletion of the gene encoding the IGF1 receptor in the conceptus decreases abundance of placental system X^-_{AG} anionic amino acid transporters EAAT 2 and 3 but increases y^+ cationic amino acid transporter abundance.[133] Since insulin does not cross the placenta, it may have differing effects at the maternal and fetal face of the placenta. Indeed experimental manipulation of fetal insulin concentration alters fetal plasma amino acid concentrations,[135,136] although it is not known whether fetal insulin regulates placental amino acid transport capacity directly.

In contrast to insulin and IGF1, circulating IGF2 levels in the pregnant mother have little effect on placental amino acid transport.[132,137] Genetic deficiency of IGF2 in the mouse fetus, due to null mutation of the *Igf2* gene, reduces placental weight-specific system A amino acid transport, *Slc38a2* gene expression, and EAAT1-4 anionic amino acid transporter abundance, concomitant with fetal growth restriction near term.[133,138] Conversely, when *Igf2* expression is specifically ablated in the placenta but not the fetus, placental system A-mediated MeAIB uptake is transiently elevated in late gestation, sustaining normal fetal growth despite reduced placental size.[138–140] Indeed, fetal overproduction of IGF2 as a consequence of deletion of the negative regulator, *H19*, sustains elevated system A amino acid transport to term in the placenta lacking local expression of the *Igf2* gene.[141] This suggests that circulating IGF2 in the fetus positively regulates placental amino acid transport in an endocrine manner. This action of IGF2 may provide an alternative explanation for the upregulation of amino acid transport in late gestation when the placenta is constitutionally small or developmentally impaired by genetic or nutritional manipulations in early gestation.[97,142,143] Overall, insulin and the IGFs appear to stimulate placental amino acid transport, irrespective of whether they are maternal or fetal in origin.

3.3 Adipokines

Adipokines including leptin, adiponectin, tumor necrosis factor α, and interleukin-6 are expressed and secreted from white adipose tissue in adult life. These adipose-derived factors modulate insulin sensitivity, metabolism, and energy homeostasis[144] and have an emerging role in fetal growth and development.[57,59,60,145] Leptin levels are elevated in obese pregnant women, who are more likely to deliver overgrown babies, and in experimental animals made obese by high fat feeding, therefore leptin may underlie increased syncytiotrophoblast system A amino acid transporter activity at term in maternal obesity.[54,74,146] Consistent with this, incubation of primary human placental villous fragments with leptin increases system A-mediated uptake of MeAIB in a concentration dependent manner (Fig. 2). Tumor necrosis factor α, interleukin-6, and monocyte chemoattractant protein-1 are similarly often elevated in the plasma of obese pregnant women[146,147] and stimulate system A amino acid transport in human trophoblast cells in vitro (Table 1). These stimulatory effects appear to be confined to adipocyte-derived cytokines, since macrophage-derived interleukin-1β downregulates insulin-stimulated MeAIB uptake in primary trophoblast cells (Table 1).

In contrast to other adipokines, adiponectin concentrations are inversely related to fat mass in adulthood, including in the pregnant mother.[74,148] A negative correlation between maternal adiponectin concentration and offspring birth weight[74] has also been described. Pregnant mice that are genetically deficient in adiponectin bear fetuses that are heavier than fetuses of the same genotype carried by a wild-type dam.[59] The inhibition of fetal growth by maternal adiponectin may be mediated by downregulated placental amino acid transport, as evidenced by reduced system A-mediated MeAIB uptake and SNAT1 and 2 protein abundance in cultured trophoblast cells in the presence of physiological insulin concentrations and increasing adiponectin (Table 1). Similarly, infusing mice with adiponectin for 4 days during late pregnancy reduces system A and L activity, SNAT and LAT protein abundance in the placental transporting membrane and fetal weight (Table 2). Furthermore, when a similar treatment is given to pregnant mice made obese by provision of a high calorie diet, the increased placental amino acid transport and fetal growth associated with maternal obesity are abolished (Fig. 2).[95] Taken together, these observations suggest that cytokines that positively indicate maternal fat mass stimulate placental amino acid transport, whereas maternal adiponectin may exercise an inhibitory effect on placental amino acid transport to limit fetal growth when maternal fuel reserves are low. However, to date, there has been no experimental investigation of whether fetal adipokines influence placental amino acid transport, although cord blood concentrations of both leptin and adiponectin are correlated with birth weight in humans.[149,150] In experimental models, manipulations of these hormones are known to influence fetal metabolism.[60,151]

3.4 Steroid Hormones

During pregnancy, the placenta is exposed to steroid hormones from both the maternal and fetal circulations. Maternal plasma concentrations of progestogens, androgens, estrogens, and glucocorticoids are elevated from the first trimester, in large part due to the steroidogenic and aromatizing activity of the placenta itself.[152] In the fetus, gonadal steroidogenesis is detectable from the point of sexual differentiation, while adrenal glucocorticoid secretion is elevated closer to term.[153,154] Unlike peptide hormones, steroids are lipid soluble and therefore able to diffuse passively across the syncytiotrophoblast plasma membranes. However, placental expression of the glucocorticoid metabolizing enzyme, 11β-Hydroxysteroid dehydrogenase type II, together with multidrug resistance transporters (e.g., phosphoglycoprotein), which mediate cellular efflux of a wide range of substrates including hormones, results in a barrier to free movement of these hormones between mother and fetus.[155,156] In addition to diverse roles in the regulation of metabolism and reproductive function in adulthood, steroid hormones have developmental effects in the fetus. Androgens influence development of the gonads, while glucocorticoids induce differentiation in fetal tissues such as the lung, gut, and liver that prepare the fetus for birth.[153,154] However, both maternal and cord blood concentrations of cortisol and testosterone are negatively correlated with size at birth in human infants and tend to be elevated in intrauterine growth-restricted fetuses.[157–162] Furthermore, although synthetic glucocorticoid administration to pregnant women expected to deliver prematurely has beneficial effects on maturity of the neonate, multiple or inappropriate doses of the steroid have been linked to indices of reduced growth in utero.[163]

The growth-inhibitory effects of fetoplacental glucocorticoid overexposure may be linked to placental system A amino acid transport. Placentae from women given the synthetic glucocorticoid, betamethasone, between 22 and 32 weeks of pregnancy and delivering at term exhibit lower sodium-dependent MeAIB uptake capacity than placentae from untreated women.[164] However, incubation with either dexamethasone or cortisol increases system A-mediated amino acid transport in choriocarcinoma cells and placental explants in vitro (Table 1). In villous explants at least, these changes in amino acid transport may be related to the concomitant stimulatory effect of dexamethasone on trophoblast morphological differentiation and microvilli formation.[91] In vivo, transplacental MeAIB transfer is suppressed near term when pregnant mice are injected with a clinically relevant dose of dexamethasone or when circulating natural glucocorticoid concentrations are elevated through oral corticosterone administration (Table 2). Testosterone administration to pregnant rats over a similar period also reduces placental system A activity and SNAT expression (Table 2). These placental transport changes are associated with reduced fetal weight.[106,107] Glucocorticoids act in a dose-dependent manner over the physiological range, as evidenced by the observation that placental system A amino acid transport is increased near term when maternal adrenal secretion is suppressed by transient glucocorticoid treatment earlier in pregnancy (Table 2). Moreover, glucocorticoid effects differ depending on whether bioavailability is increased in the maternal or fetal compartment, as genetic deficiency of placental 11β-hydroxysteroid dehydrogenase type II in the fetus increases system A amino acid transport in the mouse placenta on day 15 of gestation.[165]

Taken together, the evidence suggests that placental amino acid transport is regulated by opposing maternal anabolic or catabolic signals that up- or downregulate transport, respectively (Fig. 3). System A accumulative amino acid transporters, in particular, are subject to regulation by these signals. Maternal signals may interact with indicators of fetal nutrient availability and metabolic demand to support fetal growth. Perturbations in the levels of maternal nutritional status due to diet or lifestyle may inappropriately increase or decrease amino acid transport in the placenta, leading to fetal growth restriction or overgrowth, respectively.

4. MOLECULAR MECHANISMS UNDERLYING REGULATION OF PLACENTAL AMINO ACID TRANSPORT

Active transplacental transport of amino acids depends upon the abundance of amino acid transporters at the trophoblast apical plasma membrane. In turn, rates of uptake of nutrients depend on transcription, translation, membrane trafficking, and degradation of the transporter genes and proteins. There is evidence that these processes are regulated by nutritional and hormonal cues, but, in a number of studies, alterations in placental amino acid transporter activity are not associated with changes in either gene or protein expression (Table 2). Rather, regulation of amino acid transport activity appears to involve cellular signaling events that control membrane protein localization through both pre- and posttranslational modifications (Fig. 3).

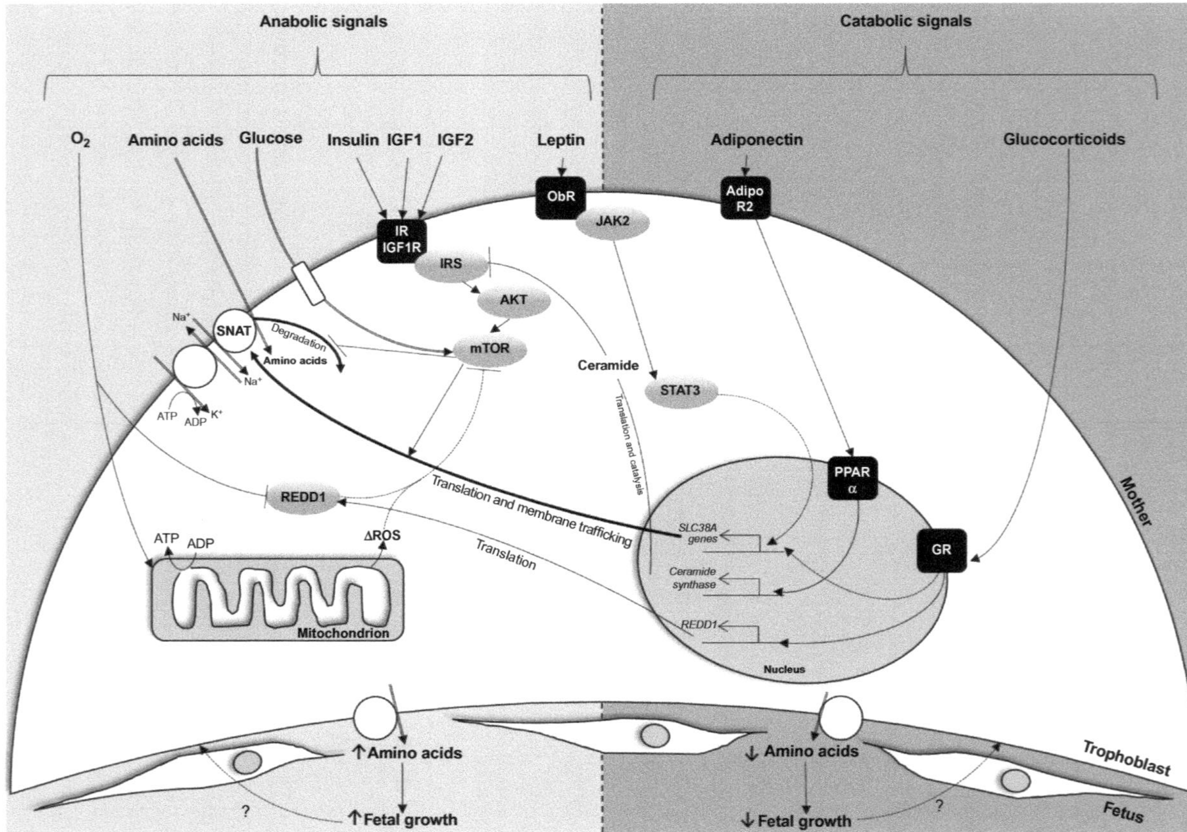

Fig. 3 See legend on opposite page.

4.1 Nutrient and Oxygen Availability

The mechanistic target of rapamycin (mTOR) complexes of signaling proteins regulate trophoblast apical membrane amino acid transporter abundance with respect to nutrient availability.[166] mTOR is a cytosolic serine/threonine kinase associated with a number of accessory proteins, including either raptor (mTOR complex 1) or rictor (mTOR complex 2). In the placenta, mTOR positively regulates activity of amino acid transporter systems A, L, and β through phosphorylation of the translational regulators ribosomal S6 kinase (S6K1 and 2), eukaryotic initiation factor 4E-binding protein (4EBP1 and 2) and eukaryotic initiation factor (eIF4G).[166–168] When mTOR complex 1 or 2 are inhibited through silencing of raptor or rictor in primary human trophoblast cells, SNAT2 abundance is reduced specifically in the microvillous membrane.[168] Moreover, mTOR complex 1 inhibits internalization and degradation of membrane-bound SNAT2 and LAT1 by inhibiting translation of NEDD4-2, which ubiquitinates these transporters and targets them to the proteasome.[169] Placental activity of mTOR complex 1 is downregulated, and NEDD4-2 abundance is increased concomitant with decreased amino acid transport capacity in pregnancies complicated by fetal growth restriction.[170,171] Conversely, placental mTOR signaling is activated in overgrown fetuses of obese mothers.[54]

Availability of nutrients such as glucose and amino acids directly increases mTORC1 activity, in part through interaction with the upstream Ras-related small G-protein, Rheb.[166,172] Culturing human primary trophoblast in low glucose concentrations reduces phosphorylation of ribosomal S6 kinase, downstream of mTORC1.[79] Meanwhile, amino acid supplementation increases mTOR activity in nonhuman trophoblast cells in vitro.[173,174] In vivo, placental mTOR complex 1 activity is increased in experimentally

Fig. 3 Regulation of placental amino acid transport by anabolic and catabolic signals. Schematic diagram showing the regulation of placental amino acid transport by maternal anabolic and catabolic signals. Regulation occurs at the level of amino acid transporter transcription, translation, membrane translocation, degradation, and ATP availability. →, activates; ⊣, inhibits; *dashed line*, hypothetical mechanism; *double line*, nutrient flux; ↑, increased; ↓, decreased; *AdipoR2*, adiponectin type II receptor; *AKT*, protein kinase B; *GR*, glucocorticoid receptor; *IGF*, insulin-like growth factor; *IGF1R*, insulin-like growth factor type I receptor; *IR*, insulin receptor; *IRS*, insulin receptor substrate; *mTOR*, mechanistic target of rapamycin; *ObR*, leptin receptor; *PPARα*, peroxisome-proliferator activated receptor α; *REDD1*, regulated in development and DNA damage 1; *ROS*, reactive oxygen species; *SNAT/SLC38A*, system A, sodium-dependent neutral amino acid transporter protein/gene; *STAT3*, signal transducer and activator of transcription 3.

induced maternal hyperglycaemia or obesity in mice and rats[25,175,176] and decreased with dietary protein restriction in rodents and nonhuman primates.[26,102] Fatty acid availability may regulate trophoblast amino acid transport through modulating mTOR activity, and this effect depends upon the chain length and degree of saturation of the fatty acid.[78] However, fatty acids may also signal through inflammatory pathways distinct from mTOR to increase system A activity.[77] Nutrient availability is also detected by the placenta through the independent amino acid response, glycogen synthase-kinase, and hexosamine signaling pathways,[67] or through microRNA-dependent mechanisms,[114] although their role in regulating placental amino acid transport remains unclear.

Oxygen availability also regulates placental amino acid transport through mTOR, as evidenced by placentae from pregnant women exposed to hypobaric hypoxia by residing at high altitude, which exhibit reduced phosphorylation of 4EBP1 and protein kinase B (AKT), downstream targets of mTORC1 and 2.[116] Similarly, Akt phosphorylation is reduced in placentae of mice exposed to normobaric hypoxia only in the last 5 days of pregnancy,[104] although mTOR activity is increased when the insult occurs throughout pregnancy.[117] Mechanistically, results of studies exposing BeWo choriocarcinoma cells to severe hypoxia (1% vs 21% normobaric) suggest that mTOR activity is downregulated as a consequence of stabilization of the hypoxia inducible factor 1α protein and increased expression of the mTOR suppressor REDD-1 (regulated in development and DNA damage response-1).[177] It has been suggested that placental mTOR–AKT signaling is indirectly downregulated during hypoxia as a result of oxidative stress and increased production of mitochondrial reactive oxygen species.[116] Indeed, placental system A amino acid transport capacity is reduced with increased reactive oxygen species production in pregnant mice with poor uterine artery conversion due to genetic deficiency of endothelial nitric oxide synthase.[46] Placental amino acid transport may also be reduced as a consequence of reduced cellular energy availability during hypoxia, when activity of specific mitochondrial complexes is suppressed resulting in reduced ATP availability for active transport.[178] Phosphorylation of AMP-activated protein kinase (AMPK) constitutes one possible mechanistic link between decreased ATP availability in the hypoxic mouse placenta and inhibition of mTOR signaling,[179] ensuring that amino acid transporter activity is aligned with cellular energy status. Conversely, inhibition of AMPK has been proposed to contribute to activation of placental mTOR and amino acid transport activity in maternal overnutrition.[54,168,176]

4.2 Insulin and Insulin-Like Growth Factors

The stimulatory effects of insulin, IGF1 and IGF2, on placental amino acid transport are mediated via the insulin receptor and IGF type I receptor, which are transmembrane tyrosine kinases. Recruitment and phosphorylation of adaptor substrates such as the insulin receptor substrates (IRS) subsequently allow activation of an array of downstream pathways. In particular, phosphoinositide 3-kinase (PI3K) activation leads to enhanced signaling through the mTOR–AKT pathway,[180] which increases amino acid transporter translation and membrane trafficking in trophoblast cells. Consistent with a role for placental insulin signaling in fetal growth, IGF type I receptor, IRS-1, PI3K, and AKT protein abundance have been shown to be decreased in intrauterine growth-restricted pregnancies,[181,182] whereas the insulin receptor, IRS-I, and PI3K are overexpressed in gestational diabetes,[183,184] when system A amino acid transport is down- and upregulated, respectively. The in vivo effects of nutritional manipulations on system A amino acid transport may be mediated by the insulin/IGF receptor cascade, as elements of the placental insulin signaling pathway are perturbed in the same direction as transport capacity in experimental animals during under-[102] or overnutrition.[93,185] However, there appears to be a complex interplay between the insulin signaling pathway and other signals intrinsic and extrinsic to the placenta, since deletion of the p110α subunit of PI3K decreases *Slc38a* gene expression but increases system A transporter capacity in the mouse placenta.[143] This demonstrates the multiple levels of placental regulation, both pre- and posttranslational, that determine overall amino acid delivery to the fetus.

4.3 Adipokines

Adipokines bind to their membrane-bound receptors and either signal through cytosolic mitogen-activated protein kinase (MAPK) cascades or alter gene expression via transcription factors such as signal transducer and activator of transcription 3 (STAT3), nuclear factor κB (NFκB), and peroxisome-proliferator-activated receptor α (PPARα). Stimulation of placental system A amino acid transport activity by leptin receptor signaling appears to depend upon transcriptional activation by Janus kinase 2 (JAK2) and STAT3, which are increased in abundance in human villous fragments or explants following incubation with leptin.[85,184] Pharmacological inhibition of either JAK2 or STAT3 abolishes the leptin-stimulated increase in MeAIB uptake in trophoblast cells.[85] However, leptin may also affect transporter translation via p38 MAPK and by impinging upon AKT–mTOR signaling, given that the activity of these cytosolic proteins is increased in trophoblast cells following leptin incubation in vitro, while transfection with kinase-null forms of either MAPK or AKT prevents leptin-stimulated protein synthesis.[186] Elements of the leptin receptor signaling pathway, including the leptin receptor,[184] p38 MAPK and STAT3,[146] are often increased in abundance or activity in gestational diabetes. Similarly, interleukin-6 induces SNAT gene expression in a STAT3-dependent mechanism,[87] whereas tumor necrosis factor α and monocyte chemoattractant protein-1 appear to increase SNAT protein translation via p38 MAPK.[86,146] STAT3 phosphorylation is also increased in the mouse placenta as a consequence of experimentally induced obesity.[185]

Adiponectin acts through its type 2 receptor to activate PPARα and p38 MAPK in the placenta.[89,105,187] However, the adiponectin-mediated suppression of insulin-stimulated system A amino acid transport appears to be linked to PPARα activity specifically, since knockdown of PPARα but not p38 MAPK prevents this effect in human trophoblasts in vitro.[89,187] In trophoblast, adiponectin activation of PPARα induces expression of the enzyme ceramide synthase, which catalyzes conversion of the bioactive sphingolipid metabolite, sphingosine 1-phosphate into ceramide.[187] Ceramide in turn inhibits IRS-1 and therefore limits insulin-AKT–mTOR-mediated amino acid transporter translation and membrane localization. Consistent with this, components of the insulin receptor signaling pathway and downstream targets of mTOR are hypophosphorylated in the placenta of pregnant mice infused with adiponectin.[105]

4.4 Steroid Hormones

In comparison to nutritional and other hormonal signals, little is known about the mechanism by which steroid hormones, in particular the glucocorticoids, regulate placental amino acid transport. Generally, lipid-soluble steroid hormones enter into the cell to induce changes in transcription directly by binding to a nuclear receptor, for example, the androgen receptor or glucocorticoid receptor.[188] While testosterone reduces rat placental MeAIB uptake in a manner dependent on *Slc38a2* gene expression,[106] the changes in transporter activity induced by glucocorticoids do not always coincide with alterations in transporter gene expression.[29,91,107,108] Glucocorticoids, such as cortisol, have transcription-independent effects in modulating inflammatory pathways and potentiating catecholamine signaling,[189] but such nongenomic actions have not been described in the placenta or fetus. In part, cortisol may regulate translocation of SNAT transporters to the trophoblast plasma membrane through serum and glucocorticoid inducible kinase 1.[90,190] Glucocorticoids may also reduce placental system A amino acid uptake by suppressing Akt–mTOR signaling. Elevated corticosterone levels in pregnant mice lead to increased placental gene expression of the stress-inducible mTOR suppressor, *Redd1* concomitant with reduced activity of the downstream targets of mTOR.[191] Ultimately, adaptations in placental amino acid transport in response to glucocorticoid concentrations in vivo are likely to reflect the interaction of their direct effects on the trophoblast with their systemic effects on insulin sensitivity, glucose tolerance, and amino acid metabolism.[192,193]

5. CONCLUSIONS AND TRANSLATIONAL PERSPECTIVES

Amino acid transport across the syncytial epithelium of the placenta is a highly regulated process that is dependent on a number of nutritional and hormonal signals (Fig. 3). In general, anabolic signals indicating abundant maternal fuel reserves stimulate placental transport of amino acids, thereby increasing their availability for fetal growth and optimizing offspring fitness. When nutritional conditions are poor and resources are scarce, catabolic signals in the mother downregulate placental amino acid transport in an attempt to preserve her own nutrient reserves and thus ensure maternal survival, fitness, and continued reproductive potential. Fetal growth and nutrient demand signals may influence placental transport of amino acids in an attempt to maintain their supply in the face of decreased availability, resulting in competition over resource allocation at the level of the placenta. Interaction of these signals occurs within the trophoblast, where cellular nutrient availability, energy status, cytosolic signaling cascades, and nuclear factors converge to influence amino acid transporter transcription, translation, localization, and activity (Fig. 3). Studies to date have focussed on the regulation of the system A sodium-dependent, neutral amino acid transporters, along with sodium-independent system L transporters, because they are required for the net accumulation of essential amino acids within the syncytiotrophoblast and are known to be essential for normal fetal growth.

Abnormal fetal growth, in particular intrauterine growth restriction or fetal overgrowth, may be a result of altered placental amino acid transport capacity in response to signals generated by maternal metabolic disease, suboptimal nutrition, or other perturbations. Development of effective and noninvasive therapies for fetal under- and overgrowth will be aided by increased understanding of the changes in placental transport that link them to maternal environmental and lifestyle factors, such as adiposity, exercise, smoking, diet, and micronutrient intake.[194,195] If decreased placental transport of amino acids is a key mechanism causing fetal growth restriction, increasing the fetal supply through modulation of maternal signaling factors may represent a novel intervention in this condition.[196,197] Likewise, increasing the abundance of anabolic factors in the uteroplacental tissues may alleviate fetal growth restriction by increasing placental amino acid transport. For example, delivery of either IGF1 or IGF2 to the mother or directly to the uteroplacental tissues has been shown to be effective in increasing amino acid transport in several animal models of fetal growth restriction.[125,198–200] Another example of targeting the placenta to modify fetal growth outcomes is maternal supplementation of adiponectin to prevent fetal overgrowth by suppressing placental amino acid transport, which is activated in cases of maternal obesity.[95] Ultimately, such therapies will be critical to improve neonatal health and reduce the risk of metabolic disease throughout the lifetime of the child.

REFERENCES

1. Pallotto EK, Kilbride HW. Perinatal outcome and later implications of intrauterine growth restriction. *Clin Obstet Gynecol.* 2006;49(2):257–269.
2. Henriksen T. The macrosomic fetus: a challenge in current obstetrics. *Acta Obstet Gynecol Scand.* 2008;87(2):134–145.
3. Barker DJ. The developmental origins of chronic adult disease. *Acta Paediatr Suppl.* 2004;93(446):26–33.
4. Battaglia FC, Meschia G. Fetal nutrition. *Annu Rev Nutr.* 1988;8:43–61.
5. Meier PR, Peterson RG, Bonds DR, Meschia G, Battaglia FC. Rates of protein synthesis and turnover in fetal life. *Am J Physiol.* 1981;240(3):E320–E324.
6. Schaefer AL, Krishnamurti CR. Whole body and tissue fractional protein synthesis in the ovine fetus in utero. *Br J Nutr.* 1984;52(2):359–369.
7. Loy GL, Quick Jr AN, Hay Jr WW, Meschia G, Battaglia FC, Fennessey PV. Fetoplacental deamination and decarboxylation of leucine. *Am J Physiol.* 1990;259(4 pt 1):E492–E497.
8. Regnault TR, de Vrijer B, Battaglia FC. Transport and metabolism of amino acids in placenta. *Endocrine.* 2002;19(1):23–41.
9. Economides DL, Nicolaides KH, Gahl WA, Bernardini I, Evans MI. Plasma amino acids in appropriate- and small-for-gestational-age fetuses. *Am J Obstet Gynecol.* 1989;161(5):1219–1227.
10. Cetin I, Corbetta C, Sereni LP, et al. Umbilical amino acid concentrations in normal and growth-retarded fetuses sampled in utero by cordocentesis. *Am J Obstet Gynecol.* 1990;162(1):253–261.
11. Anderson AH, Fennessey PV, Meschia G, Wilkening RB, Battaglia FC. Placental transport of threonine and its utilization in the normal and growth-restricted fetus. *Am J Physiol.* 1997;272(5 pt 1):E892–E900.

12. Ross JC, Fennessey PV, Wilkening RB, Battaglia FC, Meschia G. Placental transport and fetal utilization of leucine in a model of fetal growth retardation. *Am J Physiol.* 1996;270(3 pt 1):E491–E503.
13. Paolini CL, Marconi AM, Ronzoni S, et al. Placental transport of leucine, phenylalanine, glycine, and proline in intrauterine growth-restricted pregnancies. *J Clin Endocrinol Metab.* 2001;86(11):5427–5432.
14. Regnault TRH, de Vrijer B, Galan HL, Wilkening RB, Battaglia FC, Meschia G. Umbilical uptakes and transplacental concentration ratios of amino acids in severe fetal growth restriction. *Pediatr Res.* 2013;73(5):602–611.
15. Lofthouse EM, Perazzolo S, Brooks S, et al. Phenylalanine transfer across the isolated perfused human placenta: an experimental and modeling investigation. *Am J Physiol Regul Integr Comp Physiol.* 2016;310(9):R828–R836.
16. Chung M, Teng C, Timmerman M, Meschia G, Battaglia FC. Production and utilization of amino acids by ovine placenta in vivo. *Am J Physiol.* 1998;274(1 pt 1): E13–E22.
17. Holm MB, Bastani NE, Holme AM, et al. Amino acid transfer in the fetal-placental unit: a human in vivo study. *PLoS One.* Submitted. 2016.
18. Vaughan OR, Sferruzzi-Perri AN, Coan P, Fowden A. Environmental regulation of placental phenotype: implications for fetal growth. *Reprod Fertil Dev.* 2012;24:1–17.
19. Vaughan OR, Fowden A. Placental metabolism: substrate requirements and response to stress. *Reprod Domest Anim.* 2016;51(suppl 2):1–11.
20. Cetin I, Ronzoni S, Marconi AM, et al. Maternal concentrations and fetal-maternal concentration differences of plasma amino acids in normal and intrauterine growth-restricted pregnancies. *Am J Obstet Gynecol.* 1996;174(5):1575–1583.
21. Moe AJ. Placental amino acid transport. *Am J Physiol.* 1995;268(6 pt 1):C1321–C1331.
22. Cleal JK, Lewis RM. The mechanisms and regulation of placental amino acid transport to the human foetus. *J Neuroendocrinol.* 2008;20(4):419–426.
23. Cleal JK, Glazier JD, Ntani G, et al. Facilitated transporters mediate net efflux of amino acids to the fetus across the basal membrane of the placental syncytiotrophoblast. *J Physiol.* 2011;589(pt 4):987–997.
24. Desforges M, Mynett KJ, Jones RL, et al. The SNAT4 isoform of the system A amino acid transporter is functional in human placental microvillous plasma membrane. *J Physiol.* 2009;587(pt 1):61–72. 11/1708/12/received, 11/11/accepted.
25. Rosario FJ, Kanai Y, Powell TL, Jansson T. Increased placental nutrient transport in a novel mouse model of maternal obesity with fetal overgrowth. *Obesity.* 2015;23(8): 1663–1670.
26. Kavitha JV, Rosario FJ, Nijland MJ, et al. Down-regulation of placental mTOR, insulin/IGF-I signaling, and nutrient transporters in response to maternal nutrient restriction in the baboon. *FASEB J.* 2014;28(3):1294–1305.
27. Novak D, Lehman M, Bernstein H, Beveridge M, Cramer S. SNAT expression in rat placenta. *Placenta.* 2006;27(4–5):510–516.
28. Novak DA, Beveridge MJ, Malandro M, Seo J. Ontogeny of amino acid transport system A in rat placenta. *Placenta.* 1996;17(8):643–651.
29. Audette MC, Challis JR, Jones RL, Sibley CP, Matthews SG. Antenatal dexamethasone treatment in midgestation reduces system A-mediated transport in the late-gestation murine placenta. *Endocrinology.* 2011;152:3561–3570.
30. Gaccioli F, Aye IL, Roos S, et al. Expression and functional characterisation of System L amino acid transporters in the human term placenta. *Reprod Biol Endocrinol.* 2015;13:57.
31. Panitchob N, Widdows KL, Crocker IP, et al. Computational modelling of amino acid exchange and facilitated transport in placental membrane vesicles. *J Theor Biol.* 2015;365:352–364.

32. Sengers BG, Please CP, Lewis RM. Computational modelling of amino acid transfer interactions in the placenta. *Exp Physiol.* 2010;95(7):829–840.
33. Widdows KL, Panitchob N, Crocker IP, et al. Integration of computational modeling with membrane transport studies reveals new insights into amino acid exchange transport mechanisms. *FASEB J.* 2015;29(6):2583–2594.
34. Roos S, Powell TL, Jansson T. Human placental taurine transporter in uncomplicated and IUGR pregnancies: cellular localization, protein expression, and regulation. *Am J Physiol Regul Integr Comp Physiol.* 2004;287(4):R886–R893.
35. Kusinski LC, Jones CJ, Baker PN, Sibley CP, Glazier JD. Isolation of plasma membrane vesicles from mouse placenta at term and measurement of system A and system beta amino acid transporter activity. *Placenta.* 2010;31(1):53–59.
36. Ayuk PT, Theophanous D, D'Souza SW, Sibley CP, Glazier JD. L-arginine transport by the microvillous plasma membrane of the syncytiotrophoblast from human placenta in relation to nitric oxide production: effects of gestation, preeclampsia, and intrauterine growth restriction. *J Clin Endocrinol Metab.* 2002;87(2):747–751.
37. Radaelli T, Cetin I, Ayuk PT, Glazier JD, Pardi G, Sibley CP. Cationic amino acid transporter activity in the syncytiotrophoblast microvillous plasma membrane and oxygenation of the uteroplacental unit. *Placenta.* 2002;23(suppl A):S69–S74.
38. Noorlander CW, de Graan PN, Nikkels PG, Schrama LH, Visser GH. Distribution of glutamate transporters in the human placenta. *Placenta.* 2004;25(6):489–495.
39. Malandro MS, Beveridge MJ, Kilberg MS, Novak DA. Effect of low-protein diet-induced intrauterine growth retardation on rat placental amino acid transport. *Am J Physiol.* 1996;271(1 pt 1):C295–C303.
40. Glazier JD, Cetin I, Perugino G, et al. Association between the activity of the system A amino acid transporter in the microvillous plasma membrane of the human placenta and severity of fetal compromise in intrauterine growth restriction. *Pediatr Res.* 1997;42(4):514–519.
41. Mahendran D, Donnai P, Glazier JD, D'Souza SW, Boyd RD, Sibley CP. Amino acid (system A) transporter activity in microvillous membrane vesicles from the placentas of appropriate and small for gestational age babies. *Pediatr Res.* 1993;34(5):661–665.
42. Shibata E, Hubel CA, Powers RW, et al. Placental system A amino acid transport is reduced in pregnancies with small for gestational age (SGA) infants but not in preeclampsia with SGA infants. *Placenta.* 2008;29(10):879–882.
43. Mando C, Tabano S, Pileri P, et al. SNAT2 expression and regulation in human growth-restricted placentas. *Pediatr Res.* 2013;74(2):104–110.
44. Nüsken E, Gellhaus A, Kühnel E, et al. Increased rat placental fatty acid, but decreased amino acid and glucose transporters potentially modify intrauterine programming. *J Cell Biochem.* 2016;117(7):1594–1603.
45. Jansson T, Persson E. Placental transfer of glucose and amino acids in intrauterine growth retardation: studies with substrate analogs in the awake guinea pig. *Pediatr Res.* 1990;28(3):203–208.
46. Kusinski LC, Stanley JL, Dilworth MR, et al. eNOS knockout mouse as a model of fetal growth restriction with an impaired uterine artery function and placental transport phenotype. *Am J Physiol Regul Integr Comp Physiol.* 2012;303(1):R86–R93.
47. Glazier JD, Sibley CP, Carter AM. Effect of fetal growth restriction on system A amino acid transporter activity in the maternal facing plasma membrane of rat syncytiotrophoblast. *Pediatr Res.* 1996;40(2):325–329.
48. Cramer S, Beveridge M, Kilberg M, Novak D. Physiological importance of system A-mediated amino acid transport to rat fetal development. *Am J Physiol Cell Physiol.* 2002;282(1):C153–C160.
49. Jansson T, Scholtbach V, Powell TL. Placental transport of leucine and lysine is reduced in intrauterine growth restriction. *Pediatr Res.* 1998;44(4):532–537.

50. Norberg S, Powell TL, Jansson T. Intrauterine growth restriction is associated with a reduced activity of placental taurine transporters. *Pediatr Res*. 1998;44(2):233–238.
51. Aiko Y, Askew DJ, Aramaki S, et al. Differential levels of amino acid transporters System L and ASCT2, and the mTOR protein in placenta of preeclampsia and IUGR. *BMC Pregnancy Childbirth*. 2014;14:181.
52. de Vrijer B, Regnault TR, Wilkening RB, Meschia G, Battaglia FC. Placental uptake and transport of ACP, a neutral nonmetabolizable amino acid, in an ovine model of fetal growth restriction. *Am J Physiol Endocrinol Metab*. 2004;287(6):E1114–E1124.
53. Jansson T, Ekstrand Y, Bjorn C, Wennergren M, Powell TL. Alterations in the activity of placental amino acid transporters in pregnancies complicated by diabetes. *Diabetes*. 2002;51(7):2214–2219.
54. Jansson N, Rosario FJ, Gaccioli F, et al. Activation of placental mTOR signaling and amino acid transporters in obese women giving birth to large babies. *J Clin Endocrinol Metab*. 2013;98(1):105–113.
55. Jones HN, Woollett LA, Barbour N, Prasad PD, Powell TL, Jansson T. High-fat diet before and during pregnancy causes marked up-regulation of placental nutrient transport and fetal overgrowth in C57/BL6 mice. *FASEB J*. 2009;23(1):271–278.
56. Sferruzzi-Perri AN, Vaughan OR, Forhead AJ, Fowden AL. Hormonal and nutritional drivers of intrauterine growth. *Curr Opin Clin Nutr Metab Care*. 2013;16(3):298–309.
57. Forhead AJ, Fowden AL. The hungry fetus? Role of leptin as a nutritional signal before birth. *J Physiol*. 2009;587(pt 6):1145–1152.
58. Fowden AL, Valenzuela OA, Vaughan OR, Jellyman JK, Forhead AJ. Glucocorticoid programming of intrauterine development. *Domest Anim Endocrinol*. 2016;56(suppl): S121–S132.
59. Qiao L, Wattez JS, Lee S, et al. Knockout maternal adiponectin increases fetal growth in mice: potential role for trophoblast IGFBP-1. *Diabetologia*. 2016;59:2417–2425.
60. Qiao L, Yoo HS, Madon A, Kinney B, Hay WW, Shao J. Adiponectin enhances mouse fetal fat deposition. *Diabetes*. 2012;61(12):3199–3207.
61. Fowden AL, Forhead AJ, Sferruzzi-Perri AN, Burton GJ, Vaughan OR. Review: endocrine regulation of placental phenotype. *Placenta*. 2015;36(suppl 1):S50–S59.
62. Tavare JM, Holmes CH. Differential expression of the receptors for epidermal growth factor and insulin in the developing human placenta. *Cell Signal*. 1989;1(1):55–64.
63. Desoye G, Hartmann M, Blaschitz A, et al. Insulin receptors in syncytiotrophoblast and fetal endothelium of human placenta. Immunohistochemical evidence for developmental changes in distribution pattern. *Histochemistry*. 1994;101(4):277–285.
64. Fang J, Furesz TC, Lurent RS, Smith CH, Fant ME. Spatial polarization of insulin-like growth factor receptors on the human syncytiotrophoblast. *Pediatr Res*. 1997;41(2): 258–265.
65. Bodner J, Ebenbichler CF, Wolf HJ, et al. Leptin receptor in human term placenta: in situ hybridization and immunohistochemical localization. *Placenta*. 1999;20(8):677–682.
66. Dimasuay KG, Boeuf P, Powell TL, Jansson T. Placental responses to changes in the maternal environment determine fetal growth. *Front Physiol*. 2016;7:12.
67. Diaz P, Powell TL, Jansson T. The role of placental nutrient sensing in maternal-fetal resource allocation. *Biol Reprod*. 2014;91:82.
68. Metzger BE, Lowe LP, Dyer AR, et al. Hyperglycemia and adverse pregnancy outcomes. *N Engl J Med*. 2008;358(19):1991–2002.
69. Schaefer-Graf UM, Graf K, Kulbacka I, et al. Maternal lipids as strong determinants of fetal environment and growth in pregnancies with gestational diabetes mellitus. *Diabetes Care*. 2008;31(9):1858–1863.
70. Ortega-Senovilla H, Alvino G, Taricco E, Cetin I, Herrera E. Enhanced circulating retinol and non-esterified fatty acids in pregnancies complicated with intrauterine growth restriction. *Clin Sci (Lond)*. 2010;118(5):351–358.

71. Moore LG, Rounds SS, Jahnigen D, Grover RF, Reeves JT. Infant birth weight is related to maternal arterial oxygenation at high altitude. *J Appl Physiol Respir Environ Exerc Physiol.* 1982;52(3):695–699.
72. Araújo JR, Correia-Branco A, Ramalho C, et al. L-Methionine placental uptake: characterization and modulation in gestational diabetes mellitus. *Reprod Sci.* 2013;20(12):1492–1507.
73. Ditchfield AM, Desforges M, Mills TA, et al. Maternal obesity is associated with a reduction in placental taurine transporter activity (2005). *Int J Obes (Lond).* 2015;39(4):557–564.
74. Jansson N, Nilsfelt A, Gellerstedt M, et al. Maternal hormones linking maternal body mass index and dietary intake to birth weight. *Am J Clin Nutr.* 2008;87(6):1743–1749.
75. Farley DM, Choi J, Dudley DJ, et al. Placental amino acid transport and placental leptin resistance in pregnancies complicated by maternal obesity. *Placenta.* 2010;31(8):718–724.
76. Farley D, Tejero ME, Comuzzie AG, et al. Feto-placental adaptations to maternal obesity in the baboon. *Placenta.* 2009;30(9):752–760.
77. Lager S, Gaccioli F, Ramirez VI, Jones HN, Jansson T, Powell TL. Oleic acid stimulates system A amino acid transport in primary human trophoblast cells mediated by toll-like receptor 4. *J Lipid Res.* 2013;54(3):725–733.
78. Lager S, Jansson T, Powell TL. Differential regulation of placental amino acid transport by saturated and unsaturated fatty acids. *Am J Physiol Cell Physiol.* 2014;307(8): C738–C744.
79. Roos S, Lagerlof O, Wennergren M, Powell TL, Jansson T. Regulation of amino acid transporters by glucose and growth factors in cultured primary human trophoblast cells is mediated by mTOR signaling. *Am J Physiol Cell Physiol.* 2009;297(3):C723–C731.
80. Jones HN, Ashworth CJ, Page KR, McArdle HJ. Expression and adaptive regulation of amino acid transport system A in a placental cell line under amino acid restriction. *Reproduction.* 2006;131(5):951–960.
81. Novak D, Quiggle F, Haafiz A. Impact of forskolin and amino acid depletion upon System A activity and SNAT expression in BeWo cells. *Biochimie.* 2006;88(1):39–44.
82. Nelson DM, Smith SD, Furesz TC, et al. Hypoxia reduces expression and function of system A amino acid transporters in cultured term human trophoblasts. *Am J Physiol Cell Physiol.* 2003;284(2):C310–C315.
83. Jansson N, Greenwood SL, Johansson BR, Powell TL, Jansson T. Leptin stimulates the activity of the system A amino acid transporter in human placental villous fragments. *J Clin Endocrinol Metab.* 2003;88(3):1205–1211.
84. Karl PI, Alpy KL, Fisher SE. Amino acid transport by the cultured human placental trophoblast: effect of insulin on AIB transport. *Am J Physiol.* 1992;262(4 pt 1): C834–C839.
85. von Versen-Hoynck F, Rajakumar A, Parrott MS, Powers RW. Leptin affects system A amino acid transport activity in the human placenta: evidence for STAT3 dependent mechanisms. *Placenta.* 2009;30(4):361–367.
86. Aye ILMH, Jansson T, Powell TL. TNF-α stimulates System A amino acid transport in primary human trophoblast cells mediated by p38 MAPK signaling. *Physiol Rep.* 2015;3(10), e12594.
87. Jones HN, Jansson T, Powell TL. IL-6 stimulates system A amino acid transporter activity in trophoblast cells through STAT3 and increased expression of SNAT2. *Am J Physiol Cell Physiol.* 2009;297(5):C1228–C1235.
88. Aye ILMH, Jansson T, Powell TL. Interleukin-1β inhibits insulin signaling and prevents insulin-stimulated system A amino acid transport in primary human trophoblasts. *Mol Cell Endocrinol.* 2013;381(1–2):46–55.

89. Jones HN, Jansson T, Powell TL. Full-length adiponectin attenuates insulin signaling and inhibits insulin-stimulated amino acid transport in human primary trophoblast cells. *Diabetes*. 2010;59(5):1161–1170.
90. Jones HN, Ashworth CJ, Page KR, McArdle HJ. Cortisol stimulates system A amino acid transport and SNAT2 expression in a human placental cell line (BeWo). *Am J Physiol Endocrinol Metab*. 2006;291(3):E596–E603.
91. Audette MC, Greenwood SL, Sibley CP, et al. Dexamethasone stimulates placental system A transport and trophoblast differentiation in term villous explants. *Placenta*. 2010;31(2):97–105.
92. Shibata E, Powers RW, Rajakumar A, et al. Angiotensin II decreases system A amino acid transporter activity in human placental villous fragments through AT1 receptor activation. *Am J Physiol Endocrinol Metab*. 2006;291(5):E1009–E1016.
93. Sferruzzi-Perri AN, Vaughan OR, Haro M, et al. An obesogenic diet during mouse pregnancy modifies maternal nutrient partitioning and the fetal growth trajectory. *FASEB J*. 2013;27(10):3928–3937.
94. Musial B, Vaughan OR, Fernandez-Twinn DS, et al. A Western-style obesogenic diet causes maternal metabolic dysfunction with consequences for fetal nutrient acquisition in mice. *J Physiol*. 2016. Submitted.
95. Aye ILMH, Rosario FJ, Powell TL, Jansson T. Adiponectin supplementation in pregnant mice prevents the adverse effects of maternal obesity on placental function and fetal growth. *Proc Natl Acad Sci USA*. 2015;112(41):12858–12863.
96. Coan PM, Vaughan OR, Sekita Y, et al. Adaptations in placental phenotype support fetal growth during undernutrition of pregnant mice. *J Physiol*. 2010;588 (pt 3):527–538.
97. Sferruzzi-Perri AN, Vaughan OR, Coan PM, et al. Placental-specific Igf2 deficiency alters developmental adaptations to undernutrition in mice. *Endocrinology*. 2011;152:3202–3212.
98. Ganguly A, Collis L, Devaskar SU. Placental glucose and amino acid transport in calorie-restricted wild-type and Glut3 null heterozygous mice. *Endocrinology*. 2012;153(8):3995–4007.
99. Ahokas RA, Lahaye EB, Anderson GD, Lipshitz J. Effect of maternal dietary restriction on fetal growth and placental transfer of alpha-amino isobutyric acid in rats. *J Nutr*. 1981;111(12):2052–2058.
100. Coan PM, Vaughan OR, McCarthy J, et al. Dietary composition programmes placental phenotype in mice. *J Physiol*. 2011;589:3659–3670.
101. Jansson N, Pettersson J, Haafiz A, et al. Down-regulation of placental transport of amino acids precedes the development of intrauterine growth restriction in rats fed a low protein diet. *J Physiol*. 2006;576(pt 3):935–946.
102. Rosario FJ, Jansson N, Kanai Y, Prasad PD, Powell TL, Jansson T. Maternal protein restriction in the rat inhibits placental insulin, mTOR, and STAT3 signaling and down-regulates placental amino acid transporters. *Endocrinology*. 2011;152: 1119–1129.
103. Varma DR, Ramakrishnan R. Effects of protein-calorie malnutrition on transplacental kinetics of aminoisobutyric acid in rats. *Placenta*. 1991;12(3):277–284.
104. Higgins JS, Vaughan OR, Fernandez de Liger E, Fowden AL, Sferruzzi-Perri AN. Placental phenotype and resource allocation to fetal growth are modified by the timing and degree of hypoxia during mouse pregnancy. *J Physiol*. 2016;594:1341–1356.
105. Rosario FJ, Schumacher MA, Jiang J, Kanai Y, Powell TL, Jansson T. Chronic maternal infusion of full-length adiponectin in pregnant mice down-regulates placental amino acid transporter activity and expression and decreases fetal growth. *J Physiol*. 2012;590(6):1495–1509.

106. Sathishkumar K, Elkins R, Chinnathambi V, Gao H, Hankins GD, Yallampalli C. Prenatal testosterone-induced fetal growth restriction is associated with down--regulation of rat placental amino acid transport. *Reprod Biol Endocrinol.* 2011;9(1):1–12.
107. Vaughan OR, Sferruzzi-Perri AN, Fowden AL. Maternal corticosterone regulates nutrient allocation to fetal growth in mice. *J Physiol.* 2012;590:5529–5540.
108. Vaughan OR, Sferruzzi-Perri AN, Coan PM, Fowden AL. Adaptations in placental phenotype depend on route and timing of maternal dexamethasone administration in mice. *Biol Reprod.* 2013;89(4):80.
109. Roseboom TJ, van der Meulen JH, Ravelli AC, et al. Blood pressure in adults after prenatal exposure to famine. *J Hypertens.* 1999;17(3):325–330.
110. Carver TD, Quick AA, Teng CC, Pike AW, Fennessey PV, Hay Jr WW. Leucine metabolism in chronically hypoglycemic hypoinsulinemic growth-restricted fetal sheep. *Am J Physiol.* 1997;272(1 pt 1):E107–E117.
111. Owens JA, Falconer J, Robinson JS. Effect of restriction of placental growth on oxygen delivery to and consumption by the pregnant uterus and fetus. *J Dev Physiol.* 1987;9(2):137–150.
112. Pantham P, Rosario FJ, Weintraub ST, et al. Down-regulation of placental transport of amino acids precedes the development of intrauterine growth restriction in maternal nutrient restricted baboons. *Biol Reprod.* 2016;95:98.
113. Pantham P, Rosario FJ, Nijland M, et al. Reduced placental amino acid transport in response to maternal nutrient restriction in the baboon. *Am J Physiol Regul Integr Comp Physiol.* 2015;309(7):R740–R746.
114. Ganguly A, Touma M, Thamotharan S, De Vivo DC, Devaskar SU. Maternal calorie restriction causing utero-placental insufficiency differentially affects mammalian placental glucose and leucine transport molecular mechanisms. *Endocrinology.* 2016;157:4041–4054. en.2016–1259.
115. Wu G, Pond WG, Ott T, Bazer FW. Maternal dietary protein deficiency decreases amino acid concentrations in fetal plasma and allantoic fluid of pigs. *J Nutr.* 1998;128(5):894–902.
116. Yung HW, Cox M, Tissot van Patot M, Burton GJ. Evidence of endoplasmic reticulum stress and protein synthesis inhibition in the placenta of non-native women at high altitude. *FASEB J.* 2012;26(5):1970–1981.
117. Matheson H, Veerbeek JH, Charnock-Jones DS, Burton GJ, Yung HW. Morphological and molecular changes in the murine placenta exposed to normobaric hypoxia throughout pregnancy. *J Physiol.* 2016;594(5):1371–1388.
118. Brown LD, Hay Jr WW. Effect of hyperinsulinemia on amino acid utilization and oxidation independent of glucose metabolism in the ovine fetus. *Am J Physiol Endocrinol Metab.* 2006;291(6):E1333–E1340.
119. Limesand SW, Rozance PJ, Brown LD, Hay Jr WW. Effects of chronic hypoglycemia and euglycemic correction on lysine metabolism in fetal sheep. *Am J Physiol Endocrinol Metab.* 2009;296(4):E879–E887.
120. Philipps AF, Rosenkrantz TS, Porte PJ, Raye JR. The effects of chronic fetal hyperglycemia on substrate uptake by the ovine fetus and conceptus. *Pediatr Res.* 1985;19(7):659–666.
121. Milley JR. Ovine fetal protein metabolism during decreased glucose delivery. *Am J Physiol.* 1993;265(4 pt 1):E525–E531.
122. Sferruzzi-Perri AN, Owens JA, Pringle KG, Roberts CT. The neglected role of insulin-like growth factors in the maternal circulation regulating fetal growth. *J Physiol.* 2011;589(pt 1):7–20.
123. Fowden AL. The insulin-like growth factors and feto-placental growth. *Placenta.* 2003;24(8–9):803–812.

124. Luo ZC, Nuyt AM, Delvin E, et al. Maternal and fetal IGF-I and IGF-II levels, fetal growth, and gestational diabetes. *J Clin Endocrinol Metab*. 2012;97(5):1720–1728.
125. Sferruzzi-Perri AN, Owens J, Pringle KG, Robinson JS, Roberts C. Maternal insulin-like growth factors-I and -II act via different pathways to promote fetal growth. *Endocrinology*. 2006;147(7):3344–3355.
126. Baker J, Liu JP, Robertson EJ, Efstratiadis A. Role of insulin-like growth factors in embryonic and postnatal growth. *Cell*. 1993;75(1):73–82.
127. Gardner RL, Squire S, Zaina S, Hills S, Graham CF. Insulin-like growth factor-2 regulation of conceptus composition: effects of the trophectoderm and inner cell mass genotypes in the mouse. *Biol Reprod*. 1999;60(1):190–195.
128. DeChiara TM, Efstratiadis A, Robertsen EJ. A growth-deficiency phenotype in heterozygous mice carrying an insulin-like growth factor II gene disrupted by targeting. *Nature*. 1990;345(6270):78–80.
129. Liu J-P, Baker J, Perkins AS, Robertson EJ, Efstratiadis A. Mice carrying null mutations of the genes encoding insulin-like growth factor I (Igf-1) and type 1 IGF receptor (Igf1r). *Cell*. 1993;75(1):59–72.
130. Karl PI. Insulin-like growth factor-1 stimulates amino acid uptake by the cultured human placental trophoblast. *J Cell Physiol*. 1995;165(1):83–88.
131. Jones HN, Crombleholme T, Habli M. Adenoviral-mediated placental gene transfer of IGF-1 corrects placental insufficiency via enhanced placental glucose transport mechanisms. *PLoS One*. 2013;8(9), e74632.
132. Sferruzzi-Perri AN, Owens JA, Standen P, Taylor RL, Robinson JS, Roberts CT. Early pregnancy maternal endocrine insulin-like growth factor I programs the placenta for increased functional capacity throughout gestation. *Endocrinology*. 2007;148(9): 4362–4370.
133. Matthews JC, Beveridge MJ, Dialynas E, Bartke A, Kilberg MS, Novak DA. Placental anionic and cationic amino acid transporter expression in growth hormone overexpressing and null IGF-II or null IGF-I receptor mice. *Placenta*. 1999;20(8):639–650.
134. Copeland Jr AD, Porterfield SP. Effects of streptozotocin-induced diabetes in pregnant rats on placental transport and tissue uptake of alpha-amino-isobutyric acid. *Horm Metab Res*. 1987;19(2):57–61.
135. Fowden AL, Mao XZ, Comline RS. Effects of pancreatectomy on the growth and metabolite concentrations of the sheep fetus. *J Endocrinol*. 1986;110(2):225–231.
136. Milley JR. Effects of insulin on ovine fetal leucine kinetics and protein metabolism. *J Clin Invest*. 1994;93(4):1616–1624.
137. Sferruzzi-Perri AN, Owens JA, Standen P, et al. Early treatment of the pregnant guinea pig with IGFs promotes placental transport and nutrient partitioning near term. *Am J Physiol Endocrinol Metab*. 2007;292(3):E668–E676.
138. Constancia M, Angiolini E, Sandovici I, et al. Adaptation of nutrient supply to fetal demand in the mouse involves interaction between the Igf2 gene and placental transporter systems. *Proc Natl Acad Sci USA*. 2005;102(52):19219–19224.
139. Constancia M, Hemberger M, Hughes J, et al. Placental-specific IGF-II is a major modulator of placental and fetal growth. *Nature*. 2002;417(6892):945–948.
140. Kusinski LC, Dilworth MR, Baker PN, Sibley CP, Wareing M, Glazier JD. System A activity and vascular function in the placental-specific Igf2 knockout mouse. *Placenta*. 2011;32(11):871–876.
141. Angiolini E, Coan PM, Sandovici I, et al. Developmental adaptations to increased fetal nutrient demand in mouse genetic models of Igf2-mediated overgrowth. *FASEB J*. 2011;25(5):1737–1745.
142. Coan PM, Angiolini E, Sandovici I, Burton GJ, Constancia M, Fowden AL. Adaptations in placental nutrient transfer capacity to meet fetal growth demands depend on placental size in mice. *J Physiol*. 2008;586(pt 18):4567–4576.

143. Sferruzzi-Perri A, Lopez-Tello J, Fowden A, Constancia M. Maternal and fetal genomes interplay through phosphoinositol 3-kinase(PI3K)-p110α signalling to modify placental resource allocation. *Proc Natl Acad Sci USA*. 2016;113:11255–11260.
144. Pittas AG, Joseph NA, Greenberg AS. Adipocytokines and insulin resistance. *J Clin Endocrinol Metabol*. 2004;89(2):447–452.
145. Briana DD, Malamitsi-Puchner A. The role of adipocytokines in fetal growth. *Ann N Y Acad Sci*. 2010;1205:82–87.
146. Aye IL, Lager S, Ramirez VI, et al. Increasing maternal body mass index is associated with systemic inflammation in the mother and the activation of distinct placental inflammatory pathways. *Biol Reprod*. 2014;90(6):129.
147. Challier JC, Basu S, Bintein T, et al. Obesity in pregnancy stimulates macrophage accumulation and inflammation in the placenta. *Placenta*. 2008;29(3):274–281.
148. Aye IL, Powell TL, Jansson T. Review: adiponectin—the missing link between maternal adiposity, placental transport and fetal growth? *Placenta*. 2013;34(suppl):S40–S45.
149. Kotani Y, Yokota I, Kitamura S, Matsuda J, Naito E, Kuroda Y. Plasma adiponectin levels in newborns are higher than those in adults and positively correlated with birth weight. *Clin Endocrinol (Oxf)*. 2004;61(4):418–423.
150. Tsai PJ, Yu CH, Hsu SP, et al. Cord plasma concentrations of adiponectin and leptin in healthy term neonates: positive correlation with birthweight and neonatal adiposity. *Clin Endocrinol (Oxf)*. 2004;61(1):88–93.
151. Forhead AJ, Lamb CA, Franko KL, et al. Role of leptin in the regulation of growth and carbohydrate metabolism in the ovine fetus during late gestation. *J Physiol*. 2008;586(9):2393–2403.
152. Soldin OP, Guo T, Weiderpass E, Tractenberg RE, Hilakivi-Clarke L, Soldin SJ. Steroid hormone levels in pregnancy and 1 year postpartum using isotope dilution tandem mass spectrometry. *Fertil Steril*. 2005;84(3):701–710.
153. Quirke LD, Juengel JL, Tisdall DJ, Lun S, Heath DA, McNatty KP. Ontogeny of steroidogenesis in the fetal sheep gonad. *Biol Reprod*. 2001;65(1):216–228.
154. Fowden AL, Li J, Forhead AJ. Glucocorticoids and the preparation for life after birth: are there long-term consequences of the life insurance? *Proc Nutr Soc*. 1998;57(1):113–122.
155. Iqbal M, Audette MC, Petropoulos S, Gibb W, Matthews SG. Placental drug transporters and their role in fetal protection. *Placenta*. 2012;33(3):137–142.
156. Seckl JR, Holmes MC. Mechanisms of disease: glucocorticoids, their placental metabolism and fetal 'programming' of adult pathophysiology. *Nat Clin Pract Endocrinol Metab*. 2007;3(6):479–488.
157. Carlsen SM, Jacobsen G, Romundstad P. Maternal testosterone levels during pregnancy are associated with offspring size at birth. *Eur J Endocrinol*. 2006;155(2):365–370.
158. Whitehouse AJ, Maybery MT, Hart R, et al. Free testosterone levels in umbilical-cord blood predict infant head circumference in females. *Dev Med Child Neurol*. 2010;52(3): e73–e77.
159. Goland RS, Jozak S, Warren WB, Conwell IM, Stark RI, Tropper PJ. Elevated levels of umbilical cord plasma corticotropin-releasing hormone in growth-retarded fetuses. *J Clin Endocrinol Metab*. 1993;77(5):1174–1179.
160. Manabe M, Nishida T, Imai T, et al. Cortisol levels in umbilical vein and umbilical artery with or without antenatal corticosteroids. *Pediatr Int*. 2005;47(1):60–63.
161. Li J, Wang ZN, Chen YP, et al. Late gestational maternal serum cortisol is inversely associated with fetal brain growth. *Neurosci Biobehav Rev*. 2012;36(3):1085–1092.
162. Goedhart G, Vrijkotte TG, Roseboom TJ, van der Wal MF, Cuijpers P, Bonsel GJ. Maternal cortisol and offspring birthweight: results from a large prospective cohort study. *Psychoneuroendocrinology*. 2010;35(5):644–652.

163. Roberts D, Dalziel S. Antenatal corticosteroids for accelerating fetal lung maturation for women at risk of preterm birth. *Cochrane Database Syst Rev*. 2006;(3). CD004454.
164. Audette MC, Challis JRG, Jones RL, Sibley CP, Matthews SG. Synthetic glucocorticoid reduces human placental system a transport in women treated with antenatal therapy. *J Clin Endocrinol Metabol*. 2014;99(11):E2226–E2233.
165. Wyrwoll CS, Seckl JR, Holmes MC. Altered placental function of 11beta-hydroxysteroid dehydrogenase 2 knockout mice. *Endocrinology*. 2009;150(3):1287–1293.
166. Jansson T, Aye ILMH, Goberdhan DC. The emerging role of mTORC1 signaling in placental nutrient-sensing. *Placenta*. 2012;33(suppl 2):e23–e29.
167. Roos S, Kanai Y, Prasad PD, Powell TL, Jansson T. Regulation of placental amino acid transporter activity by mammalian target of rapamycin. *Am J Physiol Cell Physiol*. 2009;296(1):C142–C150.
168. Rosario FJ, Kanai Y, Powell TL, Jansson T. Mammalian target of rapamycin signalling modulates amino acid uptake by regulating transporter cell surface abundance in primary human trophoblast cells. *J Physiol*. 2013;591(3):609–625.
169. Rosario Fredrick J, Dimasuay Kris G, Kanai Y, Powell Theresa L, Jansson T. Regulation of amino acid transporter trafficking by mTORC1 in primary human trophoblast cells is mediated by the ubiquitin ligase Nedd4-2. *Clin Sci*. 2016;130(7):499–512.
170. Roos S, Jansson N, Palmberg I, Saljo K, Powell TL, Jansson T. Mammalian target of rapamycin in the human placenta regulates leucine transport and is down-regulated in restricted fetal growth. *J Physiol*. 2007;582(pt 1):449–459.
171. Chen Y-Y, Rosario Fredrick J, Shehab Majida A, Powell Theresa L, Gupta Madhulika B, Jansson T. Increased ubiquitination and reduced plasma membrane trafficking of placental amino acid transporter SNAT-2 in human IUGR. *Clin Sci*. 2015;129(12):1131–1141.
172. Lager S, Samulesson AM, Taylor PD, Poston L, Powell TL, Jansson T. Diet-induced obesity in mice reduces placental efficiency and inhibits placental mTOR signaling. *Physiol Rep*. 2014;2(2):e00242.
173. Gonzalez IM, Martin PM, Burdsal C, et al. Leucine and arginine regulate trophoblast motility through mTOR-dependent and independent pathways in the preimplantation mouse embryo. *Dev Biol*. 2012;361(2):286–300.
174. Kim J, Song G, Wu G, Gao H, Johnson GA, Bazer FW. Arginine, leucine, and glutamine stimulate proliferation of porcine trophectoderm cells through the MTOR-RPS6K-RPS6-EIF4EBP1 signal transduction pathway. *Biol Reprod*. 2013;88(5):113, 111–119.
175. Capobianco E, Fornes D, Linenberg I, Powell TL, Jansson T, Jawerbaum A. A novel rat model of gestational diabetes induced by intrauterine programming is associated with alterations in placental signaling and fetal overgrowth. *Mol Cell Endocrinol*. 2016;422:221–232.
176. Gaccioli F, White V, Capobianco E, Powell TL, Jawerbaum A, Jansson T. Maternal overweight induced by a diet with high content of saturated fat activates placental mTOR and eIF2alpha signaling and increases fetal growth in rats. *Biol Reprod*. 2013;89(4):96.
177. Zhou F, Guan LB, Yu P, Wang XD, Hu YY. Regulation of hypoxia-inducible factor-1alpha, regulated in development and DNA damage response-1 and mammalian target of rapamycin in human placental BeWo cells under hypoxia. *Placenta*. 2016;45:24–31.
178. Colleoni F, Padmanabhan N, Yung HW, et al. Suppression of mitochondrial electron transport chain function in the hypoxic human placenta: a role for miRNA-210 and protein synthesis inhibition. *PLoS One*. 2013;8(1), e55194.
179. Skeffington KL, Higgins JS, Mahmoud AD, et al. Hypoxia, AMPK activation and uterine artery vasoreactivity. *J Physiol*. 2016;594:1357–1369.

180. Knight ZA, Gonzalez B, Feldman ME, et al. A pharmacological map of the PI3-K family defines a role for p110α in insulin signaling. *Cell.* 2006;125(4):733–747.
181. Street ME, Viani I, Ziveri MA, Volta C, Smerieri A, Bernasconi S. Impairment of insulin receptor signal transduction in placentas of intra-uterine growth-restricted newborns and its relationship with fetal growth. *Eur J Endocrinol.* 2011;164(1):45–52.
182. Laviola L, Perrini S, Belsanti G, et al. Intrauterine growth restriction in humans is associated with abnormalities in placental insulin-like growth factor signaling. *Endocrinology.* 2005;146(3):1498–1505.
183. Alonso A, Del Rey CG, Navarro A, Tolivia J, Gonzalez CG. Effects of gestational diabetes mellitus on proteins implicated in insulin signaling in human placenta. *Gynecol Endocrinol.* 2006;22(9):526–535.
184. Perez-Perez A, Maymo JL, Gambino YP, et al. Activated translation signaling in placenta from pregnant women with gestational diabetes mellitus: possible role of leptin. *Horm Metab Res.* 2013;45(6):436–442.
185. Rosario FJ, Powell TL, Jansson T. Activation of placental insulin and mTOR signaling in a mouse model of maternal obesity associated with fetal overgrowth. *Am J Physiol Regul Integr Comp Physiol.* 2016;310(1):R87–R93.
186. Perez-Perez A, Gambino Y, Maymo J, et al. MAPK and PI3K activities are required for leptin stimulation of protein synthesis in human trophoblastic cells. *Biochem Biophys Res Commun.* 2010;396(4):956–960.
187. Aye ILMH, Gao X, Weintraub ST, Jansson T, Powell TL. Adiponectin inhibits insulin function in primary trophoblasts by PPARα-mediated ceramide synthesis. *Mol Endocrinol.* 2014;28(4):512–524.
188. Revollo JR, Cidlowski JA. Mechanisms generating diversity in glucocorticoid receptor signaling. *Ann N Y Acad Sci.* 2009;1179:167–178.
189. Anwar MA, Saleh AI, Al Olabi R, Al Shehabi TS, Eid AH. Glucocorticoid-induced fetal origins of adult hypertension: association with epigenetic events. *Vascul Pharmacol.* 2016;82:41–50.
190. Boehmer C, Okur F, Setiawan I, Broer S, Lang F. Properties and regulation of glutamine transporter SN1 by protein kinases SGK and PKB. *Biochem Biophys Res Commun.* 2003;306(1):156–162.
191. Vaughan OR, Fisher HM, Dionelis KN, et al. Corticosterone alters materno-fetal glucose partitioning and insulin signalling in pregnant mice. *J Physiol.* 2015;593(5):1307–1321.
192. Vaughan OR, Phillips HM, Everden AJ, Sferruzzi-Perri AN, Fowden AL. Dexamethasone treatment of pregnant F0 mice leads to parent of origin-specific changes in placental function of the F2 generation. *Reprod Fertil Dev.* 2015;27(4):704–711.
193. Vaughan OR, Davies KL, Ward JW, de Blasio MJ, Fowden AL. A physiological increase in maternal cortisol alters uteroplacental metabolism in the pregnant ewe. *J Physiol.* 2016;594:6407–6418.
194. Day PE, Ntani G, Crozier SR, et al. Maternal factors are associated with the expression of placental genes involved in amino acid metabolism and transport. *PLoS One.* 2015;10(12), e0143653.
195. Cleal JK, Day PE, Simner CL, et al. Placental amino acid transport may be regulated by maternal vitamin D and vitamin D-binding protein: results from the Southampton Women's Survey. *Br J Nutr.* 2015;113(12):1903–1910.
196. Ronzoni S, Marconi AM, Paolini CL, Teng C, Pardi G, Battaglia FC. The effect of a maternal infusion of amino acids on umbilical uptake in pregnancies complicated by intrauterine growth restriction. *Am J Obstet Gynecol.* 2002;187(3):741–746.
197. Brown LD, Rozance PJ, Thorn SR, Friedman JE, Hay Jr WW. Acute supplementation of amino acids increases net protein accretion in IUGR fetal sheep. *Am J Physiol Endocrinol Metab.* 2012;303(3):E352–E364.

198. Wali JA, de Boo HA, Derraik JG, et al. Weekly intra-amniotic IGF-1 treatment increases growth of growth-restricted ovine fetuses and up-regulates placental amino acid transporters. *PLoS One*. 2012;7(5), e37899.
199. Jones H, Crombleholme T, Habli M. Regulation of amino acid transporters by adenoviral-mediated human insulin-like growth factor-1 in a mouse model of placental insufficiency in vivo and the human trophoblast line BeWo in vitro. *Placenta*. 2014;35(2):132–138.
200. Charnock JC, Dilworth MR, Aplin JD, Sibley CP, Westwood M, Crocker IP. The impact of a human IGF-II analog ([Leu27]IGF-II) on fetal growth in a mouse model of fetal growth restriction. *Am J Physiol Endocrinol Metab*. 2016;310(1):E24–E31.

INDEX

Note: Page numbers followed by "*f*" indicate figures, and "*t*" indicate tables.

R

S

T

FSC
www.fsc.org
MIX
Papier | Fördert
gute Waldnutzung
FSC® C083411

Kolibri 360 GmbH
Ferdinand-Jühlke-Straße 7
99095 Erfurt, Deutschland
produktsicherheit@kolibri360.de